The series "Advances in Intelligent Systems and Computing" contains publications on theory, applications, and design methods of Intelligent Systems and Intelligent Computing. Virtually all disciplines such as engineering, natural sciences, computer and information science, ICT, economics, business, e-commerce, environment, healthcare, life science are covered. The list of topics spans all the areas of modern intelligent systems and computing such as: computational intelligence, soft computing including neural networks, fuzzy systems, evolutionary computing and the fusion of these paradigms, social intelligence, ambient intelligence, computational neuroscience, artificial life, virtual worlds and society, cognitive science and systems, Perception and Vision, DNA and immune based systems, self-organizing and adaptive systems, e-Learning and teaching, human-centered and human-centric computing, recommender systems, intelligent control, robotics and mechatronics including human-machine teaming, knowledge-based paradigms, learning paradigms, machine ethics, intelligent data analysis, knowledge management, intelligent agents, intelligent decision making and support, intelligent network security, trust management, interactive entertainment, Web intelligence and multimedia.

The publications within "Advances in Intelligent Systems and Computing" are primarily proceedings of important conferences, symposia and congresses. They cover significant recent developments in the field, both of a foundational and applicable character. An important characteristic feature of the series is the short publication time and world-wide distribution. This permits a rapid and broad dissemination of research results.

More information about this series at http://www.springer.com/series/11156

Advances in Intelligent Systems and Computing

Volume 863

Series editor

Janusz Kacprzyk, Systems Research Institute, Polish Academy of Sciences, Warsaw, Poland
e-mail: kacprzyk@ibspan.waw.pl

Suresh Chandra Satapathy
Vikrant Bhateja · Radhakhrishna Somanah
Xin-She Yang · Roman Senkerik
Editors

Information Systems Design and Intelligent Applications

Proceedings of Fifth International
Conference INDIA 2018 Volume 2

 Springer

Editors
Suresh Chandra Satapathy
School of Computer Engineering
KIIT University
Bhubaneswar, Odisha, India

Xin-She Yang
School of Science and Technology
Middlesex University
London, UK

Vikrant Bhateja
Department of Electronics and
Communication Engineering
Shri Ramswaroop Memorial Group
of Professional Colleges
Lucknow, Uttar Pradesh, India

Roman Senkerik
Faculty of Applied Informatics
Tomas Bata University in Zlín
Zlín, Czech Republic

Radhakhrishna Somanah
Universite des Mascareignes
Beau Bassin-Rose Hill, Mauritius

ISSN 2194-5357   ISSN 2194-5365 (electronic)
Advances in Intelligent Systems and Computing
ISBN 978-981-13-3337-8   ISBN 978-981-13-3338-5 (eBook)
https://doi.org/10.1007/978-981-13-3338-5

Library of Congress Control Number: 2018961687

This Springer imprint is published by the registered company Springer Nature Singapore Pte Ltd.
The registered company address is: 152 Beach Road, #21-01/04 Gateway East, Singapore 189721,
Singapore

Preface

The Fifth International Conference on Information System Design and Intelligent Applications (INDIA 2018) was successfully hosted by the Université des Mascareignes (UDM) from 19 July 2018 to 20 July 2018 at the InterContinental Resort Balaclava in Mauritius.

The Université des Mascareignes has strategically partnered with Université de Limoges, France, to offer a double degree with a strong emphasis on industrial placement, design its curriculum, introduce industry-based electives and facilitate students' internship at the industry for skill development. It has established a Centre of Excellence in collaboration with industry for various research and training purposes. The value addition training and career augmentation services prepare the students to meet expectations of industry demands.

Research focuses of the faculty of information and communication technology include algorithms and theory of computation, artificial intelligence, bioinformatics, cloud computing, database and data mining, data analytics, human–computer interaction, information and network security, Internet technology, image processing, mobile computing, pattern recognition, program analysis and testing, parallel and distributed computing, real-time systems, service-oriented architecture, soft computing, software engineering and wireless sensor networks.

The objective of this conference was to provide opportunities for the researchers, academicians and industry persons to interact and exchange the ideas, experience and gain expertise in the cutting-edge technologies pertaining to soft computing and signal processing. Research papers in the above-mentioned technology areas were received and subjected to a rigorous peer-reviewed process with the help of programme committee members and external reviewers. The INDIA 2018 conference received research articles in various domains, and after the rigorous review, only quality articles were accepted for publication in Springer AISC series. The hosting of the conference and full sponsorship approved by the director were the key factors in generating quality papers from UDM staff.

Special thanks to keynote speakers Mr. Aninda Bose, Senior Editor, Hard Sciences Publishing, Springer Nature; Dr B. Annappa, Professor, Department of Computer Science and Engineering, National Institute of Technology Karnataka,

Surathkal, Mangalore, India; and Dr. Jagdish Chand Bansal, Professor, South Asian University, New Delhi, and Visiting Research Fellow, Maths and Computer Science, Liverpool Hope University, UK.

We would like to express our gratitude to all session chairs and reviewers for extending their support and cooperation.

We would like to express our special gratitude to Organizing Committee Co-chair, Prof. Binod Kumar Pattanayak, Siksha 'O' Anushandhan University, Bhubaneswar, Odisha, and Publication Chair, Dr. Suresh Chandra Satapathy, KIIT, Bhubaneswar, for initiating this conference and for their valuable support and encouragement till the successful conclusion of the conference.

We are indebted to all the committee members for organizing the conference, namely Mohammad Reza Hosenkhan, Organizing Committee Chair, Founding Dean Faculty of Information and Communication Technology; Mrs Nundini Akaloo, Programme Committee Chair, Founding Dean Faculty of Business Management; Mrs Rubina Fareed and Dr. Swaleha Peeroo, Lecturers from the Faculty of Business Management; Mr Seeven Amic, Programme Committee Co-chair; and Mr Sanjeev Cowlessur, Mrs Shameera Lotun, Mrs Mahejabeen Peeermamode and Mrs Dosheela Ramlowat, Lecturers from the Department of Software Engineering.

Special thanks to administrative staff, namely Mrs Farzana Antoaroo, Mr Ravi Langur, Mr Deojeet Nohur, Mrs Vimi Lockmun-Bissessur and Mrs Yasoda Benoit.

We express our heartfelt thanks to our Chief Patron, the Hon. (Mrs) Leela Devi Dookhun-Luchoomun, Minister of Education and Human Resources, Tertiary Education and Scientific Research, who supported and inaugurated the conference.

We are first and foremost indebted to the General Chair Dr. Radhakhrishna Somanah, Outstanding Scientist, Director General of UDM, Commander of the Star and Key of the Indian Ocean (CSK)—one of the highest national awards received on the Republic Day of Mauritius on 12 March 2011, Chevalier dans l'Ordre des Palmes Academiques—Award by the Republic of France on 4 September 2009, Award by International Astronomical Union (IAU): minor planet Somanah 19318 in our solar system named after his name "Somanah" in 2007 for his contribution to research in astronomy award by NASA to my research group in June 2004 for excellence in Education and Discovery. We express our deep gratitude to Dr. R. Somanah for having believed in the importance and value of the conference and for having found solutions to many challenges and obstacles in Mauritius at the initial stage (and even after) to ensure that the conference materializes and becomes reality and has finally been a success.

Last, but certainly not least, our special thanks to all the authors without whom the conference would not have taken place. Their technical contributions have made our proceedings rich and praiseworthy.

Bhubaneswar, India	Suresh Chandra Satapathy
Lucknow, India	Vikrant Bhateja
Beau Bassin-Rose Hill, Mauritius	Radhakhrishna Somanah
London, UK	Xin-She Yang
Zlín, Czech Republic	Roman Senkerik

Contents

About the Editors

Suresh Chandra Satapathy is currently working as Professor, School of Computer Engg, KIIT Deemed to be University, Bhubaneswar, Odisha, India. He obtained his PhD in Computer Science Engineering from JNTUH, Hyderabad and Master degree in Computer Science and Engineering from National Institute of Technology (NIT), Rourkela, Odisha. He has more than 28 years of teaching and research experience. His research interest includes machine learning, data mining, swarm intelligence studies and their applications to engineering. He has more than 98 publications to his credit in various reputed international journals and conference proceedings. He has edited many volumes from Springer AISC, LNEE, SIST and LNCS in past and he is also the editorial board member in few international journals. He is a senior member of IEEE and Life Member of Computer Society of India.

Vikrant Bhateja is Associate Professor, Department of Electronics & Communication Engineering, Shri Ramswaroop Memorial Group of Professional Colleges (SRMGPC), Lucknow and also the Head (Academics & Quality Control) in the same college. His areas of research include digital image and video processing, computer vision, medical imaging, machine learning, pattern analysis and recognition. He has around 100 quality publications in various international journals and conference proceedings. Prof. Vikrant has been on TPC and chaired various sessions from the above domain in international conferences of IEEE and Springer. He has been the track chair and served in the core-technical/editorial teams for international conferences: FICTA 2014, CSI 2014, INDIA 2015, ICICT-2015 and ICTIS-2015 under Springer-ASIC Series and INDIACom-2015, ICACCI-2015 under IEEE. He is associate editor in International Journal of Synthetic Emotions (IJSE) and International Journal of Ambient Computing and Intelligence (IJACI) under IGI Global. He is also serving in the editorial board of International Journal of Image Mining (IJIM) and International Journal of Convergence Computing (IJConvC) under Inderscience Publishers. He has been editor of four published volumes with Springer (IC3T-2015, INDIA-2016, ICIC2-2016, ICDECT-2016) and few other are under press (FICTA-2016, IC3T-2016, ICMEET-2016).

Radhakhrishna Somanah is Director of Universite des Mascareignes. He is one of the three pioneers of professional astronomy in Mauritius (together with Dr Nalini Issur and Dr Kumar Golap who is now working for the VLA, NRAO). He is fully involved with the design, construction and research at the Mauritius Radio Telescope (MRT) project since 1989. MRT is one of the biggest scientific research projects at the University of Mauritius. The main aim of the project was to produce a map of the southern sky at 151.5 MHz (unique in the world), perform astrophysical interpretation of the map, develop new algorithms and software etc. He has been involved with the supervision of the civil, mechanical, electrical and electronics work. Because of its immensity, it took nearly 5 years (1989–1993) to build. He was also part of the team which developed the software to convert the raw data into deconvolved radio images, and astrophysical interpretation of the latter. He has many publications in reputed journals.

Xin-She Yang obtained his DPhil in Applied Mathematics from the University of Oxford. He then worked at Cambridge University and National Physical Laboratory (UK) as a Senior Research Scientist. Now he is Reader in Modelling and Optimization at Middlesex University, an elected Bye-Fellow at Cambridge University and Adjunct Professor at Reykjavik University (Iceland). He is the chair of the IEEE CIS Task Force on Business Intelligence and Knowledge Management. He is on the list of both 2016 and 2017 Clarivate Analytics/Thomson Reuters Highly Cited Researchers.

Roman Senkerik was born in the Czech Republic, and went to the Tomas Bata University in Zlin, Faculty of Applied Informatics, where he studied Technical Cybernetics and obtained his MSc degree in 2004, PhD degree in 2008 (Technical Cybernetics) and Assoc. Prof. degree in 2013 (VSB – Technical University of Ostrava – degree in Informatics). He is now a researcher and lecturer at the same university. His research interests are: Evolutionary Computation, Theory of chaos, Complex systems, Soft-computing methods and their interdisciplinary applications, Optimization, Neural Networks, Data analysis, Information retrieval, and Cyber-security. He is Recognized Reviewer for many Elsevier journals as well as many other leading journals in computer science/computational intelligence. He was a part of the organizing teams for several conferences, and special sessions/symposiums at IEEE CEC and IEEE SSCI events. He was a guest editor of several special issues in journals, editor of Springer proceedings for several conferences.

Integrating Information Technology and Marketing for Better Customer Value

Normada Devi Bheekharry and Upasana Gitanjali Singh

Abstract This paper aims to explore how advancement in digital technologies has contributed to identify, shape and execute digital opportunities in order to increase organisation competitive advantage. The basic principles of marketing have been heavily transformed by the explosion of information technology where the interaction of customers through new digital platforms has been mushrooming, making marketers' task more challenging. This is a conceptual study based on critical analysis of relevant literature reviews and discussion on digital marketing topics from the American Marketing Association. Technology breakouts have given rise to different digital platforms where organisations and customers interact, leading to a massive flow of data. Marketers are finding it challenging to identify the 'Who' and the 'What' of marketing. To better understand the demand of customers and deliver better customer values, different business analytics are developed and tested. Innovation in technology has brought new methods in which data can be sorted and analysed for organisational strategic development. The online environment is dynamic and this paper captures how marketers are using information technology know-how as a stepping stone to remain competitive. Understanding the digital behaviour of customers is of paramount importance to marketers to better create customer value. This paper draws on the importance and analysis of digitalisation, specifically the impact of social media, big data and big data analytics.

Keywords Social media analytics · Automation marketing · Digital marketing
Customer relationship management · Big data and big data analytics

N. D. Bheekharry (✉)
Universitè des Mascareignes, Pamplemousses, Mauritius
e-mail: nbheekharry@udm.ac.mu

U. G. Singh
University of KwaZulu Natal, Berea, South Africa
e-mail: singhup@ukzn.ac.za

© Springer Nature Singapore Pte Ltd. 2019
S. C. Satapathy et al. (eds.), *Information Systems Design and Intelligent Applications*, Advances in Intelligent Systems and Computing 863,
https://doi.org/10.1007/978-981-13-3338-5_1

1

1 Introduction

Customer is the most important assets of any organisations as it is the only one that generates revenue. Marketo CEO, Steve Lucas, says that we are in the age of the buyer, where the buyer definitely has a louder voice than the brand [1]. The business environment is very complex and with the development in information technologies (IT) new forces have arisen to challenge it and redefine it. The aim of this research is to draw upon the different breakthroughs in information system and to acknowledge the relationship between information technology and marketing. This involves an overview of the importance of data, customer relationship management and market insight. This paper recognises that new technology is a must and can be used as a competitive advantage however it is very much dependent on the organisation management philosophy and capabilities especially, human factors.

Over the recent years, the marketplace has been increasingly digitalised and research conducted by academicians outline that technological changes have made traditional marketing more efficient and effective in reaching and selling to markets [2]. Innovation in internet technologies can be summarised by the evolution of Web 1.0 into Web 2.0 and the emergence of the semantic web technologies and their integration into Web 3.0 [3, 4]. The web is a powerful marketing tool and has definitely changed the concept of marketing practices with the emergence of E-marketing. By definition, it is a new and modern commercial practice of transacting products whether it is a good, services, information and ideas via the Internet and other electronic means. Combining IT and marketing has given rise to new marketing jargons like martech and adtech. Martech lived among first-party data and owned software systems, like email and web analytics; and adtech generally adhered to third-party data and paid media (which-50). The Web 2.0 allows more exchanges among its users which include content sharing or social media interactions. Current examples of file-sharing sites, but not limited to, are (Flickr for photo sharing [5]), blogs (e.g. Blogger.com [6], wikis (e.g. Wikipedia [7], and social networking sites (e.g. Facebook [7]; Twitter [8]). As discussed previously, Web 2.0 technologies are easy to use and interactive that is why internal and external are using them for different purposes. We believe that thanks to these characteristics individuals are sharing on the web a lot of information about their experiences with products and services which are of great importance to organisations. Several strategies can be devised to better segment and target this audience. Therefore, firms tend to acquire that market intelligence using Web 2.0 technologies, together with experts to manage and act proactively for the interest and future of the organisation. Additionally, it has been argued that internet blog narratives can be used to determine one firm's competitive position [9]. Finally, Web 2.0 has the potential to effectively understand market intelligence within the firm.

Academicians have come to the conclusion that Web 2.0 has three distinct characteristics [10], namely: collaboration, participation and communication. Web 2.0 can be defined as web pages that use a two-way stream of communication between users allowing them to socialise online [11] and to share their own user generator content [6, 8, 12]. Socialising online through social media platforms whether by chatting, downloading or uploading videos, games and music, has basically a major hand in the changing role of the customer where the market has become forum where consumers play an active role in creating and competing for value. The term value, which was traditionally defined as creating and delivering goods and services prior to customers' demand effectively has dramatically changed. Marketers roles have changed as they are dealing with a global customer with different needs, wants and expectations. Combining with this, there is the culture element which has to be considered.

In the following parts, individual areas include discussion on IT software and technology, customer relationship management and big data (BD).

2 Literature Review

2.1 The Impact of the Web on Marketing

The changes and development in technology have created a revolution for marketers. New tools, software and platforms have arisen and among them are: search engine optimisation (SEO), content management, social platforms (Facebook, Instagram, Twitter and so on), knowledge management, mobile marketing and big data.

2.1.1 Search Engine Optimisation

Customers use the Internet to search for information, advice, a product or a service and companies are seeking to capture these interests through search engine optimisation [13, 14]. It has also been identified as a set of techniques used by websites to be more visible and increase site traffics [15]. To be closer to customers and reacting customers' demand, nowadays every organisations have a website where individuals can receive information and browse through the net at their ease. The research area [16] demonstrates that it has become important to evaluate the credibility and quality web-based information to build and strengthen customer's trust [15].

2.1.2 Content Marketing

Content marketing is a significant element in online marketing and understanding how it can be used in marketing is of paramount importance. Content marketing defined as is the 'creation and distribution of educational and/or compelling content multiple formats to attract and retain customers', [17]. This concept was reviewed and redefined [18] as 'content marketing is a strategy focusses on the creation of a valuable experience'. Many authors and commentators suggest that the some of the main objectives of content marketing are to create brand awareness or reinforcement, customer service and conversation, customer upsell and customer service [18–20]. The primary driver for the success of content marketing is the ability to build customer trust and four approaches were recognised to enhance trust [21]: shared values (with the customer), interdependence, quality communication and non-opportunistic behaviour. Content includes the static content forming web pages, as well as dynamic rich media content, such as videos.

2.1.3   Social Media

Social media functions as marketplace in which both buyers and sellers exist [22], along with various exchange facilitators [23] all interacting with each other in complex ways [24]. The significance of the social networks and social media has been reinforced [25], where it was seen that the former allows individual to (1) construct a profile which can be public or semipublic, (2) share a connection with a list of users and (3) view and navigate through their list of connections. The propagation of social media and its derivatives are thought to represent a great opportunity and at the same time threat for companies. Currently, firms have the chance to generate innovative business models and to extend CRM through social media. Active ways to engage customers through social marketing may compromise using traditional marketing instruments online, such as customer service, customising the offer reaching social media influences and employing customer creativity to innovate together with customers [26]. Additionally, there is a need to foster user-generated content [13, 27] and to identify influential users [28]. Furthermore, researchers pointed out, social networks are 'essential for knowledge creation, sharing and for learning' [29].

2.1.4   Knowledge Management

Knowledge management (KM) is defined [30] as a management practice that is gaining enormous importance nowadays. Many journals articles have been written on this subject and are defined as a systematic process for creating, sharing and implementing information. Knowledge originates in the mind of people [31–33] meaning any good or bad experience from a product and or service gathered in one's mind and will influence one's decision in product choice, brand and any other purchase. Now, this is no complex theory to understand if properly channelled knowledge management is an asset which can help any organisation to understand, create, generate and deliver value. Academicians and Information Technology (IT) specialists have described KM as an information technology (IT) system developed to facilitate and support the creation, dissemination and implementation of knowledge in organisation [34]. KM system is being widely used by organisations like Amazon (ecommerce), Federal Express, Proctor and Gamble, Walmart, Target and others. KM is a tool used to identify, share and utilise any information about the micro and macro environment to identify any problem whether be it (a) a decrease in sales, (b) a new competitor

(c) introducing a new product, (d) organisational processes and (e) any other. The Web produces petabytes of information daily through social connections and other digital medium and this information is the knowledge required which organisation can utilise to enhance the formulation and implementation of productive strategies [30]. Here again, expert in the field of data science and data analysis are needed to sort and analyse this information to knowledge.

2.1.5 Mobile Marketing

Smartphone (particularly iPhone) adoption by consumers presents marketing practitioners with many new opportunities and challenges to segment, reach and serve customers [35]. Research on mobile marketing indicates that as the technology continues to evolve, mobile marketing practices will likely to go through fundamental changes [36]. The purpose of classic mobile phones was mainly for conversation and text messaging. However, throughout time and advancement in technology, the mobile phone offers consumers with a wide array of features ranging from mobile web browsing, uploading and downloading of music, movies, photos, a variety of apps, email, GPS and much [35]. Consumers are attracted by smartphones for their ease of use, practical and entertainment applications. Mobile marketing combining with social media platforms provide marketers with a huge opportunity to offer superior customer experiences and products. The introduction of new mobile technologies would definitely change marketing practices [36].

2.1.6 Big Data (BD) and Big Data Analytics (BDA)

Marketing practitioners are realising the importance of Big Data even it is considered to be in the infancy stage. Research is done in this area and scholars as well as academicians are combining big data with different topics as competitive intelligence, business intelligence, knowledge management, and business analytics among others to enhance organisation effectiveness and productivity. The user interaction on the web and other digital tools has created a constantly growing flow of information and knowledge which can be defined as Big Data. Proactive firms can use BD as an important source to generate valuable insights which can help them to develop product and or services according to the customers need. Taking the right strategic decision will ultimately generate profits and enable organisations to gain competitive advantage [31, 37] and build an element of trust among customers. BD and big data analytics facilitate organisations to capture, analyse and exchange information about customers [38]. Analysing the volume and the different sources (social platforms, videos, forums and others) of data provide organisation with useful marketing insight to engage with customers [38]. Information and communication technology offers huge prospects for organisations to engage customers which is very important as from these engagements, marketers have the opportunity to understand and capture customer value and customer behaviour. At the same, this leads to advance competences helping firms to create and sustain value over time [39].

BD is of great importance to marketers for decision-making. Big companies like Netflix, Proctor and Gamble, Federal Express, Walmart, Target, Tesco, Google, Amazon, Facebook, Twitter, LinkedIn, Marriot and CEMEX use big data and big data analytics to segment and target customers and as a competitive strategy. It was [32] postulated that these companies use analytics not only as a competitive advantage but to better understand their customers, competitors and markets. The web is considered as a pull marketing environment [27]. A pull marketing environment is where companies pull their customers through search engine optimisation and social media. In pull, marketing companies are seeking to capture the interest of customers who are already seeking information, advice, a product or a service. A marketer's job is to try to understand: What do the customers want, need and expect is to know—and to create marketing programs and customer experiences that reflect this insight. Marketing practitioners should learn how to build and exploit customer understanding to identify, attract, engage and retain customers at each stage of their journey, from lead generation and campaign optimisation to customer experience (CX) design.

The previous section describes how BD and BDA are helping marketers face new challenges derived from digital marketing. The term used to describe the impact of the technologies and the nature of the new digital world is 'Big Data'. It is defined as [40]:

> [...] as a general process that involves the collection, analysis, use, and interpretation of data for various functional divisions to (or 'intending to') gaining actionable insights, creating business value, and establishing competitive advantage. We can say it is a more advance system compared to the traditional marketing information system where both structured and unstructured data can be captured and analysed. BD also facilitates to identify the intensity of the creation and movement of data and also the variability, duplication, inconsistency, quality and compatibility issues that are emerging on a large scale [41].

2.2   Combing IT Software with Artificial Intelligence

Customers are more interactive and companies should develop the philosophy of engaging their customers to better understand their needs and develop products accordingly. Nearly all organisations use digital marketing (1) to create brand awareness (2) to inform customers about new products (3) to reinforce their position in the market and this can be achieved by (1) brand reinforcement (2) lead conversation and nurturing (3) customer conversation (4) customer service (5) customer upsell and (6) passionate subscribers [18]. Further research needs to be conducted of how to use artificial intelligence to better segment and target the market. Artificial intelligence can be for effective content management for forecasting, optimisation, expert support, adaptive guidance (for customers/users), and fixing mistakes (detected along marketing process). In addition, artificial intelligence is a way where organisations can intelligently evaluate social media and gain marketing insight meaning a better way to track customers and users over the web.

3 Conclusion and Future Research

Integrating marketing and information technology has definitely bring success to many organisations be it the government, business or consumers. Innovation will keep on emerging and marketers should work together with these developments to better define and deliver customer value. Maintaining and adopting a culture where customers are considered as the main pillar in organisations are seen to bring success to many organisations. Customer engagement and customer relationship management have always been a massive topic of discussion for many practitioners and developing this culture of customer management takes time and determination. Just developing new product or new software is useless if it not used properly. Marketers' role is to devise strategies and successful strategies are built on data and insights. Digital disruption is an unstoppable force, driving marketers to place bets, take risks, fail fast and scale success—all while mastering new skills and technologies.

Acknowledgements The authors thank Professor Binod Kumar Pattanayak for his support and professional comments.

References

1. https://which-50.com/martech-adtech-may-not-convergein. Accessed on 20th May 2018
2. Woon, K.C., Mathew, S., Christopher, W., Lui, Vincent: B2B e-marketplace: an e-marketing framework for B2B commerce. Market. Intell. Plann. **28**(3), 310–329 (2010)
3. Berners-Lee, T., Hendler, J., Lassila, O.: The semantic web. Sci. Am. **284**(5), 35–43 (2001)
4. Garrigos-Simon, F.J., Alcamí, R.L., Ribera, T.B.: Social networks and Web 3.0: their impact on the management and marketing of organizations. Manag. Decis. **50**(10), 1880–1890 (2012)
5. Eason, J.A.: Online social networks: how travelers can benefit from using interactive sites. Healthcare Traveler **15**(5), 18–27 (2007)
6. Thackeray, R., Neiger, B.L., Hanson, C.L., McKenzie, J.F.: Enhancing promotional strategies within social marketing programs: use of Web 2.0 social media. Health Promot. Pract.ce **9**(4), 338–343 (2008)
7. Kennedy, G., Dalgarno, B., Gray, K., Judd, T., Waycott, J., Bennett, S., Maton, K., Krause, K.-L., Bishop, A., Chang, R., Churchward, A.: The Net Generation are Not Big Users of Web 2.0 Technologies: preliminary Findings (2007)
8. Lefebvre, R.C.: The new technology: the consumer as participant rather than target audience. Soc. Market. Q. **13**(3), 31–42 (2007)
9. Crotts, J.C., Mason, P.R., Davis, B.: Measuring guest satisfaction and competitive position in the hospitality and tourism industry: an application of stance-shift analysis to travel blog narratives. J. Travel Res. **48**(2), 139–151 (2009)
10. Li, X., Ren, X., Zheng, X.: Management of competition among sellers and its performance implications for business-to-business electronic platforms: dynamic analysis by VAR model. Nankai Bus. Rev. Int. **6**(2), 199–222 (2015)

11. Evans, D.: Social Media Marketing: an Hour a Day. Wiley, Indianapolis, IN (2008)
12. OECD. (2007). Participative Web and User-Created Content: Web 2.0, Wikis, and Social Networking, Organisation for Economic Co-operation and Development, Paris. Available at: www.biac.org/members/iccp/mtg/2008–06-seoul-min/9307031E.pdf. Accessed 14 May 2018
13. Smith, A.N., Fischer, E., Yongjian, C.: How does brand-related user-generated content differ across YouTube, Facebook, and Twitter? J. Interact. Market. **26**(2), 102–113 (2012)
14. Holliman, G., Rowley, J.: Business to business digital content marketing: marketers' perceptions of best practice. J. Res. Interact. Market. **8**(4), 269–293 (2014). https://doi.org/10.1108/jrim-02-2014-0013
15. Gandour, A., Regolini, A.: Web site search engine optimization: a case study of Fragfornet, Library Hi Tech News, 2011, pp. 6–13, page 7 (2011)
16. Kammerer, Y., Gerjets, P.: Effects of search interface and internet-specific epistemic beliefs on source evaluations during web search for medical information: an eye-tracking study. Behav. Inf. Technol. **31**, 83–97 (2012). https://doi.org/10.1080/0144929X.2011.599040
17. Pulizzi, J., Barrett, N.: Get Content, Get Customers. Voyager Media, Bonita Springs, FL (2008)
18. Rose, R., Pulizzi, J.: Managing Content Marketing. CMI Books, Cleveland, OH (2011)
19. Rowley, J.: Information marketing: seven questions. Libr. Manage. **24**(1/2), 13–19 (2003)
20. Michaelidou, N., Siamagka, N.T., Christodoulides, G.: Usage, barriers and measurement of social media marketing: an exploratory investigation of small and medium B2B brands. Ind. Mark. Manage. **40**(7), 1153–1159 (2011)
21. Peppers, D., Rogers, M.: Managing Customer Relationship: a Strategic Framework (2011)
22. Yadav, M.S., De Valck, K., Hennig-Thurau, T., Hoffman, D.L., Spann, M.: Social commerce: a contingency framework for assessing marketing potential. J. Interact. Market. **27**(4), 311–323 (2013)
23. Kaplan, A.M., Haenlein, M.: The fairyland of second life: about virtual social worlds and how to use them. Bus. Horiz. **52**(6), 563–572 (2009)
24. Hennig-Thurau, T., Hofacker, C.F., Bloching, B.: Marketing the pinball way: understanding how social media change the generation of value for consumers and companies. J. Interact. Market. **27**(4), 237–241 (2013)
25. Boyd, D., Ellison, N.: Social Networks Sites: definition, History and Scholarship (2007)
26. Constantinides (Ed.): Consumer Information Systems and Relationship Management: design, Implementation, and Use, pp. 51–73. IGI Global (2013)
27. Smith, P.R., Chaffey, D.: eMarketing eXcellence, 2nd edn. Butterworth Heinemann, Oxford (2013)
28. Trusov, T., Bodapati, A.V., Bucklin, R.E.: Determining influential users in internet social networks. J. Market. Res. **47**(4), 643–658 (2010)
29. Liang, T.P., Lin, C.Y., Chen, C.N.: Effects of electronic commerce models and industrial characteristics on firm performance. Indust. Manage. Data Syst. **104**(7), 538–545 (2004)
30. Intezari, A., Gressel, S.: Information and reformation in KM systems: big data and strategic decision-making. J. Knowl. Manage. **21**(1), 71–91 (2017)
31. Davenport, T.H.: Analytics 3.0. Harvard Bus. Rev. **91**(12), 64–72 (2013)
32. Davenport, T.H.: Big Data at Work: dispelling the Myth, Uncovering the Opportunities. Harvard Business Review Press, Boston, MA (2014)
33. Davenport, T.H., Prusak, L.: Working Knowledge: how Organizations Manage What They Know. Harvard Business Review Press, Boston, MA (2000)
34. Alavi, M., Leidner, D.E.: Review: knowledge management and knowledge management systems. Conceptual Found. Res. Issues **25**(1), 107–136 (2001)
35. Persaud, Ajax, Azhar, Irfan: Innovative mobile marketing via smartphones: are consumers ready? Market. Intell. Plann. **30**(4), 418–443 (2012). https://doi.org/10.1108/02634501211231883
36. Karjaluoto, H., Lehto, H., Leppäniemi, M., Jayawardhena, C.: Exploring gender influence on customer's intention to engage permission-based mobile marketing. Electron. Markets **18**(3), 242–259 (2008)

37. Delen, D., Demirkan, H.: Data, information and analytics as services. Decis. Support Syst. **55**(1), 359–363 (2013)
38. Hofacker, C.F., de Ruyter, K., Lurie, N.H., Manchanda, P., Donaldson, J.: Gamification and mobile marketing effectiveness. J. Interact. Market. **34**, 25–36 (2016)
39. Kunz, W., et al.: Customer engagement in a big data world. J. Serv. Mark. **31**(2), 161–171 (2017)
40. Akter, S., Fosso Wamba, S.: Big data analytics in e-commerce: a systematic review and agenda for future research. Electron. Markets **26**(2), 173–194 (2016)
41. Anuradha, J.: A brief introduction on big data 5Vs characteristics and Hadoop technology. Procedia Comput. Sci. **48**, 319324 (2015)

Use of Social Media for Improving Student Engagement at Université des Mascareignes (UDM)

Randhir Roopchund, Vani Ramesh and Vishal Jaunky

Abstract The research responds to the global trend of using technology in higher education for improving student engagement and student satisfaction. The research ontology adopted is that students may benefit from a more interactive approach of learning in classes. The methodology used is the use of a pre-designed questionnaire to test the readiness of students and faculty members for the use of social media. Three components were extracted based on EFA which are namely social media as a facilitator, improving learning proficiency and trust in data security. The research outlines the benefits, risks and challenges for adopting SNSs for improving customer satisfaction and loyalty. It is important to note that social media tools are part of the Web 2.0 interactive and intelligent system of communication being used in different fields.

Keywords Social media · Social networking sites (SNSs)
Informal scholarly communication · Higher education and social learning

1 Introduction

The present research seeks to explain the importance, challenges, and problems in using social media in Mauritian Higher education. Mauritius has a high internet penetration rate of almost 70% with more than 700,000 Facebook users. This proves that people of different ages use social media in their day to day lives. Mauritius ambitions to be an education hub in future and consequently, the use of technology

R. Roopchund (✉) · V. Jaunky
Université des Mascareignes, Beau Bassin-Rose Hill, Mauritius
e-mail: rar11@aber.ac.uk

V. Jaunky
e-mail: vishal.jaunky@ltu.se

V. Ramesh
Reva University, Bengaluru, India
e-mail: sarada889@yahoo.in

© Springer Nature Singapore Pte Ltd. 2019
S. C. Satapathy et al. (eds.), *Information Systems Design and Intelligent Applications*, Advances in Intelligent Systems and Computing 863,
https://doi.org/10.1007/978-981-13-3338-5_2

for enhancing learning is of paramount importance. Recently, the launch of a social media guide by the Minister of Education heralds the importance of social media for the Mauritian youth. The research relies primarily on the opinions provided based on a survey carried out with students at Université des Mascareignes (UDM).

2 Research Problem and Research Objectives

The present research addresses the research gap of using technology in higher education in the Mauritian context. There is little research on the prospects and challenges of using social media in higher education. However, Mauritius ranks first on Internet penetration in Africa and also has the highest number of social media users if compared to the population. The research is also a logical pursuit after exploring the use of customer relationship framework in higher education in prior research. However, the research will be context specific as the questionnaire will be administered to students of Université des Mascareignes which is a Public University.

The objectives of the research are to:

1. Understand the use of social media in Higher Education
2. Analyse how social media may be used to increase student engagement at UDM
3. Develop some recommendations to improve student engagement.

2.1 Hypothesis and Research Framework

It is hypothesized that students have positive a negative attitude towards the use of social networking sites. The hypothesis can be formally stated as:

H0: Students have a negative perception about the use of Social Networking Technologies in Higher education
H1: Students have a positive attitude towards the use of social networking technologies in Higher education

3 Literature Review

3.1 Understanding Social Media

Social media refers to media used to empower social collaboration. For the present study, the term social media technology (SMT) alludes to electronic and portable applications that enable people to make, connect with, and share new user-generated or existing content, in computerized situations through multiway communication.

Social networking sites are the term utilized as a non-exclusive term for every single social medium and PC interceded correspondence, including yet not restricted to Facebook, Twitter, LinkedIn and Myspace, including social networking sites of Cyworld, Bebo and Friendster. Ellison and Boyd [1] characterize interpersonal organization locales as electronic administrations that enable people to develop profiles, show client associations, and inquiry traverse inside that rundown of associations.

3.2 Increasing Use of Social Media

There is an increasing use of social media in Mauritius (more than 700,000 Facebook users). From the diagram below, the most commonly used social media are Facebook followed by Pinterest and Twitter (Fig. 1).

Social media in Mauritius has witnessed tremendous growth with the rise in the number of users and their participation spent their time on Social Networking Sites. Broadband too has seen a noteworthy development in 2012 with 400 thousand clients (57%) are educated youngsters. Mauritians who have relocated to Australia, Canada, and European nations for training and career are in constant touch with family and companions which prompted the ascent in utilization of informal communities such as Facebook. This trend has expanded much more with the process of globalization triggering Mauritian to use more technology such as mobiles, PCs, other registering gadgets like PCs, tablets. Most prevalent social networking media platform in

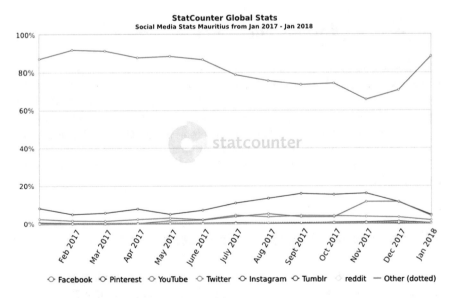

Fig. 1 Use of social networking sites in Mauritius

Mauritius is Facebook as around 58% dominated by male and for the most part, the students and professionals in the age group of 14–24 are using the social media.

3.3 Social Media in Higher Education

The Hon Minister of Education for Mauritius Mrs. Leela Devi Dookun-Luchoomun [13] stressed the growing importance of technology for Mauritians while highlighting its developmental role in the higher education sector. The Minister also highlighted that technology will entail the transformation of educational institutions into ICT-driven ones. The increasing use of tablets and other hi-tech materials may eventually contribute to making this dream a reality.

At an international level, some universities are already using podcasts, video blogs, and webcasts to share the work of students, faculty and alumni with the public at large. There is increasing use of Tweeting—the function of a status update (tweets)—is increasingly being used for discussion by students and faculty members [9]. Besides, numerous NCAA members have encouraged athletes, mentors and other staff to use Twitter and Facebook as platforms to connect with fans [20]. Faculty members have also used blogs as a pedagogical strategy. Some recent studies have investigated the use of blogs in academic disciplines including the sciences [4], language learning [8]), teacher education [7, 16] and business [21]. Faculty has also created Facebook profiles to connect with their students in a more personable and informal space [19].

3.4 Web 2.0 and Student Engagement in Learning

The advent of Web 2.0 applications, collectively known as social media, presents schools and universities with the opportunity to go beyond traditional delivery formats and develop learner-centred personalized learning environments [18]. It has significant impact for teaching and learning because they are strictly web-based and typically free, support collaboration and interaction, enhance students' learning experiences through customization and personalization and provide rich opportunities for networking. They are also highly responsive to the user and is highly student-centred approach [5].

3.5 Facebook and Blogs for Enhancing Student Engagement

The use of Facebook and blogs are increasingly being used in the higher education landscape [6]. Facebook is not only a network of contacts but may also be used for communication, sharing of videos and also chat. Gee [10] compares Facebook to affinity spaces where people may acquire social and communication skills. Facebook has additionally been utilized for formal learning with academics setting up open or private gatherings for classroom practices [14]. Through the use of blogs, lecturers have the chance to make the material open for subsequent reflection and investigation thus improving the overall learning experience. Despite the growing evidence of the use of Facebook and blogs in educating and learning, there is lack of published research linking the use of these tools for improving student engagement. Studies that exist stem to a great extent from fields outside education, for example, sociology, human sciences and communication studies [3, 11]. However, existing research in the field of education demonstrates that students use social media such as Facebook and blogs on a daily basis and trust that more utilization of such advances in scholarly settings would prompt greater planning and commitment [6].

4 Research Methodology

This research design used a survey research methodology wherein a questionnaire was used as the main research tool. Questions included demographics, challenges and perceptions. The respondents were asked to rank and comment on different questions. This was used to determine the weight or the importance of each challenge and perception. The questionnaires included a combination of both structured and semi-structured questions. The questionnaire was validated by pretesting it with a few undergraduate students and making some changes to a few questions.

5 Analysis and Results

5.1 Profile of Participants

Table 1 provides the profile of participants (58 in total) in terms of year of study and also the different age groups. We find that students from Year 1 to Year 3 students participated in the survey. The highest percentage is for the final year students (more than 40%). In terms of gender participation, there were 32 as compared to 26. Most of the students are in the age bracket of 18–25 years.

5.2 Social Media Presence

Almost all students (98.3%) are on social media based on the survey results. This is not surprising due to high internet connectivity and this corresponds to the statistics of high internet users (more than 63% overall). We also have a 50% penetration rate for Facebook users. The high rate of social media presence is also explained by the

Chart. 1 Social media landscape at UDM

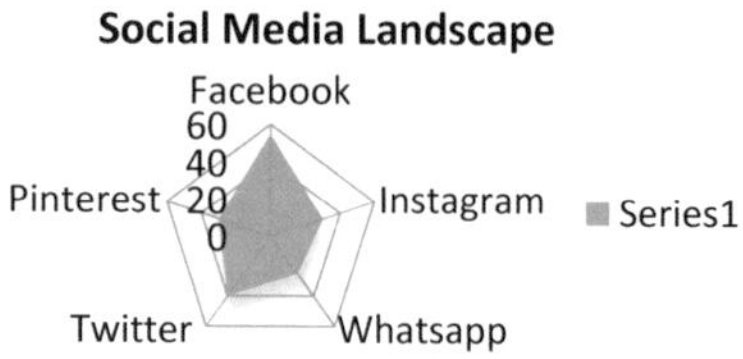

Social Media Landscape

fact that the survey has been conducted with young students who are technologically savvy. The findings corroborate with the increasing number of users as identified by the increasing internet users by global statistics. The number of Internet users in the Republic of Mauritius in the past decade has boomed. In fact, the number of internet users in the island nation has undergone a ten folded growth. Going from 30,000 users in 1998 to 290,000 12 years later with a penetration rate of over 22% [2] (Chart 1).

From the above, we find that most of the users are from Facebook, followed by Twitter and Pinterest. This is in line with the emerging trends in teaching pedagogy such as connectivism [17]. Another learning theory which suits the teaching and learning needs of digital learners is the Communities of Practice (COPs). Piktialis and Greenes [15] define Gen-Y as a person who 'values group and team learning, constructing understanding from many sources as opposed to a single authority'. A Community of Practice (COPs) is hence a natural fit to motivate and enhance the learning of Gen-Y. The statistics confirm the popularity of Facebook, Instagram and Pinterest as the most used social media. However, there may be a change in future with the problems and the data privacy issues being raised at the international level (Chart 2).

From the above diagram, we find that the students are very actively engaged on social media. They use social media on a daily basis and consequently, there is high scope for using it in the field of education. Almost 50 students claimed that they use social media on a daily basis.

Table 1 Profile of participants-year of study

Gender × year of study cross-tabulation

Count

		Year of study			Total
		Year 1	Year 2	Year 3	
Gender	Male	6	9	11	26
	Female	12	7	13	32
Total		18	16	24	58

Chart. 2 Frequency of use of social media

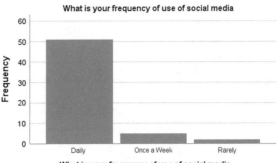

5.3 Use of Social Media for Teaching at Université des Mascareignes

The most interesting findings from the survey are that 85% of students are of the opinion that social media may be used as a teaching tool. Social media has traditionally been used as a means of making friends, networking and not as a medium of education strategies. Minocha [14] makes reference to the use of Social media in higher education for different purposes. Out of the 85% who are in favour, they believed that Facebook may be used for communicating with students and sharing resources. However, the use of Facebook as a discussion forum and Facebook page has not been selected (Chart 3).

The findings may be compared to the study carried out by [12] where students provided the following responses to the applications of social media technologies (survey carried out with 80 students). It should be explained that the sample is higher than the present study.

Chart. 3 Ways of communication

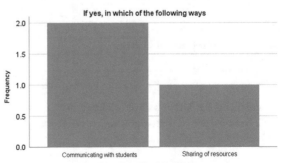

How do you use SMTS to support education	No of responses	% of responses
Assignments/project collaboration	81	97.59
Sharing of documents	76	91.57
Knowledge/information sharing	69	83.13
Activities/events updates	67	80.72
Sourcing of information	60	72.29
Communicating with professors and peers	73	87.95

5.4 Social Media and Student Engagement

The following variables have been used to assess student engagement through social media in higher education. We find that the mean is more than 3 implying that students are quite engaged and active (on a scale of 5). The grand mean for all the variables is 3.35. A factor analysis has been carried out to extract three main factors that may explain student engagement through social media. The three components will be explained later.

Total variance explained									
Component	Initial eigenvalues			Extraction sums of squared loadings			Rotation sums of squared loadings		
	Total	% of variance	Cumulative %	Total	% of variance	Cumulative %	Total	% of variance	Cumulative %
1	5.447	41.900	41.900	5.447	41.900	41.900	3.978	30.601	30.601
2	1.730	13.306	55.206	1.730	13.306	55.206	2.406	18.511	49.112
3	1.167	8.977	64.183	1.167	8.977	64.183	1.959	15.071	64.183
4	0.900	6.922	71.105						
5	0.847	6.518	77.623						

Based on the above factor loadings and eigen values, we find that three factors have been extracted with eigen values more than 1. The three components that were extracted have been named as follows:

Component 1	Social media as facilitator	Implies that social media is a facilitating or support tool. It can enhance engagement of students based on the principle of connectivism [17]
Component 2	Security and safety	Students are concerned about privacy and security issues. Students are reluctant to communicate with lecturers
Component 3	Learning proficiency	It increases interest in learning as it is more fun [12]

5.5 Components of Social Media Engagement in Education

6 Recommendations

This study provides a positive view about the perception of students for the use of social media for improving student engagement at UDM. Most of the students are using Facebook as social media platform on a daily basis. Based on the EFA, three components have been extracted which may help to improve engagement which are namely social media as facilitator and learning and proficiency. However, students are concerned about the safety and security issues with the use of social media.

References

1. Boyd, D., Ellison, N.B.: Social network sites: definition, history, and scholarship. J. Comput.-Mediated Comm. **13**, 211–230 (2007)
2. Bertrand, A.: Using Social Networks and Media to Fight for Market Supremacy in Mauritius. http://socialmedia-mauritius.com/2011/07/social-networksmedia-market-supremacy-mauritius/. Accessed 5th April 2018 (2011)
3. Bosch, T.E.: Using online social networking for teaching and learning: Facebook use at the university of Cape Town. Communication **35**(2), 185–2009 (2009)
4. Brownstein, E., Klein, R.: Blogs. J. College Sci. Teach. **35**(6), 18–22 (2006)
5. Bryant, T.: Social software in academia. EDUCAUSE Quarterly **29**(2), 61–64 (2006)
6. DeGennaro, D.: Learning Designs: an Analysis of Youth-Initiated Technology Use. http://net.educause.edu/ir/library/pdf/ERM0837.pdf. Accessed 4th March 2018 (2008)
7. Deng, L., Yuen, H.K.: Connecting Preservice Teachers with Weblogs: design Issues. Paper presented at the World Conference on Educational Multimedia, Hypermedia and Telecommunications (Ed-Media 2007), Vancouver (2007)
8. Ducate, N.A., Lomicka, L., Lord, G.: Using computer-mediated communication to establish social and supportive environments in teacher education. CALICO J. **22**(3), 537–566 (2005)
9. Dunlap, J.C., Lowenthal, P.R.: Tweeting the night away: using Twitter to enhance social presence. J. Inf. Syst. Educ. **20**(2), 129–135 (2009)
10. Gee, J.P.: Social linguistics and literacies: ideology and discourse. London: Falmer. Opportunity to learn: a language-based perspective on assessment. Assess. Educ. **10**(1), 27–46 (2003)

11. Greenhow, C., Robelia, B., Hughes, J.E.: Learning, teaching, and technology in a digital age: web 2.0 and classroom research: what path should we take now? Am. Educ. Res. Assoc. http://edr.sagepub.com/cgi/content/abstract/38/4/246. Accessed on 24th May 2018 (2009)
12. Lim, J.S.: Investigating the Use and Perceived Effectiveness of Social Media for Informatics Programs in the Malaysian Higher Education Context. Doctor of Philosophy thesis, School of Electrical, Computer and Telecommunications Engineering, University of Wollongong (2015)
13. Luchoomun, L.: Education Minister Launches Book Promoting Proper Usage of Social Media. http://www.govmu.org/English/News/Pages/Education-Minister-launches-book-promoting-proper-usage-of-social-networks.aspx. Accessed on 24th April 2018 (2017)
14. Minocha, S.: A study of the effectiveness use of social software to support student learning and engagement. JISC. http://www.jisc.ac.uk/whatwedo/projects/socialsoftware08.aspx. Accessed on 15th Jan 2018 (2009)
15. Piktialis, D., Greenes, K.A.: Bridging the Gaps: how to Transfer Knowledge in Today's Multi-generational Workforce. The Conference Board. http://tech.tac-atc.ca/private/education/pdfs/Multigenerational.pdf. Accessed on 30th April 2018 (n.d.)
16. Ray, B., & Coulter, G.: Reflective practices among language arts teachers: the use of weblogs. Contemp. Issues Technol. Teacher Educ. 8(1), 6–26 (2008)
17. Siemens, G.: Connectivism: a learning theory for a digital age. Int. J. Instr. Technol. Distance eLearning. http://www.itdl.org/journal/jan_05/article01.htm (2005)
18. Sigala, M.: Integrating web 2.0 in e-learning environments: a socio-technical approach. Int. J. Knowl. Learn. 3(6), 628–648 (2007)
19. Sturgeon, C.M., Walker, C.: Faculty on Facebook: confirm or deny? Paper presented at the Annual Instructional Technology Conference, Lee University, Cleveland, TN (2009)
20. Watson, G.: Coaches, Colleges Explore New Frontier (Commentary). http://sports.espn.go.com/ncaa/news/story?id=4308218. Accessed on 18th March 2018 (2009, July 6)
21. Williams, J., Jacobs, J.: Exploring the use of blogs as learning spaces in higher education. Australas. J. Educ. Technol. 20(2), 232–247 (2004)

Conceptualising an Online Resource Platform for Business Students at Université des Mascareignes

Nirmal Kumar Betchoo

Abstract Since the creation of the Université des Mascareignes in 2012, there has been an urge to develop online learning resources that students might need for their instruction and research during their university years. So far, resources abound in the form of textbooks, CDs and online resources like EBSCO along with an online library facility from Université de Limoges (Unilim, France). Business students in human resource management, banking and related disciplines need up-to-date resource material for their immediate use and application to their studies. This research purports the need to have a well-structured learning platform targeted specifically to student needs. By creating an online and intelligent platform, students might have access to the appropriate research undertaken by former students, resource material from local newspapers and journals and also reading material to supplement their work. Such creation of an online resource platform aims at improving the access to high-quality premium research material but also help in enhancing the quality of research. Through a demanding process of requiring updating and maintenance of resource material, the database will be a valuable tool for students and alumni alike.

Keywords Online resource · Intelligent platform · Benefits · Challenges
University

1 Introduction

University students are bound to engage themselves in research during their time they stay in campus. This is an inclusive component in the activities they should undertake in their student life when they are on a campus. Research is done at all levels ranging from the presentation of work, discussion, seminars and talks, and particularly dissertations that count up as the most important personal contribution of the student at the end of his studies. It is the role for contemporary universities to

N. K. Betchoo (✉)
Université des Mascareignes, Beau Plan, Pamplemousses, Republic of Mauritius
e-mail: nbetchoo@udm.ac.mu

© Springer Nature Singapore Pte Ltd. 2019
S. C. Satapathy et al. (eds.), *Information Systems Design and Intelligent Applications*, Advances in Intelligent Systems and Computing 863,
https://doi.org/10.1007/978-981-13-3338-5_3

have a responsibility to transcend traditional disciplinary limitations and develop a culture of academic enterprise and knowledge entrepreneurship [1]. The Université des Mascareignes (UdM) created in 2012 is no exception to such a concept and has always devoted itself to research based on its previous engagement as a polytechnic institution for the past 20 years. Contemporary students use social network platforms, the Internet and related online sources to search for data prior to and during research. Pintrich and Shunk [2] state that students with high self-efficacy beliefs about their learning capability tend to seek help more frequently because they are less likely to interpret their need for help due to the lack of ability. However, at the UdM, there appears to be a certain bottleneck in their progress. For instance, the same research themes recur annually and similar sources of data are being sought. This places the problem on how to develop a system that might be of greater assistance to students and faculty while better addressing their immediate needs. For instance, material on business is available from external various sources but this seems to be far from the existing reality. Research based on business and social sciences must also have data that comes coming from local sources. This is where the initial problem arises. There are either too few local researchers undertaking targeted research and this information is not easily available. Additionally, newspapers do provide sufficient local resources for students but this information is essentially in French and needs to be translated in the subtlest English form to address the students' needs. In this context, the effective translation of educational materials is an important tool like having the materials available to teach people what they will need to know, but in their own language, can be the difference [3].

1.1 Aims and Objectives of the Research

The key aim of this research is to conceptualise an online resource platform that could particularly address the needs of business students at the UdM. This platform would enable students to have direct access to stratified and well-maintained data that would address their research needs be it for assignments, coursework and dissertations.

The objectives would be the following:

(1) The creation of a structured online research platform that is targeted to the needs of business students at the UdM.
(2) The development of databases comprising a selection of local newspaper articles, legislation, research thematic and legislation regarding selected business areas.
(3) The creation of online versions of best dissertations and coursework that new students might refer to and adapt to their new research.
(4) The implementation of suitable teaching resource materials that students and academics might benefit from using them.
(5) The implementation, updating and maintenance of data so that the resource platform could be a useful research platform for all users.

1.2 Problem Statement

Resources for students and staff need to be developed in a university so that it emerges as a reference in education and research. This is the hallmark for any university willing to excel in the provision of quality education. The UdM is no exception as it is developing as a major research-led university catering for a niche of students and offering programmes that create some contrast with other public universities. So far, the UdM has enjoyed good employability rates from its students which has subsequently developed a good image of it from the public and society. With its low student to teacher ratio, it has also been a focal point for tailor-made learning approach for its students. However, as a small and developing university, the UdM has not yet developed an online resource platform that could be of high practical use for its students. So far, in a similar way to other universities, it relies on established external resource providers like EBSCO, Emerald Insight, and recently access to the research database of Université de Limoges (Unilim) with which it partners since 2012.

The information available from the above-cited sources is extensive and quite rich. The argument is that such data is from external sources that do not adequately address the Mauritian situation that is unique. Currently, the availability of research and development (R&D) statistics for developing countries is uneven and scarce [4]. Mauritius has its own specificity regarding practices in business as well as legislation. Certain contexts and theories that are applicable externally might not really apply to the local context. It is believed that access to foreign data, though useful, limits the opening of Mauritian students to local research and the context where they are learning.

There is an ongoing argument for developing research material that focuses on Mauritius. This is a utopia as the country has its own constraints. Locally produced research is limited and might not even exist in certain areas. Publications based on the local context are either rare or inaccessible to students. Further, newspapers might provide good quality recent information on particular topics but these are in French as a majority of dailies are published in French. Translation of articles needs to be done with subtlety in order to keep them as crisp and factual. Translation, if it is to capture the essence of a text, must take into account every thread of implication, with its relative weight understood both within the message itself [5].

This is where the need for an online resource that is tailor-made for Mauritian students arises. Since the course offer is unique, one might also ask why whether the resources provided or developed by the UdM could be unique as well? The conceptualisation of such a platform might be tough and demanding but a protocol that allows for a diligent work of developing such a platform might be of immense use for students and researchers and even add value to the quality of work undertaken at the UdM.

2   Brief Literature

Universities are expected to provide comprehensive education to prospective students. The reputation of a university might be dependent on the professors servicing the courses as well as the quality of programme and course content that they deliver. Also, a university's reputation depends on its alumni that have studied over there, developed learning competencies and becoming employable. The essential argument here is the learning facilities and logistics offered to the students. Mata [6] states that unlike in previous decades, modern teachers recognise the critical importance technology plays in teaching tomorrow's leaders. In today's learning environment, it is important to have logistics and technologies that enhance learning. Berner and Boulware [7] mention that the application of technology should be indeed encouraged and even be incorporated as a routine part of students' daily activities within basic sciences. The existence of the Internet and databases are starting points for students to easily gain access to the material. Zhang et al. [8] state that e-learning has become one of the fastest moving trends in education and poses a promising alternative to traditional learning. The time for getting information in hard copy form and through books is partly gone. Further, new information and developments in curricula are taking place and likely to supersede traditional learning methods. This view is supported by Mayer [9] stating that studies have shown that people learn considerably better from a combination of both words and images which technology enables than merely from words alone.

Universities now think of providing online learning platforms to their students. The existence of learning databases like EBSCO, Emerald Insight among others, offer a quick possibility for students to have access to learning material. These are useful in supplementing classroom learning while developing the prowess for individual learning. New learning that is interactive allows students to better grasp knowledge while keeping them in pace with current changes and developments. In the current context, the institution faces a double challenge like understanding how well-designed technology improvements can enhance a student's educational experience and being willing to examine concurrent changes in pedagogical methods [10].

Large amounts of data in particular areas are essential for students today. Such information must principally address contemporary needs. As Vermeulen [11] states, constant technological disruption is the new standard while 'Old world' concepts, models, paradigms and ideas will no longer be relevant. In the field of business, students need to develop the latest knowledge on issues like data mining, management of business information, the evolution of stock markets, international concepts like globalisation and trade agreements. Such data is available in the most appropriate mode from targeted sources. Panel [12] confirms this view by saying that increased implementation of technology will increase students' comprehension of content and development of skills in such areas as analytical reasoning, problem-solving, information evaluation, and creative thinking.

2.1 The Literature Gap

Although literature mentions the importance of learning platforms through developments in information technology, little is said about targeted learning from such platforms. For example, what type of data is suited for business students and how is this organised? Is it enough to say that heaps of information on a particular topic matter? There is then the argument of developing e-learning platforms that comprise a database of prime information. Take, for instance, human resource management. Developments in the field of legislation are ongoing in Mauritius and such information needs to be provided.

Upcoming issues like social partnership might be covered by the local press and yet such information is in French. How could such data be organised in a sequential and logical way to allow students to gain access to such prime material easily? In Mauritius, it is clear that educational and informational data might be missing in certain areas of study or might be simply disorganised. How could a university develop competences in creating a database targeted to the local students and that addresses their imminent needs? It is this argument that empowers the researcher to critically ask the need to conceptualise a tailor-made platform for business students.

3 Conceptualising an Online Resource Platform

Prior to discussing on the relevance of an online resource platform, it is worth noting that the University of Mauritius (UOM) has a virtual learning platform that caters to the needs of its students. It has the intention of consolidating its platform by offering a wider range of data that could not only benefit the public but the Mauritian community at large. The Mauritius Institute of Education (MIE) has its learning platform addressing the needs of primary and secondary level students. Some academic research is deposited on the platform with the view of sustaining interest both from its own staff and external ones. Seen from this perspective, the online research platform might be an opportunity for the UdM to come forward with some similar offering but with a different perspective.

From this standpoint, the first consideration would be the creation of an online resource platform that would particularly address the needs of business students. The premise here is that the study focuses on one category of students and faculty. The online platform should be a dynamic one capable of adequately responding to the needs of the target audience. Such audience comprises all students of the faculty, lecturers and researchers as well as external students who might also benefit from it. A set of actions might be envisaged for the conceptualisation and implementation of the system.

3.1 Classification of Data

It is first important to sort out the type of data content that is needed on the platform and how this will be managed. Different databases have to be developed that cater to specific needs but might have elements to share as and when needed. Data has to be classified into broad categories. For the sake of clarification, one discipline has been mentioned; human resource management. A few possibilities are suggested below:

General Data on Human Resource Management

This concerns information on workforce statistics. They could initially be of an economic nature but since they pertain to employment, wages earned, industry standards, trade union membership, etc. such data is of prime importance to students and researchers who need updated information in research work. HR data is dynamic and subject to constant change annually. Statistics Mauritius reports and other related industry reports might help.

Local Sources of Information

Local sources of information are available from the media. Press articles abound in human resource management, business and economics and are of practical importance. They could be used for reference and case studies. Certain papers like 'Business Magazine' have special sections that could be scanned, 'Le Défi' newspaper has a weekly section 'Monde du Travail' on industrial relations and 'L'Express-Economie' offers weekly insights into employment, labour trends and interviews.

Articles on Human Resource Management

Excellent articles on human resource management may be available from other sites. Such information could be scanned for educational purposes only. Universities do refer to 'The Economist' or 'Financial Times' on excellent business or HR-related articles that serve the needs of the target audience in terms of the quality and excellent level of English language. Such sources are also referenced in dissertations while they provide insight into the latest developments in the above-mentioned disciplines.

Dissertations

Students' dissertations namely the best and the most highly rated ones should be included on this platform. Too often, students undertake very similar dissertations and might be even tempted to copy or plagiarise some of them. The selected dissertations could also serve as models for the improvement of dissertations and could also help upgrade the standards of research. Upcoming interested and quality worthy would be included in this section. According to [13], storage of data related to research projects should be taken seriously from the outset to ensure that data resources are kept safe and are formally archived.

Teaching and Learning Resource Materials

Teaching and resource materials could be useful. For instance, past examination papers could be included in this heading. Alongside, suitable literature and reading material developed by academics might be included. To preserve information integrity and overcome the risk of plagiarising such data, passwords might be specifically developed here and students could use such information until the expiry of their course at the university.

Useful Links With Related HR Websites

This concept is useful but is mainly concerned with links. Good sources like About.com, HR.com, ACAS, etc. are highly useful and important to students. An easy access to such sites on the platform might be a very good initiative in terms of time-saving and ease of navigation for reference. Dedicated websites like Statistics Mauritius, Government of Mauritius (www.gov.mu), MCB Focus, SBM Insights including legislation like the Employment Relations Act 2008 and the Employment Rights Act 2008 should be made available with relevant updates. In a nutshell, the data classified in this section, though not exhaustive, claims that an online platform is highly anticipated and becomes a reference point for all intellectuals concerned with the use and harnessing of information. It also develops an HR Information System of high pedagogical value.

4 Management of Data

An important aspect of the learning system would comprise the management of data. There must be roles allocated to different people to monitor, maintain and manage data in the system. Actually, since responsibilities are not created for such positions, the roles might be fulfilled by computer technicians, lecturers and a systems administrator called upon to see that the platform operates in an optimum way. There will be no paid positions so long as new duties are created and people are specially assigned such tasks. The general responsibilities would be the following:

Systems Administrator

A systems administrator might have to look after the effective running of the platform and the use of the system. He/she might consider security issues pertaining to the system, permission to input, retrieve and update data including the overall management of the information system. Data content and restrictions for use could be under his control.

Computer Technician

A computer technician would be responsible for systems maintenance. Updating, deleting and amending data will be his main activity. The technician must be of help to academic staff who have information to input to the system including their classification like newspaper excerpts, external websites, etc. He/she will also have the responsibility to protect the system with antiviruses.

Academic Staff

Academic staff will comprise the core of the e-platform. They will be responsible for inputting data like lecture notes, newspaper cuttings, recent changes in legislation, developments taking place in the respective fields like HRM or economics. Staff will also ensure that past data could be preserved and made available for reference in research.

4.1 Updating and Maintenance of the System

Once the system has been set up and implemented, the key argument will be in the maintenance and updating of the system. Agreed that a 6-month moratorium is given to have all the key elements within the system installed, there should be regular updating of the system. This will come from queries asked by users and technical problems that they face. These should be dealt with rapidly and meet the expectations of the users.

As an information system, the platform should be constantly updated with new articles coming, new concepts in business being developed and new author contributions from newspapers and related sites. All legislation pertaining to the subject area needs to be updated as well. Tang and Tseng [14] just emphasise that librarians as resource persons should prepare themselves to be more familiar with the advanced information technology skills involved in the e-learning environment that can help students in solving basic technology problems during reference sessions.

4.2 An Intelligent System

If the short-term viability of the system rests on the usage and frequency of visits of users, the long-term objective must be sought as well. The conversion of the e-platform to an intelligent system cannot be undertaken overnight but certain expectations are possible.

Data Mining

In the longer run, data mining could be an eventuality. The Economic Times [15] defines data mining as 'effective data collection and warehousing as well as computer processing'. Given the wealth of research themes developed from the different sources, it is possible to envisage a data mining possibility with the use of algorithms to develop the potential of using cross-sectional data and creating new topics and possibilities for research. This can bring newer ideas to the table but also prompt users and researchers to learn about new topics for research. This could be an interesting idea for prospective Master's and Ph.D. students who need to come up with propositions that are new contributions and insights into research.

Big Data

Big data management might be considered a long-term eventuality for students and researchers. Research possibilities now offered by the Mauritius Research and Innovation Council (MRIC) and the Tertiary Education Commission (TEC) along with external bodies might require big data management. Ellingwood [16] states that big data systems are exclusively suited for surfacing undecipherable patterns and providing insight into actions that are impossible to find through conventional means. This comes from a range of cross-sectional data with possibilities for meta-analysis. Although the perspective is farfetched, nothing states that such a possibility could not be transformed into real data analysis which might also leverage the quality of data treated within the learning platform. To support this paper's key issue on online learning platform, Means et al. [17] purported that their meta-analysis found that, on average, students in online learning conditions performed better than those receiving face-to-face instruction.

4.3   Using the Online Resource Platform

The biggest asset of the learning platform comes from the usage of the system. It has not been created to remain idle and not address the needs of its incumbents. Rather, students should be invited to use the system and this will create some dynamics in information use and management. Second, researchers should be invited to use the system as well as external users like external students and researchers who could benefit from it making the system a reference point. There is no pay level to be associated with the system and the cost of running will be incurred by the UdM. Care should be taken to see that proprietary rights are respected and that data is used for non-commercial purposes. Kallerberg [18], however, claims that the re-use of personally identifiable might require the consent of the research subjects. From the technical side, the risk of hacking should be estimated and backup systems must exist to ensure the perennial nature of the information system.

5   Existing Issues and Challenges

The main challenge comes from the marketing of the learning platform and efforts to create awareness. It is accepted that students will be the initial users but there might be a need to target wider audiences. Word-of-mouth technique will work but the university website must be of great help in channelling the information to potential users. Some aspects of virtual learning and taster courses might well bring in interested parties from the outside. Interest from users must be developed. To avoid idleness of the website, there should be updates sent to the users through social platforms like Facebook, Pinterest, WhatsApp, etc.

From the technical side, data duplication must be avoided. Stale information with figures dating from the past is easy prey for users to discard using a platform. Updated and recent information are essential for the survival of the system. Compilation of new data and certain statistical analysis could be attractions for users. Also, plagiarism must be scorned off. Viorica-Torii and Carmen [19] emphasise that if these technologies are being used inadequately, they may lead to negative effects, plagiarism and the copying of the themes and projects, without the implication of any intellectual learning effort.

Easy navigation, quick use of the system as well as constant feedback are direly needed to live up to the challenges of the users. It is anticipated that the system should be a trigger to encouraging other departments within the UdM to develop similar applications.

6 Conclusion

The creation of an online learning platform is a must for a new university like the Université des Mascareignes set up in 2012. If the institution boasts providing joint certification with Université de Limoges, it must also display such academic and intellectual prowess by developing an online learning platform. It is not merely speaking of a standardised information system but a targeted one aiming at addressing the immediate needs of students, researchers and external users. This virtual platform might showcase how the university stands out from others and how it addresses effectively the needs of its students and target audiences by developing material that is fully appropriate and targeted to their needs. Dunn and Griggs [20] purport that the use of such educational strategies and technologies in the teaching activity allows teachers to learn how to use learning styles as a cornerstone of their instruction and students can learn to capitalise on their learning style strengths. In a country where related local research information is missing and inadequate, this platform will serve as a long-term enviable learning database that showcases the exacting quality of teaching. Redding [21] might well sum it up by acknowledging that in a mixed-learning approach, technology is not perceived as a replacement for the traditional classroom, but rather as a powerful tool to enhance what is already proven to be an effective pedagogy.

References

1. Crow, R.: What is the role of universities in global development? Educ. Global Develop. World Bank (2014)
2. Pintrich, P., Schunk, D.: Motivation in Education, Theory, Research and Application. Merrill Prentice Hall (1996)
3. Huddleston, G.: The Importance of Accurate Translation of Training Materials in the Workplace. languageconnect.net (2016)
4. Gaillard, J.: Measuring R&D in developing countries: main characteristics and implications for the Frascati manual. Sci. Technol. Soc. **15**(1), 77–111 (2010)
5. Vasconcellos, H.: Text and Translation: the Role of Theme and Information. Muriel Pan American Health Organisation, Washington, D.C. (1992)
6. Mata, W.: The Importance of Technology in the Classroom. Centre Technologies (2015)
7. Berner, E., Boulware, D.: Medical informatics for medical students: not just because it's there. Med. Educ. Online (1996)
8. Zhang, D., Zhao, J., Zhou, L., Nunamaker Jr., J.: Can e-learning replace classroom learning? Comm. ACM **47**(5), 75–79 (2004)
9. Mayer, R.: The Cambridge Handbook of Multimedia Learning. Cambridge University Press (2005)
10. Rose, D., Cook, J.: Education and technology. Community Coll. J. **77**(2), Oct/Nov; ProQuest Central (2006)
11. Vermeulen, E.: Education in a Digital Age. www.hackernoo.com (2017)
12. Panel, H.: Digital transformation: a framework for ICT literacy. Educ. Test. Serv. (2002)
13. Given, M.: The Sage Encyclopedia of Qualitative Research Methods. A-L; Vol. 2, M-Z Index (2008)
14. Tang, Y., Tseng, H.: Distance students' attitude toward library help seeking. J. Acad. Librarianship. Elsevier (2014)
15. The Economic Times: Definition of Data Mining. Economic India Times (2018)
16. Ellingwood, J.: An introduction to big data concepts and terminology. Digital Ocean (2016)
17. Means, B., Toyama, Y., Murphy, R., Bakia, M., Jones, K.: Evaluation of Evidence-Based Practices in Online Learning: a Meta-Analysis and Review of Online Learning Studies. Project Report, Centre for Learning Technology (2009)
18. Kalleberg, R.: Guidelines for research ethics in the social sciences, law and the humanities. De nasjonale forskningsetiske komiteer (2005)
19. Viorica-Torii, C., Carmen, A.: The Impact of Educational Technology on the Learning Styles of Students. Elsevier (2013)
20. Dunn, R., Griggs, S. (eds.): Practical Approaches to Using Learning Styles in Higher Education. Bergin & Garvey, Connecticut (2000)
21. Redding, S.: Getting personal: the promise of personalised learning. Handb. Innov. Learn. (2013)

An Unsupervised Machine Learning Analysis of the FIRST Radio Sources

David Bastien and Radhakhrishna Somanah

Abstract The large availability of radio sources from the Faint Images of the Radio Sky at Twenty cm (FIRST) has inspired us to use unsupervised Machine Learning (ML) to do a morphological segmentation of 1000 radio sources. Through techniques like shapelets decomposition, we were able to decompose each radio sources into a series of 256 coefficients that were input into unsupervised ML techniques like Isometric Mapping (ISOMAP) for dimensionality reduction and density-based spatial clustering of applications with noise (DBSCAN) as clustering algorithm. Through this process we were able to identify four groups of sources and 189 outliers. After comparing the segmentation results with our human classification, we found that the method achieved an accuracy of 0.83, with an F_1 score of 0.87. Showing that unsupervised ML could be used to classify images in the radio astronomy domain.

Keywords Unsupervised machine learning · Isomap · DBSCAN · Radio astronomy · Shapelets analysis

1 Introduction

Computer Vision (CV) has found its way in radio astronomy as a result of a rise in data being produced from surveys like Faint Image of the Radio Sky at Twenty cm (FIRST). Machine-learning-based CV has found itself to be effective and less tedious. As compared to industry-driven CV which aims at classifying and detecting everyday objects, radio-astronomy-driven CV aims at the detection and classification of astronomical entities like galaxies or supernovas. CV and ML have also found

D. Bastien
Hydrus Labs Ltd, Lot 45, Armstrong Avenue, Roches Brunes, Rose Hill, Mauritius
e-mail: david@hydruslabs.io

R. Somanah
Universite des Mascareignes, Avenue de La Concorde, Roche Brunes, Rose Hill, Mauritius

R. Somanah
University of Mauritius, Reduit, Mauritius

© Springer Nature Singapore Pte Ltd. 2019
S. C. Satapathy et al. (eds.), *Information Systems Design and Intelligent Applications*, Advances in Intelligent Systems and Computing 863,
https://doi.org/10.1007/978-981-13-3338-5_4

its way in fields with various medical applications, for example, in the detection of cancer from mammograms and detection of epileptic patients where they have shown good performance [6, 8, 9, 16]. In this work, we will focus on its application of CV for source segmentation of a particular type of astronomical entity known as Radio Galaxies (RG).

These are galaxies that belong to a group known as Active Galactic Nuclei (AGN) and have Super-Massive Black Holes (SMBH) that emit observable jets that can range from a linear size of 10pc to few Mpc.

Prior to the introduction of CV to radio astronomy, astronomers had to go through the tedious process of manually classifying the radio sources. In such a process, the radio contours from the radio image of the radio source were plotted, overlaid on the optical image and classified visually. With the introduction of FIRST, new techniques were brought forward for the classification [1, 2, 11, 12].

However, one of the drawbacks of the currently available CV system is that these are still dependent on labeled images that result from a manual classification. Indeed the new methods have accelerated the classification process but they have not completely replaced our human needs. Also as compared to classification of everyday objects, labeled data is a scarce commodity in astronomy. This is why we come forward with an unsupervised method which is not dependent on any set of pre-labeled data.

Our aim is to show that unsupervised techniques could be used for the morphological segmentation of radio sources and that the same techniques could be used for outlier or novelty detection. The technique could then be used for analysis of images from telescopes like the SKA or its pathfinders MeerKAT. The paper is divided as follows: In Sect. 2, we give a description of the unsupervised machine learning technique used as well as the feature selection. In Sect. 3, we give a brief description of the selection mechanism and the analysis done. In Sect. 4, we give our results, and we finally give our conclusion in Sect. 5.

2   Unsupervised ML and Dataset

2.1   *Unsupervised Machine Learning*

ML algorithms come in different flavors but they can basically be grouped into two groups: supervised techniques and unsupervised techniques. Unsupervised learning studies how systems can learn to represent input patterns in a way that reflects the statistical structure of the overall collection of input patterns [4], as compared to supervised learning which targets at specific outputs (labels). Unsupervised learning is the cognitive system, most common in the brain, and it is the most used visual mechanism in animals, as animals are not trained to distinguish between objects.

Unsupervised learning works solely with the input data X and comes in two groups: Dimensionality Reduction (DR) and clustering. Dimensionality reduction as its name states is used to reduce highly dimensional data to lower dimensions. In such a case, if we have a high-dimensional set $X = \{x_0, x_1, x_2, \ldots, x_{n-1}\}$, where $x_i \in R^D$ $D >> 1$, we will through DR get a new set $Y = \{y_0, y_1, y_2, \ldots, y_{n-1}\}$ such that $y_i \in R^d$, where $d << D$ such that $\forall_{ij} |x_i - x_j|_n = |y_i - y_j|_n$ [14]. There are two reasons for the adoption of DR in our analysis: i. DR is used for visualization purposes, often we have data in high-dimensional space and we want to visualize it. The only way to achieve this is by plotting our data points to lower dimensions, ii. to fight the curse of dimensionality, which is an overfitted model resulting from use of high-dimensional data.

As dimensionality reducer, we make use of Isometric Mapping (ISOMAP). ISOMAP is an approach which is able to discover the nonlinear degrees of freedom that underline complex natural observations [15]. Isomap works by exploiting the geodesic paths for nonlinear dimensionality reductions. The latter will not work with distances in higher dimensions but will rather work with geodesic distances which are projected on a lower dimensional space. The isomapping works in three steps as follows:

i. The distance between the nearest point is computed for each data points.
ii. The geodesic distance between all pairs of points is evaluated.
iii. Multi-dimensional scaling is used and the distances are projected on a d-dimensional Euclidean space which preserves the manifold.

The second unsupervised ML methods that will be employed is clustering. Clustering is a method where the data is grouped in instances in such a way that similar instances are grouped together. In our case, we are interested in the grouping of sources based on the morphological structure where each group might have different physical and observational properties. DBSCAN is here used as our algorithm of choice. DBSCAN is a hierarchical algorithm, as compared to partitioning which constructs a partition into a set of k clusters, DBSCAN creates a hierarchical distribution of the database and does not need any k input. DBSCAN relies on a density-based notion of clusters and was designed to discover clusters of arbitrary shape and size [5]. Both algorithms are implemented using the Scikit-learn machine learning framework made for Python [10].

2.2 Dataset

Our primary source of data is images of radio sources that were identified by [7], which is the last catalog derived from the FIRST survey. The catalog contains 942,432 sources and 1000 sources were studied in our work. These images were downloaded from the FIRST cutout server where 10' by 10' images were downloaded with a detection sensitivity of 1 mJy at 1.4 GHz with an angular resolution of 5" [3]. The images were manually classified into two groups: 1. point-like sources and 2. extended sources (Fig. 1). This classification was used not to train the algorithm but was used to evaluate the performance of the unsupervised algorithms.

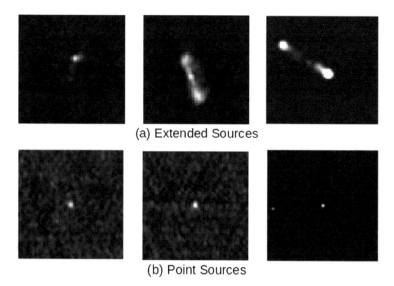

(a) Extended Sources

(b) Point Sources

Fig. 1 Sample of extended and point sources made manually

3 Feature Selection and Analysis

The shapelet decomposition technique used in [2] was used here. Through this technique each source was decomposed into a series of localized function with different shapes which are called "Shapelets" [13]. The localized basis functions are orthonormal with each other and are a set of weighted polynomial given by

$$\phi_n(x) = \left[2^n \pi^{\frac{1}{2}} n!\right]^{-\frac{1}{2}} H_n(x) e^{-\frac{x^2}{2}} \tag{1}$$

where n is a nonnegative integer and where $H_n(x)$ is the Hermite polynomial given by

$$H_n(x) = (-1)^n e^{x^2} \frac{d^n}{dx^n} e^{-x^2} \tag{2}$$

For practical purposes in imaging, a parameter β is introduced and the 1D basis function is given as

$$B_n(x; \beta) \equiv \beta^{\frac{1}{2}} \phi_n(\beta^{-1} x) \tag{3}$$

An object $f(x)$ can then be expressed as a summation of the basis functions given by

$$f(x) = \sum f_n B(n) \tag{4}$$

where f_n are the shapelets coefficients given by

$$f_n = \int_{-\infty}^{\infty} dx f(x) B_n(x; \beta) \tag{5}$$

Any source can be represented as a function $f(x)$ which is stored as fits files. To decompose the images, 2D shapelets are used and these can be obtained by taking the tensor product of two 1D basis functions, where $x = (x_1, x_2)$ and $n = (n_1, n_2)$. In that case, the dimensionless basis function is given by

$$\phi_n(x) = \phi_{n_1}(x_1) \phi_{n_2}(x_2) \tag{6}$$

and the dimensional basis functions is given as

$$B_n(x; \beta) \equiv \beta^{-1} \phi_n(\beta^{-1} x) \tag{7}$$

A radio source or object can then be expressed as a linear combination of those basis functions, where

$$f(x) = \sum f_n B(x; \beta) \tag{8}$$

where the coefficients are given as

$$f_n = \int_{-\infty}^{\infty} d^2 x f(x) B_n(x; \beta) \tag{9}$$

The coefficients can then be stored in a matrix S which will be a $max(n_1) + 1$ by $max(n_2) + 1$ matrix. In our work $max(n_1) = 15$ and $max(n_2) = 15$, thus we have a 16 by 16 matrix. So each image is given as

$$S = \begin{bmatrix} f_{(0,0)} & f_{(0,1)} & f_{(0,2)} & \cdots & f_{(0,15)} \\ f_{(1,0)} & f_{(1,1)} & f_{(1,2)} & \cdots & f_{(1,15)} \\ \vdots & \vdots & \vdots & \ddots & \vdots \\ f_{(15,0)} & f_{(15,1)} & f_{(15,2)} & \cdots & f_{(15,15)} \end{bmatrix} \tag{10}$$

Each source was then reshaped to a 256-dimensional vector which uniquely identifies the source's morphology. These vectors were then normalized using the Frobenius norm so as to preserve the information about the shape and discard the information about the source's intensity. In such a case, if two sources are similar in morphology but have different peak intensities, these once normalized will be close data points in the 256-dimensional space.

4   Results

We reduced the 256 coefficient to 3 components and made a scatter plot with marginal distribution of the first two dimensions. When compared with the manual classification, a clear separation of classes between point and extended sources was observed along the first reduced dimensions(y-axis) (Fig. 2).

Once convinced by the separation of the different classes in the reduced dimension, the reduced coefficients were input in the DBSCAN algorithm and four groups were discovered. These are described as follows:

1. Group 0 - Point sources
   These are the groups of point-like sources, these sources tend to group together as they all look similar, and hence are densely grouped together.
2. Group 1, 2, 3 - Extended source (double)
   These are the extended sources that have double components while Group 1 and 3 appear to be similar groups. Group 2 consists of two components but with one component being brighter than the other.

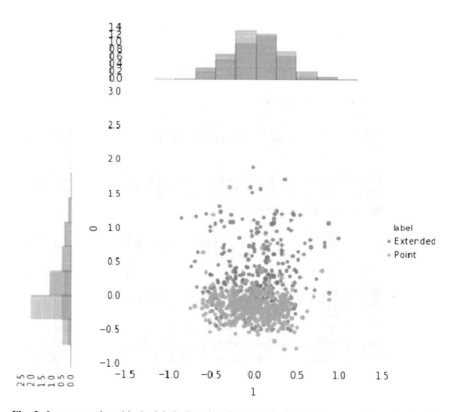

Fig. 2 Isomap results with the labels denoting the manual classification as point and extended sources

3. Outliers

These are extended sources that show no similarity with any of the above groups, and hence do not belong to any of them. They also show no similarity with any of the other outliers, and hence cannot form any group.

Figure 3 shows the reduced dimension, the clustering results and some selected sources. Point sources tend to cluster themselves around the origin of the reduced dimensional space. Also, it can be observed that highly resolved or extended sources appear in region delimited as $y > 1.0$.

To evaluate the overall performance of our method, we make use of four performance metrics (accuracy (A), precision (P), recall (R), and F_1 score.). These are found by finding the number of true positive, true negatives, false positive, and false negative, where the ground truth is the manual classification. The accuracy is not the only metric used since it is quite misleading as it depends on the number of positives and negatives in our data.

The precision defines the ratio of the total number of correct positives over the total number of positives (made by human classifiers). A low precision would imply a high number of false positive in our dataset. The recall, also known as sensitivity or true positive rate, is the ratio of correctly predicted events. A low recall implies a high number of false negative and a small number of positives. The F_1 score is the weighted average of the precision and recall. It takes both the false positive and false negative into account. It is more useful as the accuracy mainly when we have an uneven class distribution like ours.

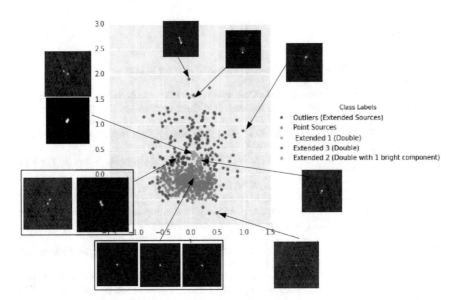

Fig. 3 Isomap results with clusters discovered through DBSCAN

Using those metrics, we got an **Accuracy of 0.83**, a **Precision of 0.85**, a **Recall of 0.92**, and an F_1 **Score of 0.87**. As compared to our previous work [2], where we achieved better accuracy (0.92) we here achieve a better F_1 score (0.87 as compared to 0.72 for supervised ML).

5   Conclusion

We have shown that unsupervised ML and shapelets analysis as feature extraction mechanism that could be used for the analysis of radio sources. These indeed achieved an accuracy close to supervised methods (0.83), while achieving better F_1 score and recall than any previously engineered method in radio astronomy. Also, the method involves no initial human classification to operate and as such this has two advantages: 1. our analysis is not affected by any human biases that can be introduced in the manual classification and as such unsupervised machine learning methods are more open to new discoveries, 2. the technique is more suited for inspecting huge amount of sources from new radio surveys. However, the full potential of the technique has not been exploited as all the 942,432 sources were not examined completely and this analysis is left as a future work. However, we have shown that the method is ideal for the incoming large radio surveys made from telescopes like the SKA or its precursor the MeerKat.

References

1. Ball, N.M., Brunner, R.J.: Data mining and machine learning in astronomy. Int. J. Mod. Phys. D **19**, 1049–1106 (2010)
2. Bastien, D., Oozeer, N., Somanah, R.: Classifying bent radio galaxies from a mixture of point-like/extended images with machine learning. IOP Conf. Ser.: Mater. Sci. Eng. **198**, 012013 (2017)
3. Becker, R.H., White, R.L., Helfand, D.J.: The FIRST survey: faint images of the radio sky at twenty centimeters. APJ, **450**, 559 (1995)
4. Dayan, P.: Unsupervised Learning. The MIT Encyclopedia of Cognitive Sciences. The MIT Press, London (1999)
5. Ester, M., Kriegel, H., Sander, J., Xu, X.: A density-based algorithm for discovering clusters. In: KDD-96 Proceedings, pp. 226–231 (1996)
6. Gautam, A., et al.: An improved mammogram classification approach using back propagation neural network. Data Engineering and Intelligent Computing. Springer, Singapore, pp. 369–376 (2018)
7. Helfand, D.J., White, R.L., Becker, R.H.: The last of FIRST: the final catalog and source identifications. Astrophys. J. **801**, 26 (2015)
8. Lay-Ekuakille, A., et al.: Multidimensional analysis of EEG features using advanced spectral estimates for diagnosis accuracy. In: 2013 IEEE International Symposium on Medical Measurements and Applications Proceedings (MeMeA). IEEE (2013)
9. Le, D.N., et al.: Optimizing feature selection in video-based recognition using MaxMin Ant System for the online video contextual advertisement user-oriented system. J. Comput. Sci. **21**, 361–370 (2017)

10. Pedregosa, F., Varoquaux, G., Gramfort, A., Michel, V., et al.: Scikit-learn: machine learning in Python. J. Mach. Learn. Res. **12**, 2825–2830 (2011)
11. Proctor, D.D.: Comparing pattern recognition feature sets for sorting triples in the FIRST database. APJS **165**, 5–107 (2006)
12. Proctor, D.D.: Morphological annotations for groups in the first database. Astrophys. J. Suppl. Ser. **194**, 31 (2011)
13. Refregier, A.: Shapelets - I. A method for image analysis. MNRAS **338**, 35–47 (2003)
14. Samudrala, S.K., Zola, J., Aluru, S., Ganapathysubramanian, B.: Parallel framework for dimensionality reduction of large-scale datasets. Sci. Program. **2015**, 1–12 (2015)
15. Tenenbaum, J.B., de Silva, V., Langford, J.C.: A global geometric framework for nonlinear dimensionality reduction. Science **5500**, 35–47 (2003)
16. Tiwari, A., et al.: ANN-based classification of mammograms using nonlinear preprocessing. In: Proceedings of 2nd International Conference on Micro-Electronics, Electromagnetics and Telecommunications. Springer, Singapore (2018)

Customer-Initiated Conversations on Facebook Pages of Tesco and Walmart

Swaleha Peeroo

Abstract Businesses are using social media as marketing tools to communicate and interact with customers. Customers initiate conversations on social networking sites of businesses but there is scant research as to why customers interact with businesses on social media in the grocery sector. This paper sets out to explore why customers initiate conversations and how do the other members of the online brand community react to those customer posts. Netnography was used to analyse the customer-initiated posts on Facebook pages of grocery stores. This study reveals that customers start discussions to express their satisfaction and gratitude, to criticise an action of the company, to complain about a product or service, to warn customers about a product or service, to recommend a product or service, or simply to engage in conversation with the community and the company. Findings show that most of customer-initiated posts are about complaints and criticisms. Also customers react more to customer complaints and criticisms than to favourable customer posts. This paper contributes to knowledge by identifying why customers post comments on the official Facebook pages of grocery stores and also by revealing the ways customers react to customer-initiated posts.

Keywords Social media · Online brand communities · Grocery stores · Facebook

1 Introduction

The ubiquity of social networking sites (SNSs) has increased social interactions and online user engagement worldwide among billions of users [1]. As proficiency with the Internet is increasing, businesses are creating online brand communities (OBCs) to engage with their customers [2]. Today, businesses and marketers need to understand that customers are shifting their attention from traditional advertisements to social media, which are perceived as the most influential marketing tool today [3].

S. Peeroo (✉)
Université des Mascareignes, Pamplemousses, Mauritius
e-mail: speeroo@udm.ac.mu

© Springer Nature Singapore Pte Ltd. 2019
S. C. Satapathy et al. (eds.), *Information Systems Design and Intelligent Applications*, Advances in Intelligent Systems and Computing 863,
https://doi.org/10.1007/978-981-13-3338-5_5

Consequently, social media have a substantial role in consumer purchasing decisions and in molding consumer behaviour [4].

Given the importance of the retail sector to the economy, it is important to understand the innovations and drivers of change. Grocery stores are harnessing social media to communicate to and engage with customers [5, 6]. However, there is scant research on the use of social media in the grocery sector. So, this paper purports to analyse why customers initiate conversations on the Facebook pages of grocery stores and how do the other customers react to these customer posts.

The following sections cover the literature relevant to the study, the methodology used and then the findings and discussion section. The paper concludes by summarising the findings, presenting the limitations of the study and proposing directions for future research.

2 Social Media

Social media are a collection of affordances sustained by an evolving and diverse technological infrastructure that allows people to communicate and collaborate in innovative ways [1]. With social media, customers and OBCs are empowered to broadcast information to a worldwide audience cheaply and instantly on the net [7]. Hence businesses have rushed to SNSs, which they use as marketing tools [8]. Owing to the substantial presence of potential customers on Facebook, businesses invest in establishing a brand community on Facebook where shoppers and fans can interact with the brand using the 'like', 'share' and 'comment' options [9].

2.1 Online Brand Communities

Individuals knowingly join like-minded persons to form groups, which interact around a focal brand [10]. These communities have been referred to as brand communities, which are 'specialized, non-geographically bound community, based on a structured set of social relationships among admirers of a brand' [11]. Corporate Facebook pages are OBCs since fans who have 'liked' the page share a common interest [12]. Within OBCs, customers not only construct knowledge from what the business communicates to them, but also from their own individually lived experiences and from conversations from other peer customers on social media platforms.

2.2 User Generated Content (UGC)

In this social media era, customers can create UGC on social media sites of businesses. UGC is the collection and leveraging of users' content on the Internet. Increasingly people create UGC about brands, companies, products and services [13]. The surge in social media usage by customers affords marketers the opportunity to shift relationships from dialogue to trialogue—in which customers have meaningful conversations and relationships with other customers and with the company [14].

Furthermore, within the realm of social media, organisations have lost total control over corporate communications as OBC members also drive conversations [15]. Businesses can no longer merely issue content they want potential customers to see; the social media ecosystem has shifted power from the organisation towards the consumer [16]. UGC with negative reviews about a brand may harm that particular brand [17]. Businesses today depend on positive UGC to maintain their legitimacy and reputation [18]. The affordances provided by social media are huge because consumers willingly contribute to posts on social media. This form of customer empowerment can be both a blessing and/or a curse for the business depending on the content created by customers [19].

2.3 Risks of Joining OBCs

There are benefits but also risks in joining and manipulating conversations in OBCs. OBCs are likened to a Pandora's box where value can be created as well as destroyed [20]. The anonymity afforded by the Internet and the open-comment platform of Facebook, create the perfect conditions for public outrage to be voiced on corporate walls [21]. When the voice of the customer carries a negative content in OBCs, it may be a threat to the organisation [22]. This may lead to co-destruction of value [23]. When businesses fail to engage customers, they have to deal with the potential risk of customer enragement [24], a state where customers can turn into value destroyers [25]. This compels businesses to be increasingly reactive, or even pro-active, to circumvent negative brand image consequences which may in turn lead to further value co-destruction.

3 Methodology

A non-participant netnography was carried out for 1 month during which data was collected from Facebook pages of Tesco and Walmart. Netnography is a 'qualitative research methodology that adapts ethnographic research techniques to study the cultures and communities that are emerging through computer-mediated communications' [26]. Netnography studies the factors and needs influencing decisions of online consumer groups by using information that is available publicly such as customer posts in online discussion groups [26].

For the coding and analysis of data, qualitative content analysis is used [26]. It is 'a research method for the subjective interpretation of the content of text data through the systematic classification process of coding and identifying themes or patterns'

[27]. Qualitative content analysis examines words intensely to group large amounts of text into a number of categories embodying similar meanings, instead of counting words [27].

To carry out the qualitative data analysis with the NVivo software, constant comparative method was used [26] in order to generate insights. Open coding was initially applied to bring meaning to the data, diligently scrutinising and comparing data for differences and similarities [28]. Each code denotes a distinct aspect of the phenomenon under study. Then axial coding was done to link the various categories identified during the open coding stage in order to identify any key underlying patterns and trends [28].

4 Findings and Discussion

For the analysis of customer reactions to customer-initiated messages, a good understanding of why customers initiate the discussion thread and factors prompting a customer to post a comment on the Facebook page of Tesco and Walmart is required. This study reveals that customers start discussions to express their satisfaction and gratitude, to criticise an action of the company, to complain about a product or service, to warn customers about a product or service, to recommend a product or service, or simply to engage in conversation with the community and the company. Analysis shows that most of these customer-initiated posts are unfavourable for the grocery stores as many customers go to the Facebook page of the store to complain, to criticise or to deter customers from buying a product or service. However, customers also post favourable comments.

4.1 Favourable Customer Posts

Customers voice out their gratitude and satisfaction for the outstanding service they had. Some customers thank the grocery store when they are satisfied with a product or service while others post comments when they are grateful to the store and its employees.

> I love shopping at Tesco too. Regardless what others say, all the staff (especially those in Customer Care) care about their customers. Services are excellent and staff are always available when required.

However, the other OBC members rarely react to such customer posts, except if another customer has had a similar experience.

When customers need more information, they initiate a conversation on Facebook. This study reveals that customers need information on prices of products, their availability, conditions attached to offers, the features and benefits of products, advice and explanation of how to use products or services. In OBCs conversations occur on prices, performance, quality and personal experiences with specific brands [29]. Other OBC members do not react to customer queries, as very often these need to be addressed by customer care officers. Interestingly, there are some cases where OBC members have provided the answer even before any customer care officer. These changes are redesigning the marketing landscape where consumer-to-consumer (C2C) marketing is replacing business-to-consumer (B2C) marketing [30].

Customers start a conversation on the Facebook page to make suggestions to the grocery store as illustrated below

> How about being responsible retailers and putting a sign by the flowers to let people know that lilies are highly toxic if a cat ingests any part. This is not well known and people should be made more aware. I think the supermarkets could do something very worthy.

Other OBC members rarely react to these suggestions unless they wish to add to the initial suggestion as depicted by the quote below

> Also on all your chocolate products please make sure there's a label saying you mustn't feed chocolate to dogs.

When OBC members make suggestions to the store, they co-create value for the company [31]. Co-created activities on the net contribute to perceived consumer empowerment, which results in higher motivation to continue co-creating value provided businesses take appropriate actions. This feeling of empowerment eases the task of businesses in satisfying customers, thereby building long term and interactive relationships [32].

4.2 Unfavourable Customer Posts

Customers initiate conversations to complain about the products and services of the grocery stores. Negative word of mouth (NWOM), in the form of complaints and criticisms, is predominant. This study shows that OBC members complain about products and services of the grocery store, behaviours of other customers within the store, its employees and its social responsibility.

OBC members tend to react differently to customer complaints depending on the content of the message. Some OBC members show empathy and provide support to the complainant when they agree with the complainant. Other customers respond to customer complaints by offering advice and helping to solve the customer's problem. For example, when a customer complained about the unavailability of Easter clothes for children at Walmart, several customers informed that customer where she could get such products at affordable prices as evidenced by the quote below

> You probably don't have one there, but if you do, JcPenny [sic] would have nice Easter attire.

Analysis of data also reveals that customers disagree with complainants and hence react by posting comments in which they criticise the complainant. For instance, a customer of Tesco complained about the misbehaviour of children accompanying parents within the store. This complaint generated many comments from OBC members who were offended and believed this complaint to be unfair and irrational. OBC members have heavily criticised the complainant as illustrated in the following quote:

> Thankfully neither of my children are [sic] big enough to wander round the supermarket on their own. Nor am I a mother who is idiotic (a generalisation which is extremely offensive to the majority of mothers!) enough to purchase £40 worth of make up

This example illustrates how social media have transformed the complaining landscape of an organisation. On social media, complaints are publicly available unlike traditional complaining by telephone, letter and email [33]. Had the complainant used traditional media to complain, the other customers would not have known of this complaint and would not have been able to defend themselves.

Moreover, this study uncovered cases when customer complaints resulted in an intense debate among OBC members who had diverging views on the issues raised. For instance, a customer complained of the poor service she had, when a cashier who while processing payment for her, was busy talking to another colleague and had refused to pack the items in a bag when asked to do so and even accused the customer of being rude. This complaint generated mixed responses from the OBC members. Some customers were highly critical of the complainant as shown below.

> Sounds like you were the one being rude as well.... pack your own shopping!

However, other customers had different reactions. Some provided support to the complainant while others criticised the other customers who disagreed with the complainant as shown below

> I am sure she is capable of packing her less than 3 items in a bag, but it is annoying and rude when someone is serving you that they are making a conversation with another oerson, [sic] wether [sic] it is a friend or a collegue, [sic] it is not professional and not good skills!!

While reacting to complaints on Facebook pages of grocery stores, customers sometimes defend the company when they believe that the business or its employees have taken the right actions. For example, a 22-year-old customer complained that the cashier refused to sell beer to her as a minor accompanied her. She was very annoyed and believed the cashier had no right to stop her buying beer. She posted a complaint and even threatened to stop patronising Walmart. OBC members took the defense of the employee by reminding the complainant of the laws about alcoholic drinks prevailing in the country.

Another major form of NWOM initiated by customers on the Facebook pages of Tesco and Walmart is criticisms about actions of the company. Customers post comments to condemn actions of organisations when they believe there is corporate misconduct. OBC members often confront businesses with their ethical, commercial and social responsibilities [34]. Tesco is heavily criticised for the sale of unlabeled

'Halal' meat and for the labelling of products containing allergens, while Walmart is criticised for the use of gestation crates, its low-wage policy and sales of genetically modified products. It was observed that some OBC members posted same criticisms repetitively.

Furthermore, when customers criticise actions of the company, these posts generate responses from OBC members who either comment on the original post or who in turn post another criticism on the same issue or sometimes simply repost the same comment. They seem to behave in this way to show to Tesco and Walmart that they support the same cause, hence coercing the organisation to take appropriate corrective action. Customers on social media can create massive waves of outrage in a short span of time when reacting to dubious statements or activities of an organisation [3].

5   Conclusion

This paper aimed to identify reasons for customer-initiated conversations and the reactions of other customers to these customer posts. Customers start conversations to complain about the products and services, to express customer satisfaction, to criticise actions of grocery stores, to request more information, to give advice to customers, to warn customers of malpractices and to make suggestions to the grocery stores. The other OBC members react to customer-initiated messages either by providing support to the customer or the grocery store. These conversations either create or destroy value for grocery stores.

This paper contributes to the literature by identifying why users initiate conversations on OBCs and how other OBC members react to these messages. A very high percentage of customer-initiated messages on Facebook pages of grocery stores are about customer complaints and criticisms. This should be a major cause of concern for businesses as such comments may damage the image of the organisation. Retailers need to monitor online conversations to enhance customer experiences. Emerging technologies such as artificial intelligence (AI) can be used to respond to customer queries instead of real employees. AI can be used to help customer service executives detect complaints and queries and determine which ones need to be addressed immediately.

The limitation of this paper is that only the official Facebook pages of Tesco and Walmart have been analysed, whereas customers post comments on various Web 2.0 platforms. Future research could be carried out to analyse various online platforms where customers of OBCs initiate conversations.

References

1. Kane, G.C.: The evolutionary implications of social media for organizational knowledge management. Inf. Organ. **27**(1), 37–46 (2017)
2. Ul Islam, J., Rahman, Z.: The impact of online brand community characteristics on customer engagement: a solicitation of stimulus-organism-response theory. Telematics. Inform (2017). http://dx.doi.org/10.1016/j.tele.2017.01.004
3. Pfeffer, J., Zorbach, T., Carley, K.M.: Understanding online firestorms: negative word-of-mouth dynamics in social media networks. J. Mark. Commun. **20**(1–2), 117–128 (2014). https://doi.org/10.1080/13527266.2013.797778
4. Mangold, W.G., Faulds, D.J.: Social media: the new hybrid element of the promotion mix. Bus. Horiz. **52**, 357–365 (2009)
5. Coelho, D., Las Casas, A.: Consumers perceptions of generation Y in the acquisition of technology products (computers) at the point of sale. Int. J. Res. IT, Manage. **3**(3), 42–57 (2013)
6. Tarnowski, J.: Social studies. Progressive Grocer **90**(7), 85–130 (2011)
7. Arnaboldi, M., Coget, J.F.: Social media and business. Organ. Dyn., 2. http://dx.doi.org/10.1016/j.orgdyn.2015.12.006 (2016)
8. Labrecque, L.: Fostering consumer–brand relationships in social media environments: the role of parasocial interaction. J. Interact. Mark. http://dx.doi.org/10.1016/j.intmar.2013.12.003 (2014)
9. Ho, C.: Consumer behavior on Facebook. EuroMed J. Bus. **9**(3), 252–267 (2014)
10. Veloutsou, C.: Brands as relationship facilitators in consumer markets. Mark. Theory **9**(1), 123–126 (2009)
11. Muniz, A., O'Guinn, T.: Brand community. J. Consum. Res. **27**(4), 412–432, 421 (2001)
12. Pöyry, E., Parvinen, P., Malmivaara, T.: Can we get from liking to buying? Behavioral differences in hedonic and utilitarian Facebook usage. Electron. Commer. Res. Appl. **12**, 224–235 (2013)
13. Vanden Bergh, B., Lee, M., Quilliam, E., Hove, T.: The multidimensional nature and brand impact of user-generated ad parodies in social media. Int. J. Advertising **30**(1), 103–131 (2011)
14. Porter, C., Donthu, N., MacElroy, W., Wydra, D.: How to foster and sustain engagement in virtual communities'. Calif. Manag. Rev. **53**(4), 80–110 (2011)
15. Colleoni, E.: CSR communication strategies for organizational legitimacy in social media. Corp. Commun.: An Int. J. **18**(2), 228–248 (2013)
16. Schulze Horn, I., Taros, T., Dirkes, S., Huer, L., Rose, M., Tietmeyer, R., Constantinides, E.: Business reputation and social media—a primer on threats and responses. IDM J. Dir., Data. Digital. Mark. Pract. **16**(3), 4. http://www.palgrave-journals.com/dddmp/index.html (2015)
17. Cheong, H., Morrison, M.: Consumers' reliance on product information and recommendations found in UGC. J. Interact. Advertising **8**(2), 1–29 (2008)
18. Jurgens, M., Berthon, P., Edelman, L., Pitt, L.: Social media revolutions: the influence of secondary stakeholders. Bus. Horiz. http://dx.doi.org/10.1016/j.bushor.2015.11.010 (2016)
19. Peeroo, S., Samy, M., Jones, B.T.: Facebook: a blessing or a curse for grocery stores? Int. J. Retail. Distrib. Manage. **45**(12), 1242–1259 (2017)
20. Jin, S.: The potential of social media for luxury brand management. Mark. Intell. Plann. **30**(7), 687–699 (2012)
21. Champoux, V., Durgee, J., McGlynn, L.: Corporate Facebook pages: when "fans" attack". J. Bus. Strategy **33**(2), 22–30 (2012)
22. Lee, H., Han, J., Suh, Y.: Gift or threat? An examination of voice of the customer: the case of MyStarbucksIdea.com. Electron. Commer. Res. Appl. http://dx.doi.org/10.1016/j.elerap.2014.02.001 (2014)
23. Smith, A.: The value co-destruction process: a customer resource perspective. Eur. J. Mark. **47**(11/12), 1889–1909 (2013)
24. Leeflang, P.S.H., Verhoef, P.C., Dahlström, P., Freundt, T.: Challenges and solutions for marketing in a digital era. Eur. Manag. J. **32**, 1–12 (2014)

25. Verhoef, P., Beckers, S., van Doorn, J.: Understand the perils of co-creation. Harvard Bus. Rev. **91**(9), 28–32 (2013)
26. Kozinets, R.V.: The field behind the screen: using netnography for marketing research in online communities. J. Mark. Res. **39**, 61–72, 62 (2002)
27. Hsieh, H., Shannon, S.: Three approaches to qualitative content analysis. Qual. Health. Res. **15**(9), 1277–1288, 1278 (2005)
28. Miles, M., Huberman, A.: Qualitative Data Analysis: a Sourcebook of New Methods. Sage, London (1984)
29. Brodie, R., Ilic, A., Juric, B., Hollebeek, L.: Consumer engagement in a virtual brand community: an exploratory analysis. J. Bus. Res. **66**(1), 105–114 (2011)
30. Kimmel, A.: Connecting With Consumers: marketing for New Marketplace Realities. University Press, Oxford (2010)
31. O'Cass, A., Viet Ngo, L.: Achieving customer satisfaction in services firms via branding capability and customer empowerment. J. Serv. Mark. **25**(7), 489–496 (2011)
32. Füller, J., Mühlbacher, H., Matzler, K., Jawecki, G.: Consumer empowerment through internet-based co-creation. J. Manage. Inf. Syst. **26**(3), 71–102 (2009)
33. Einwiller, S., Steilen, S.: Handling complaints on social network sites—an analysis of complaints and complaint responses on Facebook and Twitter pages of large US companies. Public Relat. Rev. http://dx.doi.org/10.1016/j.pubrev.2014.11.012 (2015)
34. Constantinides, E., Fountain, S.: Web 2.0: conceptual foundations and marketing issues. J. Dir., Data. Digital. Mark. Pract. **9**(3), 231–244 (2008)

Cognitive Modeling of Mindfulness Therapy by Autogenic Training

S. Sahand Mohammadi Ziabari and Jan Treur

Abstract In this paper, the effect of a mindfulness therapy based on a Network-Oriented Modeling approach is addressed. The considered therapy is Autogenic Training that can be used when under stress; it has as two main goals to achieve feeling heavy and warm body parts (limbs). Mantras have been used in therapies since long ago to make stressed individuals more relaxed, and they are also used in Autogenic Training. The presented cognitive temporal-causal network model addresses the modeling of Autogenic Training asking this into account. In the first phase a strong stress-inducing stimulus causes the individual to develop an extreme stressful emotion. In the second phase, the therapy with the two goals is shown to make the stressed individual relaxed. Hebbian learning is used to increase the influence of the therapy.

Keywords Cognitive temporal-causal network model · Hebbian learning
Extreme emotion · Mindfulness · Autogenic training

1 Introduction

Mindfulness is an approach to relaxation under an extreme emotion which has been used since many years ago. There are many mindfulness therapies to reduce the level of arousal of a stressed individual. Some are of more recent origin [1, 2], and under investigation, like Gene therapy which works by putting enzymes into the cells by using viruses as transfer. Other mindfulness therapies like music therapy and Autogenic Training have been started to work as a therapy many years ago and they have been proven to have a good effect on reducing the level of the emotion.

S. S. Mohammadi Ziabari (✉) · J. Treur
Behavioural Informatics Group, Vrije Universiteit Amsterdam, Amsterdam, Netherlands
e-mail: sahandmohammadiziabari@gmail.com

J. Treur
e-mail: j.treur@vu.nl

© Springer Nature Singapore Pte Ltd. 2019
S. C. Satapathy et al. (eds.), *Information Systems Design and Intelligent Applications*, Advances in Intelligent Systems and Computing 863,
https://doi.org/10.1007/978-981-13-3338-5_6

The first edition of Autogenic training was introduced about 70 years ago in 1932 by its founder Johannes Heinrich Schultz, who was a psychiatrist and neurologist in Berlin [3]. Autogenic training has a number of characteristics which are presented in [4]. The rate of successfulness of autogenic training is related to parameters as the clinical condition and the age of the participant, how intelligent the participant is, and his or her overall development. The time usage of an Autogenic training therapy is much less than other therapies. In Autogenic training also a group therapy is possible. Progress during the therapy can be monitored by physiological and psychological tests. The following quote from [4, p. 189] explains the use of it

> Clinical results demonstrated that autogenic training has been effective in the treatment of (a) disorders of the respiratory react; (b) disorders of the gastrointestinal tract; (c) disorders of the cardiovascular system and vasomotor disturbances; (d) disorders of the endocrine system; (e) disorders of the urogenital system; (f) disorders of pregnancy; (g) skin disorders (for example; allergic conditions, pruritus, verruca vulgaris); (h) ophthalmologic disorders and blindness (glaucoma, scotoma, certain forms of squint); (i) Neurologic disorders (certain neuromuscular disorders, brain injury, and epilepsy). It has been observed that autogenic training is very helpful in the treatment of behavior disorders and motor disturbances as for example, stuttering, writer's cramp, enuresis, hysterical dysphagia, singultus, globus hystericus, blushing, certain states of anxiety, and phobia. Patients have reported over periods ranging from a few weeks to several months that their anxiety, insecurity, and neurotic reactions are smoothing out or had gradually lost their significance.

This paper is organized as follows. In Sect. 2 the underlying neurological principles concerning the parts of the brain involved in stress and in the suppression of stress are addressed. In Sect. 3 the cognitive temporal-causal network model is introduced and illustrated by simulation of an example scenario. In Sect. 4 the simulation results of the model are discussed, and eventually in the last section a discussion is presented.

2 Underlying Neurological Principles

Mindfulness is considered as a method for emotion regulation. The alternative method called emotion suppression is declared as a maladaptive method based on keeping away or evasion from involvement in and becoming conscious of emotion of themselves [5]. Both strategies have been considered as regulation methods using different modulation systems. Both of them weaken the amygdala responses to emotional stimuli but with different pathways, on the one hand mindfulness uses operational connectivity from the medial prefrontal cortex to regulate the amygdala operation and in the other hand, suppression uses functional connectivity of other parts of the brain like dorsolateral prefrontal cortex to achieve regulation [5]. Emotion regulation is an essential tool for humans to make their sensations suitable for living in the environment.

There are many mindfulness therapies, including Music therapy [6] and Autogenic training [5, 7, 8]. In this paper, first we explain some basic mechanisms of the aforementioned therapies and then we address Autogenic training in more into detail.

In [6] it has been found that activation of amygdala is reduced by both enjoyable music and musical extemporization. In [9], it is shown how emotion regulation is done by dreaming. Dreaming is considered to be in act with four main brain parts, Amygdala, Medial Prefrontal Cortex (MPFC), Hippocampus, Anterior Cingulate Cortex (ACC).

In [10, p. 57] Autogenic training is explained as follows:

Autogenic training is a desensitization-relaxation technique developed by the German psychiatrist Johannes Heinrich Schultz by which a *psychophysiologically* determined relaxation response is obtained. [10], p. 57

In [11, p. 246] Autogenic training is explained as follows:

Autogenic training is defined as a psychophysiological self-control technique aiming at physical and mental relaxation. It uses auto-suggestions by which individuals learn to alter certain psychophysiological functions with, initially, minimal intervention by another person and, after the technique is learned, with no intervention by another person. The training uses seven short verbal standard formulas, emphasizing feelings of (1) general peace, (2) heaviness in the limbs, (3) peripheral warmth, (4) respiratory regularity, (5) cardiac regularity, (6) abdominal warmth, and (7) coolness of the forehead. The promotion of the person's capabilities to relax and to rest, (2) the reduction of overwhelming negative effects, (3) the reduction of nervousness, (4) the promotion of performance (e.g., selective attention and memory recall), (5) the self-regulation of autonomous nervous system processes (like heart rate and body temperature), and (6) the promotion of self-control and self-actualization through enhanced self-perception and self-regulation. [11], p. 246

In [12, p. 64] Autogenic bears processes as follows:

After exercise of the "simple sitting posture" (which was preferred to the horizontal training posture and the reclining chair posture because of its higher practical value), closure of eyes, and passive concentration (implying a casual and functional passivity toward the intended functional changes), and the technique of coming back to normal (flexing arms vigorously, breathing deeply, opening eyes), the standard exercises of autogenic training were introduced and trained. [12], p. 64

In [13] and [14] it has been shown that mindfulness meditation will activate and increase the functionality of bilateral insula, the rostral ACC, and the dorsomedial Prefrontal Cortex, and that during extreme stress and painful stimulation, the functionality in emotion areas in the brain like PFC, amygdala and hippocampus will be decreased and the performance in pain-processing areas of brain named ACC, Thalamus and insula will be decreased. As mentioned in [15] Autogenic training has significant influence on prefrontal cortex and insular cortex. The results in [15] have shown the performance of left precentral and postcentral cortex and also parietal cortex during the two goals considered here. It is explained that

Autogenic training has been shown to restore the balance between activity of the sympathetic and parasympathetic branches of the autonomic nervous system with important health benefits, as parasympathetic activity promotes digestion bowel movements, lowers the blood pressure, slows the heart rate and promotes the functions of the immune system. [15], p. 445

Increasing functionality in the salience network, the dorsal anterior insula and the anterior MCC, during extreme stress and pain, shows much faster neural settlement

[16]. This increment in functionality and increasing in enhanced neural habitation in the amygdala and stress-related regions of brain recommend a downregulation of processes associated with stress. Therefore, the aforementioned neurological principles are used as the basis of our model. As mentioned in [11] two goals named heavy limbs and warm limbs, which are considered as the core definition of the autogenic training, are chosen and used as the basic goals of our model in the Autogenic training therapy. As an example, a stressful individual starts first speaking a mantra for the first goal (heavy limbs) by execution and sensing the voice and the repetition in speaking the mantra preparation states for heavy limbs are activated. Then a similar process for warm limbs takes place. The extreme emotion makes the sensory representation state of the body states of body states related to the goals for heavy and warm limbs are suppressed as mentioned in [14]; the mantra's aim at overcoming the stress revealed in the body state in the context of the stressful condition as declared in [13]. The cognitive model of Autogenic training considered here is based on neurological principles as mentioned in this section; the model and its simulation results will be explained in more detail in the next section.

3   The Cognitive Temporal-Causal Network Model

First the Network-Oriented Modeling approach used to model this process is briefly explained. As discussed in detail in [7, Chap. 2] this approach is based on temporal-causal network models which can be represented at two levels: by a conceptual representation and by a numerical representation. A conceptual representation of a temporal-causal network model in the first place contains representing in an indicative manner states and connections between them that illustrate (causal) impacts of states on each other, as assumed to hold for the application domain addressed. The states are assumed to have (activation) levels that change over time. In reality, not all causal relations are equally strong, so some notion of *strength of a connection* is used. Furthermore, when more than one causal relation affects a state, some way to *aggregate multiple causal impacts* on a state is used. Moreover, a notion of *speed of change* of a state is used for timing of the processes. These three notions form the defining part of a conceptual representation of a temporal-causal network model

- Strength of a connection $\omega_{X,Y}$ Each connection from a state X to a state Y has a *connection weight value* $\omega_{X,Y}$ representing the strength of the connection, often between 0 and 1, but sometimes also below 0 (negative effect) or above 1.
- Combining multiple impacts on a state $c_Y(\ldots)$ For each state (a reference to) a *combination function* $c_Y(\ldots)$ is chosen to combine the causal impacts of other states on state Y.
- Speed of change of a state η_Y For each state Y a *speed factor* η_Y is used to represent how fast a state is changing upon causal impact.

Combination functions can have different forms, as there are many different approaches possible to address the issue of combining multiple impacts. Therefore, the Network-Oriented Modeling approach based on temporal-causal networks incorporates for each state, as a kind of label or parameter, a way to specify how multiple causal impacts on this state are aggregated by some combination function. For this aggregation a number of standard combination functions are available as options and a number of desirable properties of such combination functions have been identified; see [7, Chap. 2, Sects. 2.6 and 2.7].

In Fig. 1 the conceptual representation of the introduced temporal-causal network model is depicted. A brief explanation of the states used is shown in Table 1. The model states and their relation to the neurological principles explained in Sect. 2 are addressed in Table 2. Next, the elements of the conceptual representation shown in Fig. 1 are explained in some more detail. The states ws_c and ws_{ee} stand for world state of stimulus s from the world and internal stimulus s, respectively. The states ss_c and ss_{ee} are the sensor state of the context c and sensor state of extreme emotion. The states srs_c and srs_{ee} are the sensory representation of the context stimulus c and extreme emotion, respectively. The state srs_{ee} is a trigger affecting the activation level of the preparation state. Furthermore, ps_{ee} is the preparation state of an extreme emotional response on the sensory representation srs_c of the context c, and fs_{ee} shows the feeling state associated to this extreme emotion. The state es_{ee} represents the execution state of an extreme emotion. Two states srs_{b1}, and srs_{b2} denote sensory representation of the first goal (warm hand) and the second goal (cool head), respectively. The state $goal_{ee}$ shows the general goal of the whole mantra execution. The states ps_{b1} and ps_{b2} are the preparation states relating to the first goal and the second goal, which their connections with srs_{mantra,b_1} and srs_{mantra,b_2} are considered as the mirroring link. The states $(ss_{mantra,b_1},\ srs_{mantra,b_1},\ es_{mantra,b_1})$, $(ss_{mantra,b_2},\ srs_{mantra,b_2},\ es_{mantra,b_2})$ are the sensor states of the mantra, sensory representation and execution state of the first goal and the second goal, respectively.

The connection weights ω_i in Fig. 1 are as follows. The sensor states ss_{ee}, ss_{cc} have arriving connections from ws_{ee} and ws_c with weights ω_1, ω_2, respectively. World state of an extreme emotion ws_{ee} has an incoming connection from es_{ee} as a body-loop with weight ω_9. The sensory representation state of an extreme emotion srs_{ee} has three entering connection weights ω_{12}, ω_{13}, ω_{14} from states preparation state of an extreme emotion ps_{ee}, preparation state of the first goal ps_{b2} and preparation state of the second goal ps_{b2}, respectively. The weight ω_5 is the incoming connection weight for the feeling state fs_{ee}. The preparation state of an extreme emotion ps_{ee} has two entering connection weights ω_6, ω_7 from states srs_c and fs_{ee}, respectively. The execution state of an extreme emotion es_{ee} incoming connection weight is ω_8 from ps_{ee}.

The sensory representation states of the first and the second goal srs_{b1} and srs_{b2} both have two incoming connection weights from execution state of an extreme emotion es_{ee} (suppressing) and preparation states of them, ps_{b1} and ps_{b2}, with the values of ω_{10} and ω_{11} respectively. The general goal, $goal_{ee}$, has just one entering connection weight ω_{15} from sensory representation states of an extreme emotion srs_{ee}. The states $goal_{b1}$ has one incoming connection ω_{19} from $goal_{ee}$. The second goal, $goal_{b2}$, has two incoming connection weights ω_{16}, ω_{20} from $goal_{ee}$ and srs_{b1}, respectively. The preparations states of the first goal and the second goal has three incoming con-

Table 1 Explanation of the states in the model

X_1	ws_{ee}	World (body) state of extreme emotion ee
X_2	ss_{ee}	Sensor state of extreme emotion ee
X_3	ws_c	World state for context c
X_4	ss_c	Sensor state for context c
X_5	srs_{ee}	Sensory representation state of extreme emotion ee
X_6	srs_c	Sensory representation state of context c
X_7	fs_{ee}	Feeling state for extreme emotion ee
X_8	ps_{ee}	Preparation state for extreme emotion ee
X_9	es_{ee}	Execution state (bodily expression) of extreme emotion ee
X_{10}	srs_{b_1}	Sensory representation state of body state b_1 (heavy limbs)
X_{11}	srs_{b_2}	Sensory representation state of body state b_2 (warm limbs)
X_{12}	$goal_{ee}$	Primary goal of reducing extreme emotion ee
X_{13}	$goal_{b_1}$	Goal to go for b_1 (heavy limbs)
X_{14}	$goal_{b_2}$	Goal to go for b_2 (warm limbs)
X_{15}	ps_{b_1}	Preparation state for b_1 (heavy limbs)
X_{16}	ps_{b_2}	Preparation state for b_2 (warm limbs)
X_{17}	ss_{mantra,b_1}	Sensor state of mantra for b_1 (heavy limbs)
X_{18}	ss_{mantra,b_2}	Sensor state of mantra for b_2 (warm limbs)
X_{19}	srs_{mantra,b_1}	Sensory representation state of mantra for b_1 (heavy limbs)
X_{20}	srs_{mantra,b_2}	Sensory representation state of mantra for b_2 (warm limbs)
X_{21}	ps_{mantra,b_1}	Preparation state for mantra for b_1 (heavy limbs)

(continued)

Table 1 (continued)

X_{22}	ps_{mantra,b_2}	Preparation state for mantra for b_2 (warm limbs)
X_{23}	es_{mantra,b_1}	Execution state (speaking) for mantra for b_1 (heavy limbs)
X_{24}	es_{mantra,b_2}	Execution state (speaking) for mantra for b_2 (warm limbs)

nection weights (ω_{17}, ω_{21}, ω_{36}), (ω_{22}, ω_{25}, ω_{35}) from (srs_{b_1}, $goal_{b_1}$, srs_{mantra,b_1}) and (srs_{b_2}, $goal_{b_2}$, srs_{mantra,b_2}), respectively. Note that red lines in the model show the Hebbian learning connection (ω_{35}, ω_{36}). The mantra part of the model is the same for two goals of mantra. The states ss_{mantra,b_1} and ss_{mantra,b_2} both have two entering connection weights ω_{27} and ω_{32} from execution states of the first and the second goals, es_{mantra,b_1} and es_{mantra,b_2}, respectively. The sensory representation states of

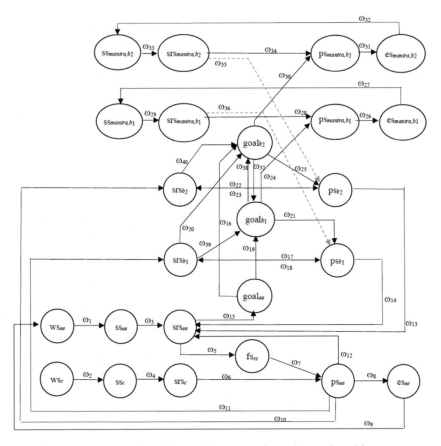

Fig. 1 Conceptual representation of the cognitive temporal-causal network model

Table 2 Explanation of the states and their relation to neurological principles

States	Neurological principles	Quotations, references
ws_{ee}	External stressor	External stress-inducing event [17]
ss_{ee}	Sensor state for perception of the stressor	'Human states can refer, for example, to states of body parts to see (Eyes), hear (ears) and fee (skin).' In [17], p. 51 and [13]
srs_{ee}	Sensory and feeling representation of stressful event	'The dACC was activated during the observe condition. The dACC is associated with attention and the ability to accurately detect emotional signals.' [5], p. 12
$goal_{b_1}$ $goal_{b_2}$ $goal_{ee}$	Executive function and manage goals (heavy and warm limbs) and emotion regulation Prefrontal Cortex (PFC)	'PFC reflects greater top-down control over pain and cognitive reappraisal of pain, and that changes in somatosensory cortices reflect alterations in the perception of noxious signals.' [18], p. 315
ss_{mantra,b_1}, ss_{mantra,b_2}	Sensor state for perception of the mantra's (hearing)	'The Autogenic training uses seven short verbal standard formulas, emphasizing feelings of (1) heaviness in the limbs, (3) peripheral warmth.' [11], p. 189
srs_{mantra,b_1} srs_{mantra,b_2}	Sensory representation of mantra's	Sensory representation state for hearing the mantra's [19–21]
ps_{mantra,b_1}, ps_{mantra,b_2}	Preparation state of mantra (preparation for speaking), PCC	'Broca area is a region in the frontal lobe of the dominant hemisphere, usually the left, of the hominid brain with functions linked to speech production.' [18], p. 318, [22]
es_{mantra,b_1}, es_{mantra,b_2}	Execution state of mantra's	Speaking the mantra's
ps_{b_1}	Preparation state for body states b_1 and b_2	'A complex mosaic of interconnected Frontal lobe areas that lie rostral to the Primary motor cortex also contributes importantly to motor functions. The medial premotor cortex, like the lateral area, mediates the selection of movements.' [23], p. 78

(continued)

Table 2 (continued)

States	Neurological principles	Quotations, references
srs_{b_1}	Sensory representation and feeling of body states (bilateral anterior temporal lobe)	'Bilateral anterior temporal lobe Task domain: emotion and affect. Core affect generation: engaging vicermotor control of the body to create core affective feelings of pleasure or displeasure with some degree of arousal.' [24], p. 2112

mantra of the first and the second goal srs_{mantra,b_1} and srs_{mantra,b_2} have two incoming connection weights ω_{29} and ω_{33} from ss_{mantra,b_1} and ss_{mantra,b_2}, respectively. The preparation states of mantra, ps_{mantra,b_1} and ps_{mantra,b_2} both have two incoming connections from the first goal and the second goal, $goal_{b_1}$, $goal_{b_2}$ with ω_{24}, ω_{39} and from sensory representation of the mantra of the first and the second goal, srs_{mantra,b_1}, srs_{mantra,b_2} with connection weights ω_{28}, ω_{34}, respectively. The execution states of the first and the second goal of the mantra, es_{mantra,b_1} and es_{mantra,b_2} both have an incoming connection weight ω_{26}, ω_{31} from preparation states ps_{mantra,b_1} and ps_{mantra,b_2}, respectively.

This conceptual representation was transformed into a numerical representation as follows [7, Chap. 2]:

- at each time point t each state Y in the model has a real number value in the interval $[0, 1]$, denoted by $Y(t)$
- at each time point t each state X connected to state Y has an impact on Y defined as $\textbf{impact}_{X,Y}(t) = \omega_{X,Y} X(t)$ where $\omega_{X,Y}$ is the weight of the connection from X to Y
- The *aggregated impact* of multiple states X_i on Y at t is determined using a *combination function* $c_Y(...)$:

$$\textbf{aggimpact}_Y(t) = c_Y\left(\textbf{impact}_{X_1,Y}(t), \ldots, \textbf{impact}_{X_k,Y}(t)\right)$$
$$= c_Y\left(\omega_{X_1,Y}X_1(t), \ldots, \omega_{X_k,Y}X_k(t)\right),$$

where X_i are the states with connections to state Y

- The effect of $\textbf{aggimpact}_Y(t)$ on Y is exerted over time gradually, depending on speed factor η_Y

$$Y(t + \Delta t) = Y(t) + \eta_Y\left[\textbf{aggimpact}_Y(t) - Y(t)\right]\Delta t \quad \text{or}$$
$$dY(t)/dt = \eta_Y\left[\textbf{aggimpact}_Y(t) - Y(t)\right]$$

- Thus, the following *difference* and *differential equation* for Y are obtained:

$$Y(t + \Delta t) = Y(t) + \eta_Y[\mathbf{c}_Y(\omega_{X_1,Y}X_1(t), \ldots, \omega_{X_k,Y}X_k(t)) - Y(t)]\Delta t$$
$$\mathbf{d}Y(t)/\mathbf{d}t = \eta_Y[\mathbf{c}_Y(\omega_{X_1,Y}X_1(t), \ldots, \omega_{X_k,Y}X_k(t)) - Y(t)]$$

For states the following combination functions $\mathbf{c}_Y(\ldots)$ were used, the identity function $\mathbf{id}(.)$ for states with impact from only one other state, and for states with multiple impacts the scaled sum function $\mathbf{ssum}_\lambda(\ldots)$ with scaling factor λ, and the advanced logistic sum function $\mathbf{alogistic}_{\sigma,\tau}(\ldots)$ with steepness σ and threshold τ.

$\mathbf{id}(V) = V$

$\mathbf{ssum}_\lambda(V_1, \ldots, V_k) = (V_1 + \ldots + V_k)/\lambda$

$\mathbf{alogistic}_{\sigma,\tau}(V_1, \ldots, V_k) = [(1/(1 + e^{-\sigma(V_1 + \ldots + V_k - \tau)})) - 1/(1 + e^{\sigma\tau})](1 + e^{-\sigma\tau})$

Here first the general Hebbian Learning is explained which is applied to ω_{35} and ω_{36} for (X_{19}, X_{15} and X_{20}, X_{16}). In a general example model considered it is assumed that the strength ω of such a connection between states X_1 and X_2 is adapted using the following Hebbian Learning rule, taking into account a maximal connection strength 1, a learning rate $\eta > 0$ and a persistence factor $\mu \geq 0$, and activation levels $X_1(t)$ and $X_2(t)$ (between 0 and 1) of the two states involved [17]. The first expression is in differential equation format, the second one in difference equation format

$$d\omega(t)/dt = \eta[X_1(t)X_2(t)(1 - \omega(t)) - (1 - \mu)\omega(t)]$$
$$\omega(t + \Delta t) = \omega(t) + \eta[X_1(t)X_2(t)(1 - \omega(t)) - (1 - \mu)\omega(t)]\Delta t$$

4   Example Simulation

An example simulation of this process is shown in Figure Table 3 shows the connection weights used, where the values for are initial values as these weights are adapted over time. The time step was $\Delta t = 0.95$. The scaling factors λ_i for the states with more than one incoming connection are also depicted in Table 3.

For goal$_{ee}$, goal$_{b1}$ and goal$_{b2}$ the alogistic combination function was considered for with the values, of $(10, 0.15)$, $(100, 0.2)$ and $(100, 0.3)$. At the first phase, an external world state of an extreme emotion-inducing context c (denoted by ws$_c$) will affect the internal state of the individual body state by triggering the emotional response es$_{ee}$ (via ss$_c$, srs$_c$, and ps$_{ee}$) leading to revealing the extreme emotion by body state ws$_{ee}$. As a result the stressed person senses the extreme emotion and the stress level increased by time, so in this step, the goal becomes to reduce this stress level by an Autogenic Training therapy goal$_{ee}$. This goal triggers two subgoals goal$_{b1}$ (for heavy limbs) and goal$_{b2}$ (for warm limbs) but in a sequential order. The goal of the first mantra starts to work by stating the autogenic therapy. The first goal goal$_{b1}$ triggers the first mantra (heavy limbs) and after that has been achieved the second goal goal$_{b2}$

Table 3 Connection weights and scaling factors for the example simulation

	1	2	3	4	5	6	7	8	9	10	11	12
Connection weight	ω_1	ω_2	ω_3	ω_4	ω_5	ω_6	ω_7	ω_8	ω_9	ω_{10}	ω_{11}	ω_{12}
Value	1	1	1	1	1	1	1	1	1	−0.01	−0.01	1
Connection weight	ω_{13}	ω_{14}	ω_{15}	ω_{16}	ω_{17}	ω_{18}	ω_{19}	ω_{20}	ω_{21}	ω_{22}	ω_{23}	ω_{24}
Value	−0.8	−1	1	0.7	1	1	0.7	1	1	1	1	1
Connection Weight	ω_{25}	ω_{26}	ω_{27}	ω_{28}	ω_{29}	ω_{30}	ω_{31}	ω_{32}	ω_{33}	ω_{34}	$\omega_{35}(0)$	$\omega_{36}(0)$
Value	1	1	1	1	1	1	1	1	1	1	0.6	0.6
Connection weight	ω_{37}	ω_{38}	ω_{39}	ω_{40}								
Value	−0.01	−0.01	−0.01	−0.1								
State	X_5	X_8	X_{10}	X_{11}	X_{15}	X_{16}	X_{21}	X_{22}				
	2	2	1	1	3	3	2	2				

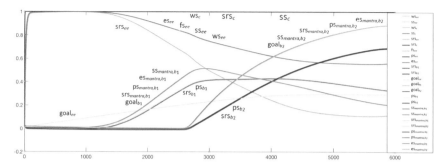

Fig. 2 Simulation results for autogenic training

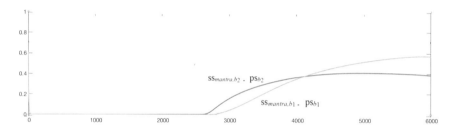

Fig. 3 Simulation results for Hebbian learning connections

for warm limbs is activated. The sensory representation of the first mantra starts working after some points of time around 1200 and after that the second goal $goal_{b2}$ starts at some point around 2700. The extreme emotion of the stressed individual starts decreasing when the first mantra starts working (around 1500). The second goal starts to have effect at time 3000 to decrease the stress of the stressed individual as the sensory representation of first mantra reached its high value. When the therapy started the stressed individual sensed their voice through both an as-if loop and an external speaking–hearing loop so she makes it stronger to become relaxed. The links from hearing and representing the mantras to the preparation states for the mantra's can be considered self-mirroring links.

So, based on the simulation results it is shown that the model for the Autogenic Training therapy works properly (Fig. 2).

As it can be seen from Fig. 3, the Hebbian learning connections which have been used from srs_{mantra,b_1}, srs_{mantra,b_2} to ps_{b1} and ps_{b1} show rocket increase from time 2700 for ps_{b1} and from 2800 for ps_{b1}. The suppression from ps_{ee} to both srs_{b1} and srs_{b2} and the mirror links to ps_{b1} and ps_{b1} prevent rocket increase and finally as can be seen at time around 4000 and 6000 this suppression take privilege and make the therapy stops working by decreasing the stress level. The parameter setting for both Hebbian Learning connections are as follows; for both connections the speed factors η are equal to 0.5, and the persistence factors μ are equal to 0.97.

5 Discussion

In this paper a cognitive temporal-causal network model was presented as a way in which mantras are used in a therapy, with two goals: heavy and warm body parts. This enables to release the stressed individual from an extreme emotion and makes two goals to affect the sensory representation of the extreme emotion. A number of simulations were performed one of which was presented in the paper. Findings from Neuroscience were taken into account in the design of the cognitive model. This literature reports experiments and measurements of therapy for emotion-inducing conditions as addressed from a computational perspective in the current paper. This model can be used as the basis of a virtual agent model to get insight in such processes and to consider certain support or treatment of individuals to handle and do the therapies of extreme emotions when they have to work in a stressful context condition and prevent some stress-related disorders that otherwise might develop. In further research, control states which happened in the brain can be added as an antecedent-focused- and response-focused emotion regulation techniques.

References

1. Sousa, N., Osborne, F.X. A.: Disconnection and reconnection: the morphological basis of (mal)adaptation to stress. Trends. Neurosci. 35(12), 742–751 (2012). https://doi.org/10.1016/j.tins.2012.08.006. Epub 21 Sep 2012
2. Masand, P.S., Gupta, S.: Selective serotonin-reuptake inhibitors: an update. Harvard. Rev. Psychiatry 7, 69–84 (1999)
3. Schultz, J.H., Luthe, W.: Autogenic Training. A Psychophysiologic Approach in Psychotherapy. Grune & Stratton, New York (1959)
4. Luthe, W.: Autogenic training: methods, research and application in medicine. Am. J. Psychother, 174–195 (1963). PMID: 13931814
5. Murakami, H., Katsunuma, R., Oba, K., Terasawa, K., Motomura, Y., Mishima, K., Moriguchi, Y.: Neural networks for mindfulness and emotion suppression. PloS One 17, 10(6), e0128005 (2015). https://doi.org/10.1371/journal.pone.0128005. eCollection 2015
6. Limb, C.J., Braun, A.R.: Neural substrates of spontaneous musical performance: an fMRI study of jazz improvisation. PloS One. https://doi.org/10.1371/journal.pone.0001679 (2008)
7. Kaixiang, Z., Minghua, B., Yu, Li., Yuman, X., Xuehua, G., Qunlin, C., Xue, D., Kangcheng, W., Dongtao, W., Huazhan, Y., Jiang, Q.: A distinction between two instruments measuring dispositional mindfulness and the correlations between those measurements and the neuroanatomical structure. Sci. Rep. 7, 652. Published online 24 July 2017. https://doi.org/10.1038/s41598-017-06599-w, PMID: 28740242 (2017)
8. Spijkerman, M.P., Pots, WT., Bohlmeijer, E.T.: Effectiveness of online mindfulness-based interventions in improving mental health: a review and meta-analysis of randomized controlled trials. Clin. Psychol. Rev. 45, 102–114 (2016). https://doi.org/10.1016/j.cpr.2016.03.009. Epub 1 Apr 2016
9. Treur, J.: Network-Oriented Modeling: addressing Complexity of Cognitive, Affective and Social Interactions. Springer Publishers, (2016)
10. Stetter, F., Kupper, S.: Autogenic training: a meta-analysis of clinical outcome studies. Appl. Psychophysiological. Biofeedback. 27(1), 45–98 (2002)
11. Gunter, K.: Evaluation of effectiveness of autogenic training in gerontopsychology. Eur. Psychol. 1, 243–254. Hogrefe Publishing (1996). https://doi.org/10.1027/1016-9040.1.4.243

12. Schultz, J.H. Luthe, W.: Autogenic methods. Autogenic therapy. In: Grune, N.Y. (ed.) Stratton, vol. 1 (1969)
13. Holzel, B.K., et al.: Mindfulness practice leads to increases in regional brain gray matter density. Psychiatry Res. **191**, 36–43 (2011). [PubMed:21071182]
14. Grant, J.A., Coutemanche, J., Rainville, P.: A no-elaborative mental stance and decoupling of executive and pain-related cortices predicts low pain sensitivity in Zen mediators. Pain **152**, 150–156 (2011). [PubMed: 21055874]
15. Schlamann, M., Naglatzki, R., de Greiff, Forsting, F., Gizewski, E.R.: Autogenic training alters cerebral activation patterns in fMRI, Int. J. Clin. Exp. Hypn, 58(4), 444–456 (2010). https://doi.org/10.1080/00207144.2010.499347
16. Lutz, A., McFarlin, D.R., Perlman, D.M., Salomons, T.V., Davidson, R.J.: Altered anterior insula activation during anticipation and experience of painful stimuli in expert mediators. Neuroimage **64**, 538–546 (2013). [PubMed: 23000783]
17. Treur, J.: Verification of temporal-causal network models by mathematical analysis. Vietnam. J. Comput. Sci. **3**, 207–221 (2016)
18. Mohr, J.P., Pessin, M.S., Finkelstein, S., Funkenstein, H.H., Duncan, G.W., Davis, K.R.: Broca aphasia: pathologic and clinical. Neurology **28**, 311 (1978)
19. Goldin, P.R., McRae, K., Ramel, W., Gross, J.J.: The neural bases of emotion regulation: reappraisal and suppression of negative emotion. Biol. Psychiat. **63**, 577–586 (2008). https://doi.org/10.1016/j.biopsych.2007.05.031
20. Craig, A.D.: How do you feel—now? The anterior insula and human awareness. Nat. Rev. Neurosci. **10**, 59–70 (2009)
21. Gusnard, D.A., Akbudak, E., Shulman, G.L., Raichle, M.: Medial Prefrontal Cortex and self-referential mental activity: relation to a default mode of brain function. Proc. Natl. Acad. Sci. U.S.A. **98**, 4259–4264 (2001)
22. Cntalpu, C., Hopkins, W.D.: Asymmetric Broca's area in great apes. Nature **414**(6863), 505 (2001). https://doi.org/10.1038/35107134.PMC2043144.PMID11734839
23. Purves, D., Augustine, G.J., Fitzpatrick, D., Katz, L.C., LaMantia, A.S., McNamara, J.O., Williams, S.M.: Neuroscience, 2nd edn, Sinauer Associates, Sunderland (MA) (2001)
24. Oosterwijk, S., Lindquist, K.A., Anderson, E., Dautoff, R., Moriguchi, Y., Barrett, L.F.: States of mind; emotions, body feelings, and thoughts share distributed neural networks. Neuroimage **62**(3), 2110–2128 (2012). https://doi.org/10.1016/j.neuroimage.2012.05.079. Epub 2012 Jun 5

Development of an Incident Prioritization Model Using Fuzzy Logic to Improve Quality and Productivity in IT Support Services

Dristesh Hoorpah, Somveer Kishnah and Sameerchand Pudaruth

Abstract Managing a high volume of incidents is a very complicated task for companies which provide support services. The application support analysts as well as managers must effectively assess the level of importance of incidents during the day to better prioritize each of them. As this process is very complex and time consuming, a lot of efforts are spent in the incident prioritization activity, which is manually carried out by the first level support team and by the analysts at the start of their shift and during the workday by going through each of the incidents and determining the order on which they need to be worked on. Bad incident prioritization leads to a decrease in quality of service as analysts fail to manage customers' expectations and this impacts productivity. To reduce this problem, a system which allows prioritization of incidents was proposed. To implement the solution, the range of factors which contributes to determine the priority of an incident was identified and a survey was conducted in multiple companies involved in ITSM to determine the importance of each of these factors. A fuzzy logic approach was formulated to determine the final priority of an incident. The results show a 19% increase in productivity and a 9% increase in quality of service.

Keywords Service management · Fuzzy logic · Incident prioritization

D. Hoorpah · S. Kishnah (✉) · S. Pudaruth
University of Mauritius, Reduit, Mauritius
e-mail: s.kishnah@uom.ac.mu

D. Hoorpah
e-mail: dristesh.hoorpah1@umail.uom.ac.mu

S. Pudaruth
e-mail: s.pudaruth@uom.ac.mu

© Springer Nature Singapore Pte Ltd. 2019
S. C. Satapathy et al. (eds.), *Information Systems Design and Intelligent Applications*, Advances in Intelligent Systems and Computing 863,
https://doi.org/10.1007/978-981-13-3338-5_7

1   Introduction

The majority of the Information Technology (IT) organizations offer support services for the software products that they develop and have adopted the Information Technology Service Management (ITSM) frameworks such as the Information Technology Infrastructure Library (ITIL). For IT companies to survive in a fiercely competitive market and to have an edge over competitors, delivering the best quality of service has become one of the topmost priority for them [1].

As the number of customers increases, the volume of incidents logged increases as well. Support teams have to deal with an increasing volume of incidents daily and are very prone to breach the service level agreement (SLA) due to bad prioritization which can eventually culminate into financial losses or drive customers away. To ensure a better alignment with the business objectives and to deliver the best customer experience, each incident needs to be resolved in a timely manner depending on its nature. The ITSM standards and frameworks define the priority of an incident as the evaluation of its impact and urgency [2].

Currently, the customer has the ability to assign a priority score to an incident and these incidents are then actioned based on the severity that customers assigned to them after being triaged and dispatched by the first level of support based on their experience and intuition or processes that have been defined by the company. The priority assigned to an incident by the customer is very subjective as an issue which is very critical and urgent to the customer might not be as urgent and critical to the support analyst and vice versa. On a daily basis, support analysts work on their inventory of incidents that they are assigned to them or they grab from queues.

The order in which they work on these incidents is very crucial to the business, so prioritization of tickets is very important. The current ad hoc methods of prioritizing incidents lead to some incidents being left out or they may not get the required attention. Prioritization has an impact on the quality of service provided and the productivity of the analyst. The priority assigned to an incident can be dynamic as it can change at any point in time. The ability to determine the priority with which the incidents should be worked to ensure that the quality factors are met while increasing the productivity is a major challenge. The development of a system which allows the automatic prioritization of incidents appear to be a proper way to tackle this problem. This will allow support analysts to work on the tickets that they have in their inventory sequentially while increasing the quality of service and productivity.

The paper proceeds as follows. In the next section, incident management is explained. The use of fuzzy logic is also described. Section 3 describes the execution of the survey and its analysis. The implementation is covered in detail in Sect. 4 while Sect. 5 discusses the results and finally, we conclude the paper in Sect. 6.

2   Background Study

2.1   Incident Management

ITIL defines an incident as an inadvertent disruption of an IT service, a degradation of the service quality or a failure of a configuration item that has not yet impacted the operations of that service [3]. Example of an incident is the inability of a user to commit payroll in a payroll application. The goal of incident management is to reestablish normal service operation in the quickest way possible to minimize impact on the operations of the business hence ensuring that the agreed levels of quality and availability are sustained.

Incidents are triggered in numerous ways. Incidents may be reported by users through emails, phone calls, or a web-based incident logging tool. Potential failures/defects are also logged by internal technical staff. All the ITSM frameworks consist of the following incident management process activities based on the ITIL best practices: incident identification, incident logging, incident categorization, incident prioritization, initial diagnosis, incident escalation, investigation and diagnosis, resolution and recovery, and incident closure.

2.1.1   Incident Prioritization

Incident prioritization is a very important aspect of the incident management lifecycle. In most ITSM frameworks, the priority defines the relative importance of an incident and is determined based on the business impact being caused by the incident and its urgency. Impact and urgency have each similar definition across the different ITSM frameworks. Impact measures the potential damage/effect of an incident on a business process before it is resolved while urgency is a measure of how fast a resolution is of the incident is required [4].

The major factor to determine the impact of an incident is the number of users affected. However, depending on the context, this factor may not enough to determine the impact level as loss of service to another user who was in the middle of a critical task can have a major business impact. Some other factors which contribute to the impact levels are listed below: number of services affected, risk to life, level of financial loss, effect on business reputation and breaches on legislations or regulations. The most popular method to derive the priority of an incident is given by the incident priority matrix in Tables 1, 2, 3 and 4. Other frameworks, for example ENISA, use color (red, yellow, orange) codes to determine priorities with red having the highest priority [5]. The Microsoft Operation Framework uses the ITIL incident priority matrix and uses categorization to determine the relative priorities [6].

To appropriately allocate priority level, support staffs are trained to be able to determine the correct urgency and impact levels. Guidance on how to use the priority levels is also provided to clients during service level negotiations. There are many cases where the priority levels will be overridden, for example: dealing with an

Table 1 Categories of urgency [7]

Category	Description
High (H)	Rapid increase in damage caused by incident The damage caused by the Incident increases rapidly High Sensitive task on hold Minor incident can translate to major incident if not actioned immediately. High number of users affected
Medium (M)	Considerable increase in damage caused by incident over time Only 1 VIP user affected
Low (L)	Marginal increase in damage caused by incident over time Work on hold but not time sensitive

Table 2 Categories of impact [7]

Category	Description
High (H)	High number of staff/task affected High financial impact caused by incident which exceeds a threshold High damage to the reputation of the company
Medium (M)	Moderate number of staff/task affected Moderate financial impact caused by incident which is likely to exceed a threshold (below the high threshold) Moderate damage to the reputation of the company
Low (L)	Minimal number of staff affected Workaround to complete affected tasks exist Minimal financial impact caused by incident which is likely to exceed a threshold (below the high threshold) Minimal damage to the reputation of the company

Table 3 Priority matrix [7]

		Impact		
		High	Medium	Low
Urgency	High	1	2	3
	Medium	2	3	4
	Low	3	4	5

Table 4 Priority code [7]

Priority code	Description	Target response time	Target resolution time
P1	Critical	Immediate	1 h
P2	High	10 min	4 h
P3	Medium	1 h	8 h
P4	Low	4 h	24 h
P5	Very low	1 Day	1 Week

uncompromising client who does not agree to the priority that the support staff has assigned to his incident as the incident might be very important to the client but not per the priority criteria. In some organizations, incidents from VIPs (politicians, high ranking executives, etc.) are handled with higher priorities. The priority of an incident may be dynamic to adapt to different circumstances. An incident which was logged as low priority can have its priority increased if it has not been resolved within the SLA target times.

2.1.2 Requirement Prioritization

Requirement prioritization is one of the most important and critical activity in the software development lifecycle. To prioritize the requirements, various techniques have emerged during the past years. However, they do not take into consideration the uncertainty and fuzziness that exists in requirements. Various research has been recently carried out to use fuzzy logic to efficiently prioritize requirements [8]. Dabbagh and Lee proposed a requirement prioritization approach which uses fuzzy logic to prioritize nonfunctional requirements based on their relationship with the functional requirements [9].

Ruby and Balkishan developed a fuzzy logic-based requirement prioritization (FLRP) approach to help with decision making [10]. The FLRP technique uses the following variables as input: customer importance, cost to implement, time to implement and risk. They used the Mamdani fuzzy system where all the input membership functions are of Gaussian type [11]. The output parameter is *Priority* which has membership function (triangular type) ranging from *Very High* to *Very Low*. Eighty-one IF-THEN rules were generated using these four input parameters. Requirements with higher weights have higher priority and those with lower weights have lower priorities.

2.1.3 Test Case Prioritization

Testing is the most important quality assurance activity in the software development process. Prioritization is needed as due to tight schedules, not all test cases can be tested. In [12], the authors proposed a fuzzy logic test case prioritization technique for a GUI-based software which considered multiple criteria for test case prioritization and prioritized the test cases that uncover the most faults. The following three factors were considered as inputs to the fuzzy system and their membership functions ranged from *Very Low* to *Very High*. The rule base consists of 125 (5^3) rules to determine the priority (output) of test cases.

3   Survey Execution and Analysis

The aim of the survey is to determine the importance of the different ITSM parameters that affect the priority of an incident. The importance of each of the different parameters may vary from one organization to the other and depending on the role of the person and his involvement in each ITSM process. The parameters were analyzed to determine their weights. A questionnaire was designed and circulated electronically in eight different software development companies, including the one where the first author is currently working. This particular company will be referred to as Company A.

3.1   Survey Analysis

Forty respondents from Company A and thirty respondents from seven other companies responded positively to the survey. Respondents had to assign a value between 1 (lowest priority) and 5 (highest priority) to each of the 21 parameters listed in the questionnaire. These parameters are: deadline, percentage of employees affected, number of escalation, status of incident, priority set by client, visibility of client, client risk category, number of days since last incident, client's revenue, number of days opened, incident category, is it a repeated incident, percentage of SLA breaches, number of days since system is live, client's employee count, size of project, client's business segment, most recent net promoter score from client, description of incident, and the number of incidents that are currently being processed for that client.

From the survey, we saw that deadline has been considered the most important factor to consider when deciding which task to work on next. Other highly rated factors are: percentage of employees affected by that requirement, the number of escalations, the status of the incident, and the priority set by the client. The description of the incident and the number of tickets opened by the clients were considered as the least important. For both Company A and all the eight companies (Global) taken together, the mean score for each parameter is very similar, which means that the parameters are rated in the same manner in the different companies.

Impact of Incident Prioritization on Productivity
The majority of the respondents reacted positively to this question. They strongly believed that an incident prioritization system which takes the parameters listed in

the survey will help to boost their productivity. The mean for company A was 4.83 and the mean for all companies combined was 4.79 on a scale of 1–5.

Impact of Incident Prioritization on Quality
Most of the respondents agreed that an incident prioritization system which considered the parameters listed in the survey will help to increase quality of service provided. The mean for company A was 4.93 and the overall mean was 4.90. This shows that respondents from all company were very confident that all these parameters should be taken into consideration when calculating the priority of an incident.

4 Implementation

The incident prioritization portal is a web-based application and provides a front-end access of the system. The portal communicates with a fuzzy logic system for calculating the priority of each parameter. All data are stored in a database server.

4.1 Fuzzy Logic System

The fuzzy inference system (FIS) was implemented using the Mamdani-type fuzzy inference. To come up with the optimal configuration of the fuzzy logic system. The design of the FIS was broken down into multiple scenarios to test different configurations in terms of the number of inputs variables and the number of rules combinations. The following scenarios were considered and the most efficient one was used for the fuzzy logic calculation in the incident portal.

Scenario 1
The first scenario consists of a basic implementation of the fuzzy logic system which takes all the 20 variables as input. For the system to behave well and provide a correct output, all rules for each combination of the factors need to be created. Generating rules with 20 variables, each having 5 factors causes a combinatorial explosion as the number of inference rules is equal to $5^{20} = 95,367,431,640,625$. Therefore, this approach is not feasible unless we have supercomputing power.

Scenario 2
The second scenario consists of a two-level fuzzy system. It involves grouping of the input variables into four different sets and generating all possible combinations of inference rules for each of them. In the first level, the input variables are grouped in sets of five in the order of their weights (descending) and the output of each set will serve as input to the second level which generates the final priority. The setup in this scenario brings down the number of inference rules to 13,125. The number of inference rules in the first set is: $5^5 = 3125$, and the total number of inference rules is therefore: $4 \times 5^5 = 12,500$. An example of a fuzzy rule is shown below.

Table 5 Variable-weight example

Variable	Weight	Parameter
Deadline	4.70	Very low (1)
Number of escalations	4.65	Very low (1)
Status	4.55	Low (2)
Number of employees affected	4.50	Low (2)
Importance of client	4.35	Moderate (3)

If *NDaysDeadline* is *Very Low* and *NEscalation* is *Very Low* and *Status* is *Low* and *EmpAffected* is *Low* and *ClientImportance* is *Moderate* then *Priority* is *Very Low*.

Table 5 shows a set of top 5 variables, their weights and parameter values. The parameters have a value from 1 to 5. In this case, the priority of the incident has been calculated as *Very Low*.

4.2 Incident Prioritization Portal

The incident prioritization portal has been built using the ASP.NET MVC (Model-View-Controller) framework and the database server is implemented using the Microsoft SQL Server. A priority is assigned to an incident as soon as it is created, and it can be recalculated on demand, giving the analyst a great span of control. The system implements a date driven concept which means that no tickets (including the lowest priority ones) will be left unattended. The usage of windows services in the implementation of the system has led to a considerable load reduction on the web-based incident portal as the priority calculations are performed in the background and are only triggered when required.

5 Results and Discussions

Five experienced support analysts were asked to prioritize 20 incidents from the incident portal. This proved to be a very complex and tedious task to complete as they had to open each incident to see their different factors and compare the 20 incidents against each other. Each application support analyst involved in this exercise took around 30–45 min to review the list of 20 incidents and eventually give a priority order index to each of them. A priority order index was used as metric as it would have been more time consuming and difficult for the individuals to rate the priority based on the weights of each parameter. An average of the priority provided by the five analysts was computed and the results were sorted to get the final priority order index.

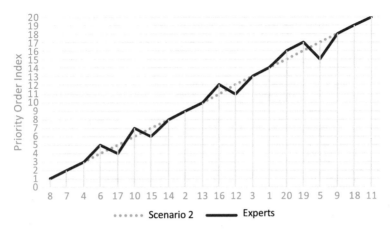

Fig. 1 Experts versus scenario 2 ranking of Incidents

The results obtained from this exercise were compared against the results from scenario 2. The results were converted into the priority order index to be able to compare them. The results from scenario 2 matches significantly the experts view results as shown in Fig. 1. Sorted by scenario 2 results (highest to lowest priority), the same priority order is given to the first 3 test runs (8, 7, 4) in both cases. However, in other cases, the priority order index given by the two cases are different, but the positions are close to each other. For example, while scenario 2 places incident #20 in the 4th position and incident #32 in the 5th position, the experts have swapped the positions of these two incidents.

To assess the impact of the proposed solution on productivity and quality, the system was implemented for a 1-week period in Company A and involved 5 application support analysts of varied experience levels. As the system has been implemented on a local machine, a copy of the database and the project was deployed on the machine of each of the 5 experts involved. The execution of this testing phase has been carried out in a parallel configuration since the incidents are logged through the incident management application (Parature) used by the support team. The execution of this testing exercise is very complex in this environment as during these 5 days of testing, the analysts should manually create a copy of all the incidents being logged on Parature as well as mirror all the updates to the incidents during the day. An average of 20 incidents was entered in the system by each analyst. To assess the impact of the system, the number of tickets closed per week was used as a productivity metric and the score of quality reviews (which is conducted internally and takes into consideration the percentage of SLA missed, responsiveness, setting expectation, etc.) was used as a quality metric. Productivity is measured as the number of tickets closed over the weekly target (10 incidents) and quality is measured in terms of a percentage.

Since the number of clients who log the incidents is very high, existing clients on the incident portal will be used to simulate the testing. The existing clients on the incident portal have similar features to existing clients in terms of the different

Table 6 Testing results—quality

Analyst	Week 4	Week 3	Week 2	Week 1	4 Weeks avg	New system
1	70	67	64	72	68	77
2	79	83	85	83	82	88
3	57	73	66	71	67	71
4	60	57	68	57	60	69
5	82	79	72	68	75	81

Table 7 Testing results—productivity

Analyst	Week 4	Week 3	Week 2	Week 1	4 Weeks avg	New system
1	14	7	12	10	11	14
2	10	13	7	10	10	12
3	15	9	12	13	12	11
4	8	12	9	10	10	13
5	4	9	11	7	8	10

variables (revenue, risk, employee count, visibility, etc.). The results from the real-life testing exercise are very promising as shown in Tables 6 and 7. However, it is also very difficult to draw a conclusion based on a week's data. To fully assess the impact of the new system on productivity and quality, a longer period of time, for example, 1–2 months, with more application analysts and testing in other companies would have been ideal. Thus, the results obtained from the 1 week of testing were then compared against previous data which is based on a period of 4 weeks.

The quality reviews score for that 1 week was above the average for all the analysts involved. The quality reviews result also shows an increase of 9% over the last 4 weeks' average. As the analysts were provided with the order in which they had to work on the incidents assigned to them, they were all able to set the correct expectations to the customers regarding when they will provide an update or resolution. Since updates to the clients were provided according to the deadline set and other factors, the percentage of SLA breaches also decreased. Two of the analysts involved also received fantastic scores from the customers whose tickets they were working on during the week for creating a better customer experience as they had been able to prioritize the incidents while assessing all the factors involved. According to the analysts, the new system has completely reduced the time that they were taking to prioritize the incidents during the day and provided a more systematic way of working thus allowing them more time for investigating the incidents. For the majority of the analysts, the productivity for that week was above the weekly average productivity. The results show an increase of 19% in productivity over the last 4 weeks average.

6 Conclusions

The implementation of this automated prioritization system was very complex due to the high number of variables which needed to be considered. The scenario used to implement the system consists of a two-level fuzzy logic system which groups the variables into four sets of five variables and generates all the possible combinations for each set. A web-based incident portal which allows incident management and implements the fuzzy logic model has also been developed. The incident portal allows its users to view all the incidents in the queue as well as the incidents assigned to them. The system provides dynamic functionalities as priorities are calculated as soon as incidents are created, and the priorities can be recalculated on demand for one or for all incidents in real-time. This helps the support analysts to always have an updated list of tickets to work on, sorted by priority. The analysts involved in the testing confirmed that the system helped them to increase their quality of service and productivity. The results show an increase of 9% in the quality of service and an increase of 19% in productivity over the previous 4 weeks' average. The system allowed the application support analysts to better plan their work and allowed them to set the correct expectation to the client which helps to create a better customer experience.

References

1. Schütze, R.: An intuitionistic fuzzy approach for service level management. In: Improving Service Level Engineering. Fuzzy Management Methods. Springer, Cham (2018)
2. The HP IT Service Management (ITSM) Reference Model. ftp://ftp.hp.com/pub/services/itsm/info/itsm_rmwp.pdf
3. van Bon, J.: Foundations of ITIL® V3 (Best Practice IT Management). Van Haren Publishing, The Netherlands (2007)
4. Sarnovsky, M., Surma, J.: Predictive models for support of incident management process in IT service management. Acta Electrotechnica et Informatica 18(1), 57–62 (2018)
5. European Network and Information Security Agency: Good Practice Guide for Incident Management. ENISA, Greece (2010)
6. Microsoft Operations Framework. https://technet.microsoft.com/en-us/library/bb497036.aspx (2018)
7. Kempter, S.: Checklist Incident Priority. IT Process Wiki. https://wiki.en.it-processmaps.com/index.php/Checklist_Incident_Priority (2018)
8. Zadeh, L.A.: A very simple formula for aggregation and multicriteria optimization. Int. J. Uncertainty Fuzziness Knowl.-Based Syst. 24(6), 961–962 (2016)
9. Dabbagh, M., Lee, S.: An approach for prioritizing NFRs according to their relationship with FRs. Lecture Notes on Software Engineering, vol. 3, issue 1, pp. 1–5 (2015)
10. Ruby, Balkishan: Fuzzy logic based requirement prioritization (FLRP)—an approach. Int. J. Comput. Sci. Technol. 6(3), 61–65 (2015)
11. Mamdani, E., Assilian, S.: An experiment in linguistic synthesis with a fuzzy logic controller. Int. J. Man Mach. Stud. 7(1), 1–13 (1975)
12. Chaudhary, N., Sangwan, O., Singh, Y.: Test case prioritization using fuzzy logic for GUI-based software. Int. J. Adv. Comput. Sci. Appl. 3(12), 222–227 (2012)

Automatic Text-Independent Kannada Dialect Identification System

Nagaratna B. Chittaragi, Asavari Limaye, N. T Chandana, B Annappa
and Shashidhar G. Koolagudi

Abstract This paper proposes a dialect identification system for the Kannada language. A system that can automatically identify the dialects of the language being spoken has a wide variety of applications. However, not many Automatic Speech Recognition (ASR) and dialect identification tasks are carried out in majority of the Indian languages. Further, there are only a few good quality annotated audio datasets available. In this paper, a new dataset for 5 spoken dialects of the Kannada language is introduced. Spectral and prosodic features have captured the most prominent features for recognition of Kannada dialects. Support Vector Machine (SVM) and neural networks algorithms are used for modeling text-independent recognition system. A neural network model that attempts for identification dialects based on sentence level cues has also been built. Hyper-parameters for SVM and neural network models are chosen using grid search. Neural network models have outperformed SVMs when complete utterances are considered.

Keywords Dialect identification · Sentence level · Complete utterance level
MFCCs · Support vector machines · Neural networks

N. B. Chittaragi (✉) · A. Limaye · N. T. Chandana · B. Annappa · S. G. Koolagudi
Department of Computer Science and Engineering,
National Institute of Technology Karnataka, Surathkal, Mangalore, India
e-mail: nbchittaragi@gmail.com

A. Limaye
e-mail: asavari.limaye@gmail.com

N. T. Chandana
e-mail: chandanant58@gmail.com

B. Annappa
e-mail: annappa@nitk.edu.in

S. G. Koolagudi
e-mail: koolagudi@nitk.edu.in

N. B. Chittaragi
Department of Information Science and Engineering,
Siddaganga Institute of Technology, Tumkur, India

© Springer Nature Singapore Pte Ltd. 2019
S. C. Satapathy et al. (eds.), *Information Systems Design and Intelligent
Applications*, Advances in Intelligent Systems and Computing 863,
https://doi.org/10.1007/978-981-13-3338-5_8

1 Introduction

The dialect of any language represents a pattern of pronunciation followed by the community of native speakers belonging to the particular geographical region. The dialect of the language spoken is largely influenced by his/her area of origin and also by the region(s) in which the person has lived. Further, geographical location, social class, and ethnicity can also influence dialects.

An automatic dialect identification can be used by spoken information retrieval systems to provide more relevant and personalized results to their users. Certain words those are quite often and vary across dialects can model dialects effectively. Dialect detection systems thus can improve the performance of dictation software as well as personal assistants. Dialect recognition can also be useful in preparing speaker profiles, verifying records (for background screening, etc.), investigating forensic cases, and making processes like address verification more robust and efficient. The dialects and accents of a person largely depend on their geographical location and native information. Performance of ASR systems is highly improved with dialect recognition systems [1].

This paper proposes a model that is capable of capturing phonological variations followed among dialects and classified appropriately. Dialect-specific spectral features are captured through modeling vocal tract information. Variations in pronunciations can be clearly recognized through the varying shapes of the vocal tract. Features such as spectral centroid, spectral flux, spectral density, spectral roll-off, and popular MFCCs are extracted. Similarly, prosody differences among dialects are explored using short-term energy and pitch features. Further, a combined feature vector is postprocessed statistically and derived feature vector is created. SVM and neural networks machine learning techniques are used for identification of five dialect classes. Optimized hyper-parameters selection has produced better dialect recognition with neural networks. Performance analysis is also conducted with sentence and utterance level speech samples.

Remaining part of the paper is organized as follows. Section 2 discusses briefly the existing works in the area of dialect processing. Section 3 presents the proposed dialect recognition system covering the individual steps. Section 4 describes experiments carried out using two machine learning models. Section 5 presents the discussion of obtained results and analysis. The conclusion of the present work and future directions are given in Sect. 6.

2 Literature Review

This section briefly explains various approaches adopted for dialect identification tasks. Although extensive research has been found for language identification, limited work has been observed for dialects. It is a challenging task over language identification due to unavailability of clear boundaries between dialects. Majority of the systems have considered dataset properties such as a speaker, the transcription, gender information and type of spoken text, either spontaneous or read. Indeed, text independent (spontaneous) is most preferred for accurate identification of dialects [2, 3].

Two approaches namely, acoustic-phonetic and/or phonotactic models are more commonly followed for addressing dialect identification. Phonotactic approaches such as Phone Recognition Language Modeling (PRLM) and few variants namely, parallel PRLM (PPRLM), Parallel Phone Recognition (PPR) are language dependent models suggested for dialect processing [4].

From acoustic-phonetic perspective, a system is proposed by considering the collection of common words used across dialects and HMMs trained for each word. Later, a majority vote over all the words in an utterance predicts dialect for a whole utterance [3, 5]. A study is carried out by combining prosodic features with MFCC features. Prosodic features like intonation and rhythm are extracted from larger units of speech such as syllable-like units instead from phonetic segments (vowels and consonants) [2, 6]. Existing systems have stated that knowing gender of the speaker beforehand greatly influences dialect identification performance [7].

Experiments on Indian languages and dialects have also been observed. From five distinct Hindi dialects prosodic duration, pitch and energy contours and spectral MFCC features are extracted from syllable units. Applied auto-associative neural networks model for implementation and shown better results [8, 9]. Very few studies can be found for addressing dialect recognition problem for other Indian languages including Kannada. Proposed studies have referred commonly used spectral and prosodic features for classifying three dialects of Kannada [10]. Even though research is found for dialect classification, it still remains active due to the following reasons:

1. Significant research is limited for regional Indian languages like Punjabi, Malayalam, Tamil, Kannada, Marathi, etc.
2. The average accuracy is lingering around 70% which is quite low.
3. Existing systems consider whole utterance instead the phoneme, word, or sentence level may be selected.
4. An audio signal contains a lot of information and a wide variety of dialect-specific features can be extracted.

3 Proposed Dialect Identification System

Various processes involved in the development of proposed system are discussed briefly in this section. A workflow diagram describing all the significant stages is represented in Fig. 1. The system is implemented in Python.

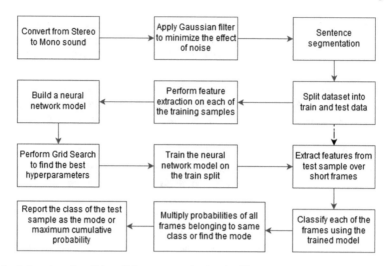

Fig. 1 A flowchart describing all the processes in the classification system

3.1 Dataset Used

The dataset used in this work consists of five classes each representing a dialect of Kannada. These five dialects are unique pronunciation patterns of Kannada spoken across the Karnataka. The dataset used in this work has been recorded from native speakers in an interview style by asking random questions from a questionnaire with 15 specific questions. While selecting participants sufficient care is taken. Only native speakers who have stayed in the same place since childhood and their education qualification is lesser are selected. Recording involved natural interaction between the interviewer and speaker as the interviewer asked them questions on their daily routine, family, occupation and so on. Hence, the speech available is spontaneous and text independent. The recording was done with Sony Voice recorder with a sampling frequency of 44,100 Hz. Approximately, 10 hours of data is available for training and testing.

3.2 Data Cleaning and Data Preparation

1. **Converting from Stereo to Mono** The recorded audios from various speakers were stereo sounds and are converted to mono for easier processing.
2. **De-noising of audio samples** Few of the recordings are contained faint background noises. Gaussian filters and Savitzky–Golay smoothing filters are used to de-noising. Gaussian filters gave better Signal-to-Noise Ratio (SNR) and hence were used in the experiments.

3. **Sentence segmentation algorithm** In this paper, classification of dialects at the sentence level is done. An segmentation algorithm is designed using period of silence and a silence threshold features to obtain sentence level data from the recordings. A silence period of 350 ms and a threshold of −32 dB is found heuristically for best splitting of majority of the audio clips into sentences. However, few exceptions are found and these are handled by considering the time duration of each chunk. If the duration is too small (less than 1 s) it is discarded, if it lies between 1 and 5 s, its retained and if its big (more than 5 s), then it is recursively split into further smaller chunks to get sentence level data.

3.3 Feature Extraction

Spectral as well as prosodic features are extracted from the audio samples for effective dialect discrimination. These features are extracted from short-term processing of a frame size of 20 ms and an overlap of 10 ms. Spectral features such as 39 RASTA processed MFCCs (13 coefficients each for MFCC, Delta and Delta Delta), energy entropy, spectral centroid, spectral spread, and spectral flux are extracted from each frame. These features are known to capture the unique dialectal information. Dialects of a language also show variations at supra-segmental or prosodic level. Hence, features such as energy, pitch, and loudness are capable of capturing the stress, rhythm, and intonation patterns exist among dialects. All the features are extracted using a toolbox [11]. The final feature vector contains 46 features. The final feature vector is further postprocessed statistically by computing mean and standard deviation of two consecutive frames from the original feature vector thereby giving rise to a new derived feature vector. The derived feature vector helped to improve performance significantly [12].

4 Implementation of Models

In this paper, dialect identification systems are designed using two different machine learning techniques namely SVMs and Neural Networks. Scikit-Learn was used to implement the SVM and Neural Network models.

SVM: In the beginning, SVM is used for classification of five Kannada dialects using the derived feature vectors. Since MFCC is a critical feature, and SVMs are also trained with feature vectors comprising of MFCC feature and its variations. SVMs for five class classification are implemented using the one-versus-rest approach with Radial Basis Function (RBF) kernel. The SVM model has been trained and tested in 80:20 ratio with fivefold cross-validation configuration to make the model more stable and to involve every part of the dataset for training and testing. For SVMs, the optimal value of hyper-parameter C(cost of mis-classification) is chosen with grid search and 5-fold cross-validation.

Neural Networks: Neural network based models are built for classifying the dialects at the sentence level as well as complete utterance. Neural networks are capable of learning more complex nonlinear relationships as compared to SVMs and hence are expected to outperform SVMs. A train test split of 80% and 20% is used for training the model and testing the model, respectively. Fivefold stratification is used to ensure that there is equal representation from each dialect class in the train and test split.

Hyper-parameter Optimization: Neural networks involve many hyper-parameters like activation function, number of hidden layers, solver, etc. Grid Search has been used to configure the hyper-parameters for the neural network model. For each hyper-parameter, the values which are to be tried out are provided as lists. Grid Search then tries out all possible combinations of these hyper-parameter values provided by constructing and evaluating one model for each combination and returns the best combination. Fivefold cross-validation has been used to evaluate each individual model. Precision and Recall have been optimized for. The activation function, solver, and structure of the neural networks are chosen using the following three steps:

1. Choosing the activation function: The activation functions "logistic," "tanh," and "relu" have been tried out along with the network hidden layer sizes (40, 40, 40, 40, 40) and (200, 200, 200).

2. Choosing the solver: Once the activation function is chosen, the solvers "adam" and "lbfgs," are tried out, along with the other hyper-parameters. The activation function chosen from the previous step is used, and the hidden layer sizes used are (200, 200, 200).
3. Choosing the structure of the network: The hidden layer sizes (20, 20, 20), (40, 40, 40), (200, 200, 200, 200), (40, 40, 40, 40), (100, 80, 40, 30) are tried out with the selected activation function and solver from the previous two steps.

Similar procedures are followed to perform predictions at the sentence level and the complete utterance. Features are extracted over short windows of the test sample and then the dialect class for each window is predicted using the trained model. Subsequently, two methods are followed to predict the dialect of the entire sample.
1. Multiplying the probabilities from each window of the sample for each class, and choosing the class with maximum probability.
2. Finding the class which is predicted the most often (mode) for all the windows in the sample.

5 Experiments and Results

This section describes the significant results of the experiments carried out. Accuracy has been used as a performance metric to compare various classification models built.

5.1 Performance with SVM

Table 1 shows the optimal value of C, train and test accuracies, and the difference between them for the different feature combinations that have been experimented with. For evaluation of the dialect identification system, feature sets are explored from test speech samples and are verified with five individually trained SVM models. Using the derived feature vector for training and testing, the SVM model produced a train accuracy of 46% and a test accuracy of 38% . The best result with a train accuracy of 71% and a test accuracy of 66% was obtained with the feature vector comprising of MFCC and it's variations.

5.2 Performance with Neural Networks

Optimized neural networks with ReLU as the activation function, the Adam solver, and use of four hidden layers with 200 neurons each consistently gave higher train and test accuracies for complete utterance level classification as compared to SVMs. Experiments are conducted to analyze neural network performance at both sentence and complete utterance level. The results obtained after training the neural network using the best hyper-parameters chosen are given in Table 2 for sentence and Table 3 for complete utterance level.

Neural networks are better at classifying dialects at the complete utterance level as compared to sentence level. The test accuracy for classification of the complete

Table 1 Training and test accuracies for various feature combinations using SVMs

Feature combination	Best value of C	Train accuracy (%)	Test accuracy (%)	Difference (%)
Derived feature vector (spectral+prosodic)	0.0001	46	38	8
MFCC	0.5	66	71	5
MFCC delta	25	84	48	36
MFCC double delta	50	59	59	0
MFCC + MFCC delta + MFCC double delta	0.5	71	66	5

Table 2 Dialect recognition accuracies with neural network model for sentence level

Accuracy/Method	Neural network prediction	Multiplied probabilities	Mode of predicted classes
Train accuracy (%)	91.90	99.63	98.95
Test accuracy (%)	74.93	83.09	83.09

Table 3 Dialect recognition accuracies with neural network model for utterance level

Accuracy/Method	Neural network prediction	Multiplied probabilities	Mode of predicted classes
Train accuracy (%)	92.71	100	100
Test accuracy (%)	78.91	99.24	99.24

utterance is 99.24% whereas the test accuracy for classification at the sentence level is significantly lower at 83.09%. The reason for this behavior is that dialectal cues are extracted more effectively from longer utterances when compared to shorter utterances like sentences.

6   Conclusion

This paper presented a method to build an automatic dialect recognition system for five dialects of Kannada. Text-independent dialect dataset is collected and used for evaluation. Spectral and prosodic features are extracted by frame level processing of the speech signal. Among all the features, MFCC features alone are most suitable for classifying dialects. MFCC features model the shape of the vocal tract which determines the type of sounds generated. Hence, they can help in capturing the phonological variations among the dialects. Experiments are conducted with two popular machine learning techniques namely, SVMs and neural networks. Hyper-parameters for the neural networks are chosen using grid search and they outperformed SVMs. The idea of classifying small frames of the input audio sample and then multiplying the probabilities of each class for each of the frames or finding the mode of the classes gave higher accuracy as opposed to classifying the entire input sample as a whole. This work can be further extended by attempting classification at the word level and phoneme level. Few more hyper-parameters can be fine-tuned and deeper neural networks can be used to enhance the performance of the system.

7   Declaration

We have taken the prior permission from the respective people and the dataset used in this study. We shall be responsible for any kind of conflicts in future.

References

1. Harris, M.J., Gries, S.T., Miglio, V.G.: Prosody and its application to forensic linguistics. LESLI: Linguist. Evid. Secur. Law Intell. **2**(2), 11–29 (2014)
2. Rouas, J.L.: Automatic prosodic variations modeling for language and dialect discrimination. IEEE Trans. Audio Speech Lang. Process. **15**(6), 1904–1911 (2007)

3. Huang, R., Hansen, J.H.L., Angkititrakul, P.: Dialect/accent classification using unrestricted audio. IEEE Trans. Audio Speech Lang. Process. **15**(2), 453–464 (2007)
4. Zissman, M.A.: Comparison of four approaches to automatic language identification of telephone speech. IEEE Trans. Speech Audio Process. **4**(1), 31–44 (1996)
5. Chittaragi, N.B., Koolagudi, S.G.: Acoustic features based word level dialect classification using SVM and ensemble methods. In: 2017 Tenth International Conference on Contemporary Computing (IC3), pp. 1–6 (2017)
6. Lei, Y., Hansen, J.H.L.: Dialect classification via text-independent training and testing for Arabic, Spanish, and Chinese. IEEE Trans. Audio Speech Lang. Process. **19**(1), 85–96 (2011)
7. Clopper, C.G., Smiljanic, R.: Effects of gender and regional dialect on prosodic patterns in American English. J. Phon. **39**(2), 237–245 (2011)
8. Sinha, S., Jain, A., Agrawal, S.S.: Speech processing for Hindi dialect recognition. In: Advances in Signal Processing and Intelligent Recognition Systems, pp. 161–169 (2014)
9. Rao, K.S., Koolagudi, S.G.: Identification of Hindi dialects and emotions using spectral and prosodic features of speech. Int. J. Syst. Cybern. Inform. **9**(4), 24–33 (2011)
10. Soorajkumar, R., Girish, G.N., Ramteke, P.B., Joshi, S.S., Koolagudi, S.G.: Text-independent automatic accent identification system for Kannada language. In: Proceedings of the International Conference on Data Engineering and Communication Technology, pp. 411–418. Springer, Berlin (2017)
11. Giannakopoulos, T., Pikrakis, A.: Introduction to Audio Analysis: A MATLAB Approach. Academic, Amsterdam (2014)
12. Chittaragi, N.B., Prakash, A., Koolagudi, S.G.: Dialect identification using spectral and prosodic features on single and ensemble classifiers. Arab. J. Sci. Eng. **43**, 4289–4302 (2017)

Identifying the Reusable Components from Component-Based System: Proposed Metrics and Model

Neelamadhab Padhy, Rasmita Panigrahi and Suresh Chandra Satapathy

Abstract Reusability is the key component from the software development prospective. This job describes a measurement which is popularly called as reprocess measurement to discover and investigate the static activities of the module. This paper proposed a set of reusability metrics especially partly adaptable, completely changeable and moderately capable modules. The process of reusability can be measured to degree of module in component-based system. This paper provides the novel model as well as the proposed metrics. We propose reusability-metric for all categories of components including partially modifiable, fully modifiable as well as for off-the-shelf components. Using reusability-metric, we draw a reusability-matrix containing the reusability ratios of all the different classes of components. This paper introduces selection criteria for components by using the reusability features of component-based software.

Keywords CBS (Component-Based System) · Reusability · Reusability-matrix

1 Introduction

Software is assured because one of the most important developments finished till today, which eventually has motivated person culture with wide-ranging applications like as science and technology, industry, investment, medicinal, and defence. Human life has become more competent and lavish, due to high rate surfing of software

N. Padhy (✉) · R. Panigrahi
Rasmita Panigrahi Gandhi Institute of Engineering and Technology, GIET (Autonomous), Gunupur, India
e-mail: neelamahdabphd@gmail.com

R. Panigrahi
e-mail: rasmi.mcamtech@gmail.com

S. C. Satapathy
Kalinga Institute of Industrial Technology, Bhubaneswar, Odisha, India
e-mail: sureshsatapathy@gmail.com

© Springer Nature Singapore Pte Ltd. 2019
S. C. Satapathy et al. (eds.), *Information Systems Design and Intelligent Applications*, Advances in Intelligent Systems and Computing 863,
https://doi.org/10.1007/978-981-13-3338-5_9

applications and its associated technologies. This leads to non-changeable object of the contemporary person life would not be astounding, since it has made easier plus comfortable in each useful action used for human existent in upcoming events and plans. These significances symbolize the importance of software in human life, science, and technologies. Recently, the growing price of software-based computations has initiated academia-corporate/industries to progress best assured software design solutions that may ensure cost-effective software systems deprived of cooperating with its reliability, scalability, stability, and overall quality. Accomplishing software quality, while making sure the lowest cost of design can make software industry to remain noteworthy in market and to attain higher market share. Currently, ensuring reliability (survivability, aging-resilient processing) can be of supreme significance for the manipulators and industries to make software as a fragment of individual efficacies.

With is new component we will get several advantages these are as below.

- Condensed charge of growth,
- More consistency,
- Required small expenditure for continuance.

The CBSD "Component-Based Software Design" system has driven importance all over the sphere, due to reusability feature, low cost, and reliable software solution. This type of a software development has provided novel measures of the price effectual and excellence software plan.

2   Literature Survey

Taking into consideration of the significance of literature study and analysis, in this paper, a number of literatures discussing various techniques for software reusability and especially the fault. With help of the literature survey, we got the research questions as well as developed the hypothesis. Different models have been developed in order to get the optimal reusability as well as cost-effective one. The major benefits of the systematic literature survey are given below:

- To know the effectiveness of the research
- To recognize the alive work in that particular problem statement
- Be aware of the flaws of the research
- To know the improvization of the research.

The abovementioned points help us to further deep research in the future.

Diamantopoulos et al. [1] explored the software reusability and they have focused the procedure of collecting the reusable components from online repositories. They have further integrated the different sources of files into single one and created one recommendation system for others researchers to follow. Hinkel et al. [2] present the model where a document can be designed. Properly as well as it analyzes, run and generate the code. Hudaib et al. [3] explored that how software reusability possible by using novel soft computing self-organizing maps. They also discussed how time will reduce by using software reusability concept. Irshad et al. [4] studied and presented the literature about the software reuse necessity and importance of quality. They have done the systematic literature survey and used the snowball sampling method to investigate the papers from the database. They also focused industrial investigation about reusability. Sarro et al. [5] published a paper related to evolutionary algorithms in software project management. They pointed out how software engineer delivered the poor quality products. To avoid the cited problem, they presented novel multi objective decision system to equilibrium the development. Al Dallal et al. [6] conducted a review where primarily 76 relevant articles discussed where mainly focused software refactoring. A cost estimation model is derived to assess cost-efficacy of each algorithm. Singh and Tomar [7] present the reusability hierarchy as well as mentioned the factor influences related to software reusability estimation. Padhy [8] discussed the complexity parameter through the Multiparadigm approach. They have taken a sample of 20 programs on the java script and use their proposed metrics and estimate the complexity. They have developed their prototype models as well as new metrics. Padhy [9] suggested the combination of metrics which can perform the better result. They have taken the two sets of metrics and exhibited using C++ constructor program. They have created own data set and plot the graph between the proposed metrics. In a program, if we obtain added sufficient amount of methods then the collision will more. These finds they proposed in their paper. Padhy [10] defined the set of reusable properties.

They have done the literature survey from the decades 2010–2017 and found most reusable factors from the software code. They have identified a list of assets and plot a graph author and year-wise publication. Padhy et al. [11] examined both OOMCK "Object-Oriented Metrics and CK metrics suite". Numerous researchers have calculated software reuse. However, there exists a gap between the literature review and its corresponding reusability. This summary obviously mentioned the requirement of reusability in software projects.

Padhy [12] discussed aging system. The first time presented the paper related to aging system. If we excess reuse the existing classes into new module, then the system may be safe or unsafe system. In their paper, they have observed the dissimilar reusability forecasting models. Their primary target was the OO-SM "object-oriented software metrics" and taken as key elements, i.e., "cohesion, coupling, and complexity ". They have proposed the model which is called as CERM "cost-efficient reusability prediction model" which exhibits the novel techniques out of that reusability threshold estimation is one of the important characteristics. Different machine learning algorithms have been discussed to achieve the task.

Padhy et al. [13] designed a prototype and it is proposed to make use of a new evolutionary computing-based artificial intelligence or machine learning scheme for regression tests to be used for reusability assessment. Padhy [14] discussed the algorithms and model for estimating the metrics from the mobileApps. They have discussed several facts about the MobileApps and taken the case study to access the software quality metrics.

Padhy et al. [15] focused the reusability cost estimation techniques by using novel machine learning algorithms. They presented how an evolutionary computing algorithm works to estimate software cost-estimation. They presented Enhanced evolutionary computing based artificial intelligence model for web-solutions software reusability estimation.

Padhy et al. [16] presented a set of 20 programs each and demonstrated the complexity of the metrics. They have proposed the model and algorithms to estimate the metrics set from object oriented programs.

3 Proposed Reusability-Metric from Component-Based System

3.1 The Complete Function Tip of a Module

It is stated as the summing up of reprocessing task points and fresh function points.

$$|\text{FPC}_i| = \left|\text{FPC}_{i\text{-Reprocessed}}\right| + |\text{FPC}_{\text{Fresh}}| \tag{1}$$

where FPCi be the entire amount of task point of a module, - is to calculate of reusable function points, and FPCFresh is the sum of fresh task points. The task point is commonly known as the functional points.

3.2 The Total Number of Reprocess Task Points of a Module

It takes an account of off-the-shelf task points and flexible task points of a module.

$$\left|\text{FPC}_{i\text{-Reprocessed}}\right| = |\text{FPC}_{i\text{-Off-the-shelf}}| + \left|\text{FPC}_{i\text{-complete\_capable}}\right|$$
$$+ \left|\text{FPC}_{i\text{-Fraction\_capable}}\right| \tag{2}$$

where FPCi-Reprocessed be the adding up of reusable task points, FPCi-Off-the-shelf is the count of off-the-shelf task points, FPCi-complete_capable is the amount of completely capable task points, and FPCi-PFraction_Capable is the amount of partly capable task points.

3.3 Flexible Task Points of a Module

Flexible module consists of the partly capable at the same time completely capable works. Partly capable task points are accomplished as of the partly capable tasks which require a key quantity of revision to acquire reused in the existing claim. Completely capable task points are accomplished beginning from the completely capable works which require a negligible amount of revision to acquire reprocess in the present relevance.

$$|FPC_{i\text{-Flexible}}| = |FPC_{i\text{-Complete\_capable}}| + |FPC_{i\text{-Fraction\_capable}}| \quad (3)$$

where Ci-Flexible is the whole number of reusable task points, FPCi-Complete_Capable is the numeral of completely capable task points, and FPCi-Fraction_capable is the amount of partly capable task points.

3.4 Novel Task Points of a Module:

It is the task point which can be obtained by using restricted of reused task points from the entire task points of the module

$$|FPC_{Fresh}| = |FPC_i| - |FPC_{i\text{-Reprocessed}}| \quad (4)$$

where FPCFresh symbolizes the fresh task points, and FPCi-Reprocessed is the add up of reprocessed task points.

4 Calculation of Function Points

In this part of an article, the task points are considered as the base measurement for the reusability. When we identify the key components from the database, then their task points must attach with their module. Calculation of task point is a just once movement and be able to set aside with the task as a routine characteristic in the database for further utilize others we should use our traditional approach, i.e., TLOC(Total Lines of Code Metrics). This paper present the novel reusability metrics approach at several layers that may be the module level and component-based system(CBS) level.

4.1 Factor Influenced in Reprocessed Measurement at Module Stage (FIRMCi)

By means of task point, we describe reprocessed measurement at the module stage when the relative amount among the full amount number of task points of a module and the total amount of reprocessed task points of the matching module.

$$\text{FIRMC}_i = \frac{\left|\text{Task}_{i\text{-Reprocessed}}\right|}{\text{Task}_i} \tag{5}$$

4.2   Factor Influenced in Reprocessed—Measurement at Task Stage (FIRMCCBS)

In this stage, reprocessed metrics may be computed by using the total number of fresh, partly and completely capable in task-based software system (TBSC). While task points are obtainable then the measurement can be stated proportion between the sum of reused task points of every module and the sum of task points of the TBSC.

$$\text{RMC}_{\text{TBSC}} = \frac{\left|\text{TPC}_{\text{TBSC-Reused}}\right|}{\left|\text{Task}_{\text{TBSC}}\right|} \tag{6}$$

This article the reusability factors can be calculated in different ways. These are as below.

4.3   Flexible Module

This is the module where we can obtain the reusability, where fully qualified adaptable and partially qualified adaptable component plays a vital role.

Off-the-shelf components
This is the module where we can achieve the reusability without any modification.

5   Adaptable Reusability-Metric (RMAdaptable)

The term flexible modules are the modules which primarily contains the active necessities, code, design pattern as well some test cases with negligible or major change. As per requirements flexible tasks can be broadly classified into two dissimilar groups these are as below:

- Completely capable module (Slightly required to change in the software code)
- Moderately capable modules (More required to change in the software code).

At the beginning of the scenario the amount of alteration, we have two scenarios

(a)   Flexible reusability-metric at module level (FRMM)
(b)   flexible reusability-metric at system level (FRMS).

6 Completely Capable and Flexible Reusability-Metric at Component Level (CCFRMi-Full-Qualified)

It required a small modification to stable in the new frameworks. At this point, the reprocess assessment can be done along with the function points. There are various ways to measure these are as below:

RQ 1: What will happen when only flexible parts of the component are involved? Solution: In this part of the research question can be elaborately defined.

It is the ratio between sums of total numbers of completely capable task divided by the total numbers of flexible task points of the module. By using the above, we can define the reusability ratio:

$$\text{CCFRM}_{\text{i-Completely-Capoable}} = \frac{\left|\text{FPC}_{\text{i-Completely-Capoable}}\right|}{\left|\text{FPC}_{\text{i-flexible}}\right|} \tag{7}$$

RQ 2: What will happen when every parts associated with the module?

It completely depends on the two ratios and these are completely capable component as well as sum of all the FP (Function Points) of a particular module. This can be represented as below:

$$\text{CCFRM}_{\text{i-Completely-Capoable}} = \frac{\left|\text{FPC}_{\text{i-Completely-Capoable}}\right|}{\left|\text{FPC}_{\text{i}}\right|} \tag{8}$$

RQ 3: What will happen when some of the parts are reused during the component selection?

To answer this,
The mathematical equation has been developed and it is mentioned below:

$$\text{CCFRM}_{\text{i-Completely-Capoable}} = \frac{\left|\text{FPC}_{\text{i-Completely-Capoable}}\right|}{\left|\text{FPC}_{\text{i-Reprecessed}}\right|} \tag{9}$$

i.e., the ratio between completely capable module and completely reused module.

7 Proposed Model

From the below mentioned proposed model (Fig. 1) which indicates the following steps:

(1) Reliability measurement can be done through the module repositories
(2) Reliability measurement can be defined through the context model
(3) Measure the component reliability in the current context

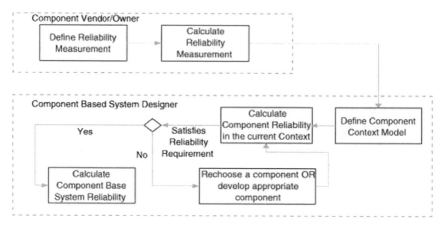

Fig. 1 CBS dependability process

Table 1 Matrix for fresh and reprocessed FP (Functional points)

Reusability-matrix	Task1	Task2	Task3	Task4
CCFRMi	0.88	0.79	0.77	**0.99**
CCFRMi-partially-capable	0.54	0.58	0.36	**0.51**
CCFRMi-Completely-capable	0.36	0.29	0.36	**0.42**
CCFRMCi-Moderately-capable	0.27	**0.39**	0.08	0.27
Total CCFRM For all functional points	**0.98**	0.74	0.58	**0.99**

(4) Does it Satisfies Reliability Requirement? If yes, then calculate the Component Base System Reliability. Otherwise, develop the suitable module else rebuild the module.

In order to represent the fully capable and partially and moderately capable, we have taken one case study where four tasks are identified and the values we have obtained using the previous mentioned equations. In Table 1, the data we have derived by using the other two tables which has not mentioned here.

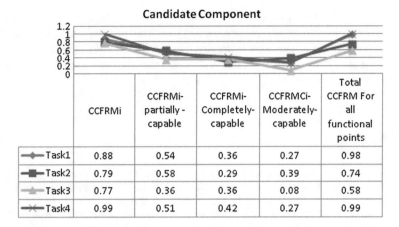

Fig. 2 For candidate component

	CCFRMi	CCFRMi- partially - capable	CCFRMi- Completely- capable	CCFRMCi- Moderately- capable	Total CCFRM For all functional points
Task1	0.88	0.54	0.36	0.27	0.98
Task2	0.79	0.58	0.29	0.39	0.74
Task3	0.77	0.36	0.36	0.08	0.58
Task4	0.99	0.51	0.42	0.27	0.99

8 Reusability-Matrix

It is one type of a matrix surrounding the calculations of reusability-metric of every module in the component-based software. Information surrounding reusability-matrix working as an assortment and confirmation criterion for the candidate components.

The below-mentioned figure obtained from the above Table 1. The values obtained from the different equations. In this part of this article, we have taken the case study which analyzes the reusability metrics from the component-based system. There are initially four tasks are considered which is further known as tasks. At this point, hypothesis is that the function points are accessible with each candidate component.

In the above Fig. 2 clearly indicates that, four tasks are considered for the same purpose. In this stage, we have taken one case study which is one kind of project. That project consists of the different modules where student registration is one of the them where the user must log in as a candidature. This module verifies the username, password, and Rollno as well as registration number. In the above mentioned Fig. 2, four tasks are identified for candidate component generation.

9 Conclusion

This article presents the novel models as well as the reusability estimation metrics from the component-based system. There are different types of estimation modules done like fresh, partly capable, completely capable components. The acquired consequence brings to a close that the reusability-metric be able to be used at three dissimilar stages for the apparatus selection as well as confirmation. The distinct reusability-metric be capable to help while apparatus are selected at organization stage. There are different values are obtained which are primarily used for selection or elimination of the component. Primarily, these estimation values are important for both the reason which is to be stored in the database like component-based system and performance coefficient.

References

1. Diamantopoulos, K., Thomopoulos, Symeonidis, A.: QualBoa: reusability-aware recommendations of source code components. In: IEEE/ACM 13th Working Conference on Mining Software Repositories (MSR) Austin, TX, pp. 488–491 (2016)
2. Hinkel, G., Kramer, M., Burger, E., Strittmatter, M., Happe, L.: An empirical study on the perception of metamodel quality. In: Proceedings of the 4th International Conference on Model-Driven Engineering and Software Development (MODELSWARD). IEEE, Rome, Italy, 16990400 (2016). Electronic ISBN: 978-989-758-2
3. Hudaib, A., Huneiti, I.: Othman: software reusability classification and predication using self-organizing map (SOM). Commun. Netw. **8**, 179–192 (2016). http://dx.doi.org/10.4236/cn.2016.83018
4. Irshad, M., Petersen, K., Poulding, S.: A systematic literature review of software requirements reuse approaches. Int. Softw. Technol. (2018)
5. Sarro, F., Ferrucci, F., Harman, M., Manna, A., Ren, J.: Adaptive multi-objective evolutionary algorithms for overtime planning in software projects. IEEE Trans. Softw. Eng. **43**(10), 898–917 (2017). https://doi.org/10.1109/TSE.2017.2650914
6. Al Dallal, J., Abdin, A.: Empirical evaluation of the impact of object-oriented code refactoring on quality attributes: a systematic literature review. IEEE Trans. Softw. Eng. **44**(1), 44–69 (2018)
7. Singh, A.P., Tomar, P.: Estimation of component reusability through reusability metrics. World Academy of Science, Engineering and Technology. Int. J. Comput. Inf. Eng. **8**(11) (2014)
8. Padhy, N., Singh, R.P., Satapathy, S.C.: Complexity estimation by using Multiparadigm approach: a proposed metrics and algorithms. International Journal of Networking and Virtual Organisations, Vol. X, No. Y, Inderscience, Switzerland (Press) (2018)
9. Padhy, N., Singh, R.P., Satapathy, S.C.: Estimation of complexity by using an object oriented metrics approach and its proposed algorithm and models. Int. J. Netw. Virtual (Press) (2018)
10. Padhy, N., Satapathy, S., Singh, R.P.: State-of-the-art object-oriented metrics and its reusability: a decade review. In: Satapathy, S., Bhateja, V., Das, S. (eds.) Smart Computing and Informatics. Smart Innovation, Systems and Technologies, vol. 77, pp. 431–441. Springer, Singapore (2018). https://doi.org/10.1007/978-981-10-5544-7_42
11. Padhy, N., Panigrahi, R., Baboo, S.: A systematic literature review of an object oriented metric: reusability. In: 2015 International Conference on Computational Intelligence and Networks, pp. 190–191, Bhubaneswar (2015)
12. Padhy, N., Singh, R.P., Satapathy, S.C.: Cluster Computing, "Cost-Effective and Fault-Resilient Reusability Prediction Model by Using Adaptive Genetic Algorithm Based Neural Network for Web-of- Service Applications". Springer US (2018). Print ISSN 1386-7857, Online ISSN 1573-7543, https://doi.org/10.1007/s10586-018-2359-9
13. Padhy, N., Singh, R.P., Satapathy, S.C.: Software reusability metrics estimation: Algorithms, models and optimization techniques. Elsevier, Comput. Electr. Eng. vol. 69, pp. 653–668. (2017)
14. Padhy, N., Singh, R.P., Satapathy, S.C.: Utility of an object-oriented metrics component: Examining the feasibility of .Net and C# object-oriented program from the perspective of mobile learning. Int. J. Mobile Learn. Organ. (IJMLO) (2018). 10011924 Issue: 3, 263–279. Inderscience Publishers (IEL). https://doi.org/10.1504/ijmlo.2018

15. Padhy, N., Singh, R.P., Satapathy, S.C.: Cluster Computing, "Enhanced evolutionary computing based artificial intelligence model for web-solutions software reusability estimation". Online ISSN 1573-7543, Springer, Cluster Comput. (2017). https://doi.org/10.1007/s10586-017-1558-0
16. Padhy, N., Singh, R.P., Satapathy, S.C.: Utility of an object-oriented metrics component: Examining the feasibility of .Net and C# object-oriented program from the perspective of mobile learning. Int. J. Mobile Learn. Organ. vol.12, issue: 3 (2018). https://doi.org/10.1504/IJMLO.2018.092777

Dr. Neelamadhab Padhy received his M.C.A. in 2003 from the BPUT, Raurkela, and his M.Tech. (CS) in 2009 and Ph.D. in 2018. Currently, he is working as an Associate Professor in Gandhi Institute of Engineering and Technology, GIET, Gunupur (Autonomous). He has published many articles in SCI Indexing as well as Scopus Indexing like Springer, Elsevier, Inderscience, etc. He has also published many conference papers and book chapters. He got the best faculty award in 2013, best researcher award in 2016, and best paper presenter in the conference at GIET, Gunupur. Apart from this, he got the Dr. Sarvepalli Radhakrishnan Lifetime Achievement Award in 2018 and is nominated as Young Researcher in Computer Science and Engineering and will be awarded in January 2019. He is a senior member of the CSI and a Life Member of the Computer Society of India, IE and Soft Computing Society. In addition, he also serves on the editorial board of several international journals.

Dr. Suresh Chandra Satapathy Senior IEEE Member, received his Ph.D. (Computer science and Engg) from JNTU, Hyderabad and M.Tech (CSE) from NIT Rourkela. He has published more than 80 publications to his credit in both national and International level. He has 27 yrs of teaching and 10 yrs of research experience. He has been involved/organized as program chairs, conveners, etc., of number of programs related to computer science engg in his campus and other campuses. As a senior member of IEEE, he also actively engaged in various students activities under IEEE. He is a Reviewer for Information Science Journal, published by ELSEVIER and a reviewer for IET.

Information Security Journal (formerly known as IEE proceedings for Information Security). He is presently working as a professor at school of Computer Engineering department in KIIT Odisha.

Cloud Forensics in Relation to Criminal Offences and Industrial Attacks in Mauritius

Dhanush Leelodharry

Abstract The cloud can be referred as an infrastructure with large data centres that gives the possibility to run and share resources on your computer, mobile phones or any other device connected with the Internet and in. In Mauritius, many companies are moving towards the cloud and cloud forensics have become important components in the field of investigation. Cloud forensics means how to investigate cybercrimes in cloud environments. Emphasis on concept of the cloud, its mode of forensics and difficulties encountered during an investigation on the cloud. Further, law implications and legislations in Mauritius for computer misuse are also being described. It is obvious that the cloud provides many advantages as described in this paper. However, along with this service, there are several challenges before extracting evidence in case of attacks. These include acquisition of evidence and its preservation, chain of custody and legislation.

Keywords Cloud forensics · Cybercrime · Cloud computing · SLA
Chain of custody

1 Introduction

Have you ever thought how data used to be stored in ancient civilisations? Everyone at a point of time must have thought about this question. Today, information age has penetrated in every aspect, concept of modern society and talking about petabytes of data is not a new thing—even the highly reputable social media Facebook stores 100+ petabytes. Cave paintings and drawings are the first evidence of data being stored 40,000 BC ago. It took another 37,000 years before Papyrus was used as a writing material by the Egyptians, and it was only around 100 BC before China introduced and produced paper as storage for writing and drawing. In the mid 1960s, people started to hear more about computer storage. However, it is only in the mid-2000 that

D. Leelodharry (✉)
Université des Mascareignes, Beau Bassin-Rose Hill, Mauritius
e-mail: dleelodharry@udm.ac.mu

© Springer Nature Singapore Pte Ltd. 2019
S. C. Satapathy et al. (eds.), *Information Systems Design and Intelligent
Applications*, Advances in Intelligent Systems and Computing 863,
https://doi.org/10.1007/978-981-13-3338-5_10

world have witnessed a real game changer in storage terms with the introduction of cloud storage [1].

The cloud provides the structure and convenience of being able to access files from anywhere in the world provided one has an Internet access. It is a physical infrastructure consisting of numerous computers housed in data centres spread over the world. In Mauritius, several companies providing such cloud services such as Orange Business Services, Emtel Ltd, Linkbynet, Harrel Mallac and Data Communication LTD.

Any storage device, given that they store data and information, are subject to crimes, vulnerabilities, threats and other malwares or attacks. In order to protect the stored data, it is imperative to establish proper frameworks so as to counter these attacks and to be able to recover data in case of loss. This is where people have to resort to suitable and effective forensics measures in order to remediate the situation as early as possible. In addition, referring to Mauritius law, there is the CMCA (Computer Misuse and Cyber Security Act), which make provision for penal servitude. Forensics are scientific tests which are associated with crimes and which will lead to some results. These tests are usually carried out in forensics labs.

National Institute of Standards and Technology (NIST) defines digital forensics as 'an applied science for the identification, collection, examination and analysis of data while preserving the integrity of the information and maintaining a strict chain of custody for the data' [2].

Computer forensics can be simply defined as the application of technology to obtain digital evidence which can be presented and shall be admissible in a court of law. Cloud forensics, on the other hand, can also be defined as applying all the processes of digital forensics and using right knowledge in the cloud environment. In Mauritius, people are lucky to have different Acts regarding the abuses and wrong doings in the Information Technology (IT) field. It can be said that have a full fledge legislation to cover these pertinent issues. For instance, the existence of the updated Data Protection Act (DPA) 2017 approved in the Cabinet since 1st December 2017 (Data Protection Bill, 2017) [3, 4]. In this context, malicious people are warned of the penalties associated with criminal offences. Victims are sensitised and channelled to the right instances in case of cyberattacks. Besides the CMCA. Mauritius has also a special unit known as the cybercrime unit in the police division for laws enforcement.

2   Background Study

2.1   Cloud Storage and Cloud Computing

Basically, cloud storage is a platform where any types of files or media can be stored and which can also be accessed and retrieved from multiple devices anytime, anywhere one needs them. In simple words, this infrastructure allows the subscriber to save their information on the Internet, the same way that anyone would have saved the same information on any workstation or secondary storage such as pen drives. In Mauritius, the commonly used cloud storage is Google Drive, DropBox and ICloud. This new kind of repository is fast replacing physical storage systems in organisations as it is cheaper than traditional storage mediums.

Cloud computing implies the presence of a cloud storage somewhere. In fact, they are interconnected somehow.

2.2 Types of Cloud Computing Services in Mauritius

Cloud computing is an on-demand delivery of IT capabilities which can provide IT infrastructure and applications to subscribers as a metered service over a network. Three popular types of cloud computing service are Infrastructure-as-a-Service (IaaS), Platform-as-a-Service (PaaS) and Software-as-a-Service (SaaS).

IaaS is based on virtualisation of resources such as hardware and operating systems which are controlled through an Application Programming Interface (API) for executing services. Amazon Elastic Cloud Computing (EC2), Amazon S3, GoGrid are examples of such infrastructure [5].

AfrAsia Bank is exploring contemporary technologies for the banking industry in Mauritius and the move to cloud is a first step in their digital journey. The elasticity in the cloud pricing enables banks to achieve scalability and agility in an unprecedented manner. Migration to cloud offers the opportunity to redesign the operating processes to standardise and align these with global best practices. The AfrAsia bank has an advanced, flexible and agile digital banking platform in the local financial sector. By taking this recent step into utilising cloud-based IT infrastructure, the bank anticipates a lucrative return on investment due to increased costs savings, quicker turn-around and improved efficiency [6].

With PaaS, clients can create and customise their host applications through the API. In this context, the PaaS offers development tools, configuration management, middleware, operating systems and deployment platforms on-demand that can be used by subscribers to develop custom applications. The 'Google App engine' is an example of such platform [7].

Linkbynet which is a Microsoft cloud service provider in Mauritius offers such kind of service, for instance, Azure technology.

With SaaS, customers need not bother about installation, update and maintenance of software applications. However, the SaaS vendors have to shoulder the responsibility of deploying all applications as per the requirement of the subscriber. This service is offered via a web browser or an application to subscribers over the Internet [8]. Agileum in Mauritius provides such facility.

2.3 Cloud Deployment Models

- Public Cloud—These clouds are entirely hosted and managed by the providers who has the sole responsibility to manage. (E.g., Microsoft Azure)
- Private Cloud—Expensive and more secure comparing to public clouds. Subscribers and providers have optimised control over the infrastructure. (E.g., Eucalyptus systems)
- Community Cloud—Like Facebook, community cloud is a social platform used to communicate among employees or people sharing same objectives and concerns.
- Hybrid Cloud—Comprises of both public and private cloud deployment models that remain unique entities but are bound together, offering the benefits of multiple deployment models [9].

2.4 Cybercrime Strategy of Mauritius (2017–2019)

Based on Cybercrime Strategy 2017–2019 of Mauritius, the Government has come up with a public and private partnership to involve different stakeholders to develop a more effective legal framework in order to tackle and prosecute cybercrime. Some of the stakeholders are the Ministry of Information and Communication Technology (ICT), National Cybercrime and Security, National Computer Emergency Response Team (CERT), Mauritius Police Force, Data Protection Office (DPO), IT Security Unit and another regulators office. Their mission is to enhance the Government efforts and together to tackle cybercrime by providing effective law enforcement and criminal justice response. To educate the community on the risks of cybercrime is one of the key goal of this recent strategy. The government of Mauritius is focusing in this crucial field, as technology when wrongly used leads to a lot of damage and negative impact on individuals, companies and states. Thus, regular awareness, campaigns, or workshops are carried by the stakeholders to pursue with this objective [10].

3 Literature Review

There have been several researches regarding the challenges forensics investigators met during an investigation, but till now, there has been no reliable solutions to meet the gap of cloud forensics and computer forensics [11].

Ruan et al. [12] conducted a survey among 257 international digital forensic experts and practitioners. Their survey includes key questions on cloud forensics ranging from definitions, challenges, opportunities, implications and missing capabilities. According to the results, more than 80% of the respondents strongly agreed in the following four challenges: (1) Jurisdiction—90%; (2) Lack of international collaboration and legislative mechanism in cross-nation data access and exchange—85%; (3) Lack of law/regulation and law advisory—81%; and (4) Investigating external chain dependencies of the cloud provider—80%. It can be seen that cloud forensic practitioners consider legal challenges the biggest issue in cloud forensics.

Biggs [13] stated that 'Criminal users deliberately rent a cloud storage and exploit all possible loopholes in order to apply information warfare. Software Licence Agreement (SLA) should always be well defined, so as to obtain a proper roadmap and acts as an insurance cover when there are problems in the cloud environment. Legislation is a significant matter as data centres are located in different segments in the world with their different respective judicial prosecutions. Unlike computer forensics, cloud computing has still a long way before forensic readiness'.

O'Shaughnessy and Keane [14] explained that the aspect of cloud computing changes the traditional way of carrying digital forensics. In cloud computing, data and evidences are stored in different geographical location on servers. For example, while multiple subscribers may have access to these data, these servers may be under the control of different providers having different legislation to cope with. This situation summarises the complexities associated with forensic investigation.

Suresh Chandra Satapathy et al. describe the security concept of cloud computing as a worrisome aspect as users do not have control over the devices proposed to them. Many information is confidential; thus, cloud providers must ensure that this information are safe. Implementation of security in a cloud reveals to be challenging, as there will be always a black hat hacker trying to have access to user's privacy by disabling the security measures [15].

4 Research Methodology and Findings

Qualitative research was done for this study whereby interviewees were asked questions regarding cloud forensics in Mauritius. Only two companies responded as it is a very confidential and sensitive issue. However, anonymity was requested for the purpose of protecting the reputation of the company.

Case I: Jay is the sole owner of an online shopping hosted in Mauritius. Customers visit this website regularly and make their transactions as per their needs and requirements. Jay generates high income from his respective website business. With an aggressive marketing strategy and high reputation of the website, there should be no risk of downtime else this will affect Jay's goodwill. Jimmy, a black hat hacker launched a Distributed Denial of Service (DDoS) attack on Jay's website. The former subscribed himself to the service of a particular CSP and using the Virtual Machines, he could successfully put down the shopping platform. Result, there was no online access for approximately 1 h. Jay retained the services of a forensic investigator and the latter found the logs which revealed IP addresses of clients who accessed the website. Further investigation showed that the website was flooded by IP addresses attributed by the local CSP. The investigator issued a warrant to procure him the network logs. However, the CSP supported Jimmy malicious acts and provide a fake log. The investigator could not check the authenticity and integrity of the records while Jimmy remained anonymous. Assuming that the CSP was honest, there could be no trace, if Jimmy terminated his contract at the right time and thus leaving no trace hence there will be again no logs at all.

Case II: Mauritius Police learnt about a criminal who sells stolen articles on a website hosted in a foreign CSP. The police need to gather maximum evidence to prosecute this criminal. The investigator needs to enquire about the location of the hosted CSP so as to confirm if he has jurisdiction to pursue his investigation. In that case, the fact that the server is abroad, the investigator need international cooperation to collect relevant data. During preliminary investigation, the police finds that the CSP was in a foreign country X and the servers was in another foreign country Y. The extraterritorial jurisdiction is a significant challenge in the cloud forensics phases. If investigators do not get enough support in terms of jurisdiction from foreign states, this case may be dropped. It is crucial to apply digital cloud forensic to extract relevant evidence so as to prosecute the attacker.

Unfortunately, cloud environment is not forensic friendly and have some limitations regarding the forensic part. It is not uncommon that forensic investigation is an intensive process, whether data is in the cloud or not. The obstacles seen in the case clearly imply that there is still area for improvements, research and developments. Based on the case study above, Cloud subscribers need to negotiate and query about support from the CSP in case of data breaches or any other kind of attacks. Drafting a proper Service Level Agreement (SLA) is crucial. This legal document provides an assurance and clearly defined support from their CSP during an investigation.

In IaaS environment, VM hosting can be in a different location and this is time consuming element in the analysis phase. It can be a tedious task which involves different countries jurisdiction.

Ultimately, chain of evidence remains cumbersome as several people have access to evidences. In a cloud environment, there is no physical hardware and data is the unique evidence.

4.1 Why Cloud Forensic Is Complex

In a cloud environment, files and data of different users from different parts of the world are stored. Thus, seizing servers from a data centre in Mauritius and accessing the files of different users is considered to be a violation of privacy according to DPA and Information and Communication Technology Act. In many circumstances, investigators have to rely on the CSP's word. This issue clearly makes the cloud forensics a new challenging task.

The fact that in a cloud environment there are so many subscribers, the suspect subscriber may refute ownership of data during investigations. The suspect can deny his malicious act by claiming that his instance was compromised.

Moreover, termination of a VM lead to a complete loss of logs and relevant evidence as data residing in a VM are volatile.

Chain of custody is a chronological documentation that records the sequence of interactions, control, transfer and custodianship during an investigation process. In traditional computer forensic, maintaining a proper chain of custody is easier as all evidence, tools are physical and local, such as hard disk, mobile phones. On the other hand, in a cloud environment, this kind of documentation reveals to be a complicated task for the following reasons:

- Lack of information on location of VM
- Different countries with their own time zones.

4.2 Recommendations

Cloud forensics prove to be challenging especially due to inaccessibility, remote VM and numerous reasons as listed in this paper. However, still mitigate the damage by adopting some precautionary measures. Risks are everywhere in all fields and all sectors. With the evolution of technology, even the cloud which is a dominating area is not spared. To ensure people are in a safer zone, the cloud can be isolated. Isolation means keeping systems separate from each other. In the context of a cloud environment, there should be segregation of customers' applications and separate program instances. Else, hackers could tunnel from subscriber X's external system into Y's cloud and then jump from Y's cloud to Z's cloud. Cloud information breaches could occur mainly through file systems and social engineering. Tunnelling from system to system is an open door to hackers.

Clients should be restricted from the network traffic of other clients through isolation of the network traffic itself. Allowing this visibility would mean inviting a hacker to exploit any vulnerabilities. In addition, hacker might escalade privileges and then use them to spy on other enterprises with stronger security.

Need to encrypt all data using algorithms like RSA. Similarly, crypto-shredding should be used to delete data.

Perform regular penetration test to assess vulnerabilities and implement remedial actions. There is a need to implement a cloud disaster recovery, ensure proper SLA with clearly defined guidelines and bring awareness to users.

In poorly configured cloud environment, vulnerabilities are subjected to exploitation and attacks. Configurations need to be strengthened with skilled, experienced and qualified security staff.

5 Conclusion and Future Works

The small island of Mauritius is moving along with technology. Enterprises, people and government agencies should ensure that they are in line with the cloud technology. The government is highly focusing on the IT hub, Smart cities and numerous latest technological concept. Thus, moving to the cloud should not be just a desire but a necessity to ensure business continuity. In every society, there are risks and Mauritius is not to be spared. However, government should educate the citizens to keep pace with technology progress and associated laws. Forensic investigation proves to be a difficult task indeed, not only in Mauritius but across the world. The various recommendations proposed in this paper, if properly implemented can at least minimise risk and to be in a risk appetite zone.

Acknowledgements We undertake that we have the required permission to use images/dataset in the work from suitable authority and we shall be solely responsible if any conflicts arise in future.

References

1. Hussey, D.: From Cave Paintings to IBM Cloud (2018). https://www.linkedin.com/pulse/from-cave-paintings-ibm-cloud-declan-hussy
2. NIST: Digital Forensics in the Cloud (2013). https://pdfs.semanticscholar.org/5795/3e2dd1f572d769f1eeb4f98dc8a0ef313b4d.pdf
3. The Mauritius Assembly: The Data Protection Bill (2017). http://mauritiusassembly.govmu.org/English/bills/Documents/intro/2017/bill1917.pdf
4. DataGuidance: Mauritius: DPB Should in Principle Lead to EU Adequacy (2017). https://www.dataguidance.com/mauritius-data-protection-bill-principle-provide-eu-adequacy/
5. Satapathy, S.C., et al.: Cloud Computing: Security Issues and Research Challenges (2011). https://www.ijcsits.org/papers/Vol1no22011/13vol1no2.pdf
6. Newsroom News: AfrAsia Pioneers Successful Deployment of Cloud (2017). https://www.afrasiabank.com/en/about/newsroom/news/2017/afrasia-pioneers-successful-deployment-of-cloud-based-financial-solution-in-mauritius
7. George Grispos: Challenges of Cloud Computing in Digital Forensics (2012). https://arxiv.org/pdf/1410.2123.pdf
8. Steve Ranger: Introduction to Cloud Computing (2018). https://www.zdnet.com/article/what-is-cloud-computing-everything-you-need-to-know-from-public-and-private-cloud-to-software-as-a/
9. Satapathy, S.C., et al.: Cloud Computing: Security Issues and Research Challenges (2011). https://www.ijcsits.org/papers/Vol1no22011/13vol1no2.pdf
10. Government of Mauritius. Cybercrime strategy of Mauritius (2017). http://cert-mu.govmu.org/
11. Martini, B., Choo, K.: Cloud Forensic Technical Challenges and Solutions: IEEE Cloud Computing, pp. 20–25 (2014)
12. Ruan, K., Carthy, J., Kechadi, T., Baggili, I.: Cloud forensics definitions and critical criteria for cloud forensic capability: an overview of survey results. Digit. Investig. **10**, 34–43 (2013)
13. Biggs: Cloud Forensics and SLA (2013). http://users.cis.fiu.edu/~fortega/df/research/Cloud%20Forensics%20II/Other%20references/Cloud%20Forensics%20Issues%20and%20Opportunities.pdf
14. O'Shaughnessy, S., Keane, A.: Impact of Cloud Computing on Digital Forensic Investigations (2013). https://link.springer.com/content/pdf/10.1007%2F978-3-642-41148-9_20.pdf
15. Satapathy, S.C., et al.: Cloud Computing: Security Issues and Research Challenges (2011). https://www.ijcsits.org/papers/Vol1no22011/13vol1no2.pdf

HECMI: Hybrid Ensemble Technique for Classification of Multiclass Imbalanced Data

Kiran Bhowmick, Utsav B. Shah, Medha Y. Shah, Pratik A. Parekh and Meera Narvekar

Abstract Imbalanced data is a problem which is observed in many real-world applications. Although a lot of research is focused on achieving a solution to handle this problem, most of them assume binary classes. However, occurrence of multiple classes in most of the applications is not uncommon. Multiclass classification with imbalanced data poses additional challenges. This paper proposes a hybrid ensemble approach for classification of multiclass imbalanced data (HECMI). A hybrid of data based and algorithm based approach is proposed to deal with the imbalance and multiple classes. The ensemble created focuses on misclassified instances that are added to the partitioned dataset. HECMI proves to be more accurate than traditional algorithms.

Keywords Multiclass · Ensemble · Imbalance · Classification

1 Introduction

Data that is generated by the real-world applications contains some classes that are heavily underrepresented as compared to other classes. This imbalanced distribution of data causes difficulty in classification as the classifier is biased towards the majority

K. Bhowmick (✉) · U. B. Shah (✉) · M. Y. Shah · P. A. Parekh · M. Narvekar
Dwarkadas J. Sanghvi College of Engineering, Vile Parle, Mumbai, Maharashtra, India
e-mail: kiran.bhowmick@djsce.ac.in

U. B. Shah
e-mail: shahutsav195@gmail.com

M. Y. Shah
e-mail: shahmedha1001@gmail.com

P. A. Parekh
e-mail: pratikparekh217@gmail.com

M. Narvekar
e-mail: meera.narvekar@djsce.ac.in

© Springer Nature Singapore Pte Ltd. 2019
S. C. Satapathy et al. (eds.), *Information Systems Design and Intelligent Applications*, Advances in Intelligent Systems and Computing 863,
https://doi.org/10.1007/978-981-13-3338-5_11

class and tends to misclassify the minority class. Even with high accuracy, there are high chances of misclassification of minority class [1]. The cost of misclassification of minority class can be expensive for applications such as credit card fraud detection, network intrusion detection, medical diagnosis, and hence focus should be put on this problem. A lot of research has been directed to provide a solution but most of them assume that the classes are binary. However, there often exist more than two classes in an application. For example, in credit card transaction the fraud is always of a different type, in internet traffic data we have MAIL, HTTP, ATTACK, and CHAT which are multiple classes [2]. Multiclass tasks have been shown to suffer more learning difficulties than its binary counterpart in offline learning because multiple classes increase the data complexity and aggravate the imbalanced distribution [3].

Many researchers have proposed solutions with either new multiclass algorithms or extending the binary classification methods for multiple classes. The latter includes widely used strategies one-versus-all (OVA) and one-versus-one (OVO). The OVA decomposes an m class problem into m binary problems. The strategy involves training a single classifier per class where the samples of that class are positive samples and all other samples are negative samples. The OVO, on the other hand, creates $m(m-1)/2$ classifiers for m classes. The strategy involves pairing each class with every other class. With more number of classes, OVA suffers from the problem of imbalance classification [4] and so is not suitable for our problem.

In this paper, we propose HECMI to deal with the problem of multiclass imbalance data which has more than one majority and minority class. HECMI is generic in nature and can be used for any domain. It is a hybrid of data-based and algorithm based method for imbalanced data. The base classifier of the ensemble is selected from the existing traditional models with the best cross-validation score for the concerned dataset. This makes HECMI adaptable to different domains. The data set is initially partitioned into n parts for n iterations. A data-based approach of oversampling minority class instances is used to balance the dataset in each of the iterations. The class with the least recall is the one misclassified the most and hence should be paid attention to by the classifier. The instances of this class are oversampled and added to the next data part in training. Along with these, the instances of classes with recall less than the threshold are also added to the next data part. Final prediction is done by taking the majority of votes of the classifiers in the ensemble. HECMI is thus a hybrid of boosting as it focuses on the misclassified instances and bagging as it applies similar data splitting technique and majority voting for prediction. HECMI is tested on real-world datasets and the results are compared with traditional algorithms. We prove that it successfully overcomes the data imbalance issue for multiple classes and classifies data as good as the balanced data.

In the following sections of the paper, Sect. 2 discusses the related work in this area, Sect. 3 explains the proposed framework solution describing our novel approach of partitioning and iterative ensemble creation, Sect. 4 focuses on the implementation details and further Sect. 5 explains the results, and Sect. 6 discusses the conclusion and future work.

2 Related Work

Most of the existing research in multiclass imbalanced data focuses on effective ways to decompose the multiclass problem into many binary classification tasks. In this section, we will discuss some of the earlier work in this area.

Mikel Galar et al. in [5, 6] provided a detailed comparative study of different techniques of multiclass classification like the bagging, boosting, OVO and OVA. Fernandez et al. in their paper [7] proposed a methodology of combining OVO with oversampling instances that are imbalanced. Jeatrakul in [8] paper proposed a technique named as the One-Against-All with Data Balancing (OAA-DB) algorithm to solve the multiclass imbalanced problem. Data balancing is done by integrating the undersampling technique using Complementary Neural Network (CMTNN) and SMOTE [9].

A parallel research focused on cost-sensitive approach for multiclass imbalance problem. A cost-sensitive approach that focuses on finding appropriate cost matrix using GA was proposed in [10]. In their paper [11], Zhou has proposed a novel approach for rescaling in multiclass imbalanced problem. Their framework is especially helpful when unequal misclassification costs and imbalance occur simultaneously. Krawczyk in [12] proposed an OVO cost-sensitive neural network approach for scaling the cost function in multiclass imbalance problem. The cost is automatically detected using ROC analysis for each OVO pair and a dedicated classifier fusion approach is then used for final prediction.

Recent research also focussed on adaptive frameworks to deal with this problem. Yijing et al. in [13] proposed an adaptive multiple classifier system which uses IR, dimension and the number of classes to differentiate into different types of imbalanced data. Wei et al. [14] have proposed a hybrid adaptive framework combining undersampling, Adaboost, cost-sensitive weight modification to learn from imbalanced data. Ortigosa-Hernández et al. [15] have proposed a novel measure to detect imbalance ratio in multiclass datasets. The framework in [16] proposed by Yuan et al. uses stratified undersampling for balancing the data and regularizes the classifiers in the ensemble to improve performance. In [17], Bi and Zhang proposed a novel multiclass classifier that generates codes for sparse schemes. A codeword is generated for each class to obtain the distance between classes and a dichotomy classifier is built for each codeword. A dynamic ensemble of classifier proposed in [18] by Garcia et al. uses hybrid random oversampling random undersampling and SMOTE to balance the dataset and the classifier selected is based on its competence to classify minority instances in the local region. The authors in [19] proposed a cost-sensitive approach with an efficient boosting algorithm that uses cost matrix as a tool to learn class boundaries.

There are many solutions to this problem and each fits some application or the other. Our approach differs in the way the data is partitioned and synthetic data generated and also the way in which the ensemble is created. To the best of our knowledge, there has been no such research in this direction.

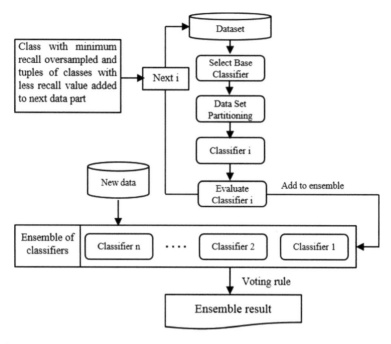

Fig. 1 Proposed model

3 Proposed Solution

An adaptive ensemble of N classifiers is built to classify the data. The process first begins by selecting a suitable base classifier for the ensemble. Partitioning of the dataset is made to create training and testing parts for each classifier in the ensemble. The classifier is evaluated on the respective testing part and added to the ensemble. In the next iteration, data part is modified for the next classifier to be trained (Fig. 1).

3.1 Selecting the Base Classifier for the Ensemble

Different traditional classification algorithms such as random forest, decision tree, Naive Bayes, logistic regression, support vector machine, etc., can be used as the base classifier of ensemble classifier for classifying imbalanced dataset. But the algorithm which will be most efficient and having high accuracy will depend on the nature and size of dataset, number of attributes, number of classes, etc. In order to make the approach adaptive and precise, a suitable classification algorithm is chosen as the base classifier based on the dataset. The advantage of this approach is that there is no fixed base classifier and will ensure the accuracy and recall rate produced will the highest among all other classification algorithms.

Algorithm: HECMI

---

1. Read the dataset.
2. Select the base classifier.
    Base classifier ⇐ classifier with max cross validation score.
3. Split dataset into training and testing dataset.
4. Split training set into N partitions.
5. For each data part i do
    - a. Split data part into training and testing part i
    - b. Create a Classifier i on the training part i
    - c. Evaluate Classifier i on testing part i for each class
    - d. Add classifier to the ensemble
    - e. Modify data part $i + 1$
        - i.   Oversample the class with the least recall value
        - ii. Collect instances of classes with recall less than the threshold
        - iii. Append the instances from i & ii to data part $(i + 1)$.
    - f. Move to next iteration $(i + 1)$
6. Read data x
7. Get class label for x: $c_i = \text{Class}(x)$.
8. Class with majority votes is the final label.
    Final $c = \text{argmax} \sum_{i=1}^{N} c_i$

3.2 Data Splitting and Creation of "N" Data Parts

In order to classify the imbalanced dataset, a three-step approach is used. In the first step, the original data set is separated into two parts: Training and Testing. In the second step, the training part is further divided into N parts called Data Part i ($i = 1$ to N). In the third step, the Data Part 1 is read and partitioned into two sets Training set 1 and Testing set 1. The classifiers are built on the training set and tested on the testing set (Fig. 2).

3.3 Development of Ensemble of Models

After the base classifier is selected, the model iteratively performs the following processes:

1. Building n classifiers iteratively.
2. Modifying data part for next iteration.

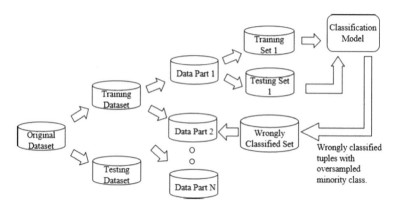

Fig. 2 Data partitioning

Building "N" classifiers iteratively. The ensemble model consists of "N" classifiers where N is the total number of different class attribute values. Each of the N classifiers is trained on N different data parts. These different classifiers are trained and tested on data part's training dataset and testing dataset respectively. In each iteration, a new classifier is developed based on the training set of data part.

Modifying the dataset. Once a classifier is trained, it is evaluated using the respective testing set. The instances of the class having the least recall are oversampled using SMOTE technique. SMOTE generates synthetic instances of the misclassified class with the number of instances generated equal to the number of instances of the majority class thus balancing data in the process. Along with the class having least recall, instances of classes having recall below a particular threshold are also sampled and added to the next data part. The threshold is the minimum of the average of recall rate of all classes and 0.7. A new classifier is then trained on this modified data part in the next iteration.

3.4 Ensemble Result

The final output class label is based on the majority voting technique. Each classifier's class label is taken and the one with the maximum vote is given as the final class label for the unseen data.

$$\text{Final } c = \arg\max \sum_{i=1}^{N} c_i \tag{1}$$

Table 1 Description of datasets

Dataset	Classes: (class distribution)	IR
Activity	7: (33,677I928I11,179I26,860I3191I2917I83,748)	1:90
Balance	3: (7840I46,080I46,080)	1:6
CMC	3: (42,780I23,574I33,646)	1:2
Dermatology	6: (30,129I16,554I19,367I14,342I13,898I5710)	1:5
Ecoli	8: (42,559I22,917I15,476I10,417I5952I1489I595I595)	1:72
Gesture	4: (27,763I10,108I21,240I40,889)	1:4
Glass	7: (32,991I35,629I9129I255I5446I3536I13,014)	1:140
Hayes-Roth	3: (37,761I38,843I23,396)	1:1.6
Thyroid	3: (69,767I13,953I16,280)	1:5
Page Blocks	5: (89,149I6347I675I1492I2337)	1:132
Sensor reading	4: (40,414I15,139I38,435I6012)	1:7
Shuttle	5: (78,497I166I280I15,691I5366)	1:473
Volcanoes	5: (90,549I2285I1766I2661I2739)	1:51
Wine (white)	7: (335I3366I29,296I45,365I17,920I3619I99)	1:458
Yeast	10: (31,199I29,908I16,442I10,984I3437I2965I2358I2022I1348I337)	1:93
Zoo	7: (38,175I22,222I5516I13,819I4560I6572I9136)	1:8

4 Implementation

HECMI is evaluated on 16 different datasets that are publicly available at the UCI repository [20]. Table 1 describes the datasets with the number of classes followed by the number of instances in each class. The imbalance ratio IR is calculated by taking the ratio of instances of the class with least number of instances to the instances of the class with maximum number of instances. As the original number of instances was very less, we generated synthetic samples of the data by using the treeEnsemble and indAttrGen method in R language. The treeEnsemble method creates a tree ensemble which is used for data generation. The indAttrGen method generates data which has the same distribution as the original data. Each of the datasets was increased to 100,000 tuples. The final implementation of the algorithm is done in python.

5 Results and Discussions

We have compared the performance of HECMI with some of the traditional classification approaches namely Logistic Regression (LR), Linear Discriminant Analysis (LDA), K Nearest Neighbors (KNN), Support Vector Machine (SVM).

Table 2 Evaluation results considering micro-average recall values of all the classes

Datasets	LR	LDA	KNN	SVM	HECMI
Activity	0.9131	0.7252	0.9105	0.7410	**0.9182**
Balance	0.9983	0.8367	0.9989	**1.0000**	0.9972
CMC	0.9989	0.5152	0.9728	**0.9998**	0.9996
Dermatology	0.9673	0.9245	0.9689	0.9792	**0.9875**
Ecoli	0.9550	0.8671	1.0000	1.0000	**1.0000**
Gesture	0.9009	0.5559	1.0000	1.0000	**1.0000**
Glass	0.9134	0.6138	0.9935	0.998	**0.9987**
Hayes-Roth	1.0000	0.5680	1.0000	1.0000	**1.0000**
Thyroid	0.9994	0.9231	0.9957	0.9972	**0.9999**
Page Blocks	**0.9873**	0.9306	0.9481	0.9035	0.9828
Sensor reading	0.9586	0.6792	0.9016	0.9993	0.9979
Shuttle	0.9942	0.9304	0.9865	0.7946	**0.9942**
Volcanoes	**0.9750**	0.9104	0.9477	0.9532	0.8113
Wine (white)	**0.9115**	0.5291	0.7314	0.7352	0.8431
Yeast	0.8443	0.5866	1.0000	1.0000	**1.0000**
Zoo	0.9677	0.9373	0.9961	**1.0000**	0.9999

Due to the imbalance and multiclass classification, we have used micro-average of recall values as an evaluation metric. Table 2 displays the micro-average of recall values calculated for each dataset under consideration.

HECMI performs at par with the traditional classification techniques when the imbalance ratio is small to moderate. For datasets with high imbalance ratio, however, HECMI has proven to give better recall rates for minority classes. Also, the recall for majority class does not fall too low for these datasets. For the datasets where HECMI performs a little lower, we have found that the minority class recall value is very high and the majority class recall also does not fall too low for HECMI as compared to the traditional algorithms. This is displayed in Table 3.

It is evident from the evaluation that HECMI is more stable and accurate than the traditional approaches.

Table 3 Recall values for dataset with slightly lower HECMI results

Class	LR	LDA	KNN	SVM	HECMI
Balance dataset					
0	1	0	1	1	0.99
1	1	0.9	1	1	1
2	1	0.91	1	1	1
CMC dataset					
1	1	0.65	0.99	1	1
2	1	0.35	0.92	1	1
3	1	0.46	0.99	1	1
Page blocks dataset					
1	1	0.99	0.99	1	1
2	0.91	0.46	0.63	0.03	0.85
3	0.58	0.53	0.4	0	0.6
4	0.75	0.76	0.57	0.26	0.83
5	0.99	0.3	0.6	0.08	0.91
Volcanoes dataset					
1	1	0.98	0.99	1	1
2	0.52	0	0.44	0.37	0.71
3	0.2	0.01	0.51	0.11	0.65
4	0.87	0.11	0.52	0.52	0.7
5	1	0.62	0.57	1	1
Wine dataset					
3	0.04	0.07	0.42	0.41	0.87
4	0.03	0.2	0.45	0.54	0.66
5	1	0.46	0.72	0.94	0.89
6	1	0.75	0.82	0.98	0.87
7	0.83	0.26	0.64	0.89	0.81
8	0.4	0	0.47	0.67	0.52
9	0	0	0.42	0.42	0.97

6 Conclusion and Future Scope

In this paper, we proposed a hybrid model to deal with the problem of imbalanced datasets with multiple classes. The main contribution of our model HECMI is that it deals with multiclass imbalanced data by creating a diverse ensemble of classifiers which is maintained by the way the data is sampled while creating the model. The number of classifiers in the ensemble is proportional to the number of classes which are not known a priori. The method is more susceptible to noise and outliers and does not depend on the size of the data.

As a future work, we intend to extend our model for classifying multiclass imbalanced data streams.

References

1. Haixiang, G., Yijing, L., Shang, J., Mingyun, G., Yuanyue, H., Bing, G.: Learning from class-imbalanced data: review of methods and applications. Int. J. Expert Syst. Appl., 220–239. Elsevier (2017)
2. Elaheh, A., Kantardzic, M., Sethi, T.S.: A partial labeling framework for multi-class imbalanced streaming data. In: International Joint Conference on Neural Networks (IJCNN), IEEE, Anchorage, AK, USA (2017)
3. Wang, S., Minku L., Yao X.: Dealing with multiple classes in online class imbalance learning. In: International Joint Conference on Artificial Intelligence (IJCAI-16), pp. 2118–2124. IEEE, New York, USA (2016)
4. Rafiez, A., Raziff, A., Sulaiman, M.N., Mustapha, N., Perumal, T: Single classifier, OvO, OvA and RCC multiclass classification method in handheld based smartphone gait identification. In: AIP Conference Proceedings (2017)
5. Galar, M., Fernandez, A., Barrenechea, E., Bustince, H., Herrera, F.: A review on ensembles for the class imbalance problem: bagging, boosting, and hybrid-based approaches. IEEE Trans. Syst. MAN Cybern. **42**, 463–484. IEEE (2011)
6. Galar, M., Fernandez, A., Barrenechea, E., Bustince, H., Herrera, F.: An overview of ensemble methods for binary classifiers in multi-class problems: experimental study on one-vs-one and one-vs-all schemes. Int. J. Pattern Recogn., pp. 1761–1776 Elsevier (2011)
7. Fernández, A., Jesus, M.J., Herrera, F.: Multi-class imbalanced data-sets with linguistic fuzzy rule based classification systems based on pairwise learning. In: Hüllermeier, E., Kruse, R., Hoffmann, F. (eds.) Computational Intelligence for Knowledge-Based Systems Design. IPMU 2010. Lecture Notes in Computer Science, vol. 6178. Springer, Berlin, Heidelberg (2010)
8. Jeatrakul, P., et al.: Enhancing classification performance of multi-class imbalanced data using the OAA-DB algorithm. In: IJCNN (2012)
9. Chawla, N.V., Bowyer, K.W., Hall, L.O., Kegelmeyer, W.P.: Smote: synthetic minority over-sampling technique. J. Artif. Intell. Res. (JAIR) **16**, 321–357 (2002)
10. Alejo, R., Sotoca, J.M., Valdovinos, R.M., Casa˜n, G.A.: The multi-class imbalance problem: cost functions with modular and non-modular neural networks. In: Wang, H., Shen, Y., Huang, T., Zeng, Z. (eds.) The Sixth International Symposium on Neural Networks (ISNN 2009). Advances in Intelligent and Soft Computing, vol. 56. Springer, Berlin, Heidelberg (2009)
11. Zhou, Z.H., Liu, X.Y.: On multi-class cost-sensitive learning. In: AAAI (2009)
12. Krawczyk, B.: Cost-sensitive one-vs-one ensemble for multi-class imbalanced data. In: IJCNN, IEEE, Canada (2016)
13. Yijing, L., Haixiang, G., Xiao, L., Yanan, L., Jinling, L.: Adapted ensemble classification algorithm based on multiple classifier system and feature selection for classifying multi-class imbalanced data. Int. J. Knowl. Based Syst. **94**, 88–104. Elsevier (2016)
14. Wei, L., Zhe, L., Chu, J.,: Adaptive ensemble under sampling-boost: a novel learning framework for imbalanced data. Int. J. Syst. Softw. **132**, 272–282. Elsevier (2017)
15. Ortigosa-Hernández, J., Inza, I., Lozano, J.A.: Measuring the class-imbalance extent of multi-class problems. Int. J. Pattern Recogn. Lett. **98**, 32–38. Elsevier (2017)
16. Yuan, X., Xie, L., Abouelenien, M.: A regularized ensemble framework of deep learning for cancer detection from multi-class, imbalanced training data. Int. J. Pattern Recogn. **77**, 160–172. Elsevier (2018)
17. Bi, J., Zhang, C.: An empirical comparison on state-of-the-art multi-class imbalance learning algorithms and a new diversified ensemble learning scheme. Int. J. Knowl. Based Syst. **94**. Elsevier (2018)
18. García, S., Zhang, Z.L., Altalhi, A., Alshomrani, S., Herrera, F.: Dynamic ensemble selection for multi-class imbalanced datasets. Int. J. Inf. Sci., vol. 445–446, pp. 22–37. Elsevier (2018)
19. Fernández-Baldera, A., Buenaposada, J., Baumela, L.: BAdaCost: Multi-class Boosting with Costs. Int. J. Pattern Recogn. **79**, 467–479. Elsevier (2018)
20. Dua, D., Taniskidou, K.E.: UCI machine learning repository (http://archive.ics.uci.edu/ml). Irvine, CA: University of California, School of Information and Computer Science (2017)

A Flexible and Reliable Wireless Sensor Network Architecture for Precision Agriculture in a Tomato Greenhouse

Vimla Devi Ramdoo, Kavi Kumar Khedo and Vishwakalyan Bhoyroo

Abstract Agriculture in the twenty-first century faces significant challenges given the ever-increasing need to produce more food to feed a growing population. Wireless Sensor Networks (WSNs) have recently emerged in agriculture to improve crop yields as well as to facilitate decision-making. Numerous environmental sensors that can sense data such as humidity, pressure, temperature, and light are deployed in the WSN that can be used both on land and underground. This paper proposes the design of a reliable and flexible WSN using heterogeneous environmental sensor data streams for precision agriculture in a tomato greenhouse. The proposed system is used to monitor the ever-changing greenhouse environmental conditions that will allow the farmer to have access to real-time as well as historical data of the greenhouse. Furthermore, the proposed system is evaluated using a qualitative approach. Finally, challenges and future works related to WSNs design for precision agriculture are explored.

Keywords Wireless sensor network · Precision agriculture · Greenhouse

1 Introduction

The Food and Agriculture Organization (FAO) of the United Nations predicts that the world will have to produce 70% more food globally by 2050 to sustain the growing population of the Earth [1]. To meet this high demand, farmers are gradually turning towards precision agriculture to increase production capabilities while minimizing cost and preserving resources. Over the next few years, sensors and WSNs

V. D. Ramdoo (✉)
Curtin Mauritius, Moka, Mauritius
e-mail: vramdoo@curtinmauritius.ac.mu

K. K. Khedo · V. Bhoyroo
University of Mauritius, Réduit, Mauritius
e-mail: k.khedo@uom.ac.mu

V. Bhoyroo
e-mail: v.bhoyroo@uom.ac.mu

© Springer Nature Singapore Pte Ltd. 2019
S. C. Satapathy et al. (eds.), *Information Systems Design and Intelligent Applications*, Advances in Intelligent Systems and Computing 863,
https://doi.org/10.1007/978-981-13-3338-5_12

have emerged at an extraordinary rate [2, 3]. With respect to precision agriculture, wireless sensors can be used to monitor environmental conditions that influence crop development such as light irradiance, soil pH, soil salinity, soil moisture, soil nutrients level (total dissolved solids, TDS and electrical conductivity, EC), NPK soil nutrients (nitrogen, phosphorus, potassium), leaf moisture/wetness, color of leaf (indicating deficiency of nutrients), air humidity, air carbon dioxide content, and air temperature [4–7]. The decision of which parameters to monitor will depend on the crop type and its environment such as field, semi-shaded or fully shaded greenhouses. The measured values from the heterogeneous sensors should be transferred reliably with minimal power consumption, regardless of their position in the greenhouse and be flexible to accommodate more sensors in the future that will allow for real-time monitoring.

This paper is structured as follows: Section 2 presents a brief literature review on WSNs for precision agriculture. Section 3 proposes a WSN architecture design for precision agriculture in a tomato greenhouse where the requirements are detailed at the sensor or node level, computing at the sink or gateway level and analytics and visualization at the remote cloud server level. Section 4 presents the evaluation of the architecture per tier. Section 5 explores further challenges and future works. Section 6 concludes the paper and highlights future directions.

2   Brief Literature Review on WSNs for Precision Agriculture

The main aim of WSNs is to optimize cost, scalability, accuracy, flexibility, ease of deployment and reliability of sensor nodes communication with the sink or gateway node, and the remote server [8]. Hamouda and Elhabil [9] implemented a WSN using Bluetooth (BT) connectivity for precision agriculture in greenhouses named Greenhouse Smart Management System (GSMS) to monitor the ambient relative humidity and temperature and automatically control the irrigation and cooling of greenhouses. However, the continuous monitoring of other environmental parameters that directly influences crop development such as carbon dioxide level, light, soil pH, soil moisture, leaf moisture among others, has not been catered in GSMS.

Mat et al. [10] implemented a Wireless Moisture Sensor Network (WMSN) named Greenhouse Management System (GHMS) that managed the greenhouse's wetness of soil using moisture sensor, wetness of air using humidity sensor and heat of the air using temperature sensor. Sensors' data enable GHMS to automatically switch devices ON and OFF such as water pumps for irrigation, fans for air flow and mist to increase humidity of the air. However, the proposed system considered only a few plants rather than the whole greenhouse and was also limited to the monitoring of moisture only.

Math and Dharwadkar [11] proposed a low cost and energy-efficient WSN framework for precision agriculture with three tiers namely field sensing, wireless communication, and actuation and monitoring, respectively. However, the field sensing was limited to temperature, humidity and moisture content of the soil, and was not tested on a large area such as a greenhouse.

Flores et al. [12] developed a Precision Agriculture Monitoring System (PAMS) using WSN to monitor humidity, temperature, moisture, luminosity, electrical conductivity and pH, and Raspberry Pi microcontroller as a local server. Despite the use of more sensors as compared to the previous systems, PAMS was not deployed and tested on a large area, therefore its reliability is still unexplored.

Regardless of the increasing emergence of WSNs in Precision Agriculture, there has been only limited research concerning the implementation of reliable and flexible WSNs to enable the real-time monitoring and control of greenhouses.

3 Proposed WSN Architecture for Precision Agriculture

This section describes the proposed WSN architecture for precision agriculture inside a greenhouse for remote monitoring of crop growth with the aim of optimizing yield. The proposed architecture is designed to accommodate more sensors in the future as well as allow reliable information flow and easy accessibility of crop growth conditions to farmers from any remote greenhouse location. Multiple heterogeneous environmental sensors are used to monitor the greenhouse's parameters in real-time for data acquisition. The sensors collect and send data to the sink node or gateway for data processing which is then transferred to the remote cloud server where the farmer can easily have access to data visualization and analytics.

3.1 Architecture Design Requirements

In the proposed WSN architecture for precision agriculture, heterogeneous environmental sensors collect the crops' environmental data for instance light irradiance, soil parameters (such as temperature, pH, salinity, moisture, nutrients), leaf parameters (such as moisture/wetness, thickness), and air parameters (such as humidity, carbon dioxide content, temperature). Thus, the quality of data acquisition is highly crucial as well as effective data processing and transfer throughout the WSN system infrastructure. Any possible errors in the network such as signal strength, latency, delay in transmissions, and failure-prone sensors could lead to erratic reliability of the system. Moreover, since sensor technology is advancing at a fast rate, the proposed system should be flexible to accommodate more sensors and be interoperable, that allows the exchange of information between different devices. Therefore, ensuring reliability, flexibility as well as interoperability are crucial requirements of the proposed WSN system for precision agriculture.

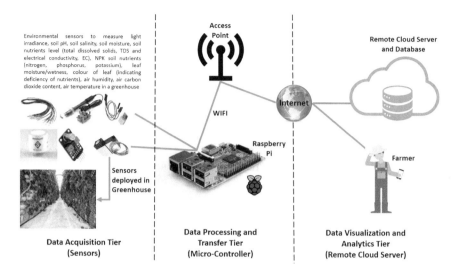

Fig. 1 High-level WSN architecture for precision agriculture

3.2 High-Level WSN Architecture

The proposed high-level WSN architecture for precision agriculture comprises of three main tiers namely the (1) Data Acquisition tier, (2) Data Processing and Transfer tier, and (3) Data Visualization and Analytics tier, as shown in Fig. 1. The data acquisition tier consists of multiple heterogeneous environmental sensors that interact with the crops' environment in a greenhouse and transfers the sensed data wirelessly to the gateway for data processing and transfer. The gateway consists of a microcontroller such as a Raspberry Pi that aggregates the sensed data, uses inference engines to bring intelligence to the aggregated data, and transfers the preprocessed data to the remote cloud server through the Internet. The third tier consists of the remote cloud server and the farmer's device such as a smartphone. The remote cloud server collects, categorizes, and records the crops' environmental data from the WSN in a database for further visualization and analytics.

The proposed architecture can sense, aggregate, and process heterogeneous data across different data sources in a flexible and extensible manner, using standardized protocols as well as open-source technologies in the domain of Internet of Things (IoT). This is primarily used as many powerful and scalable frameworks already exist for other domains such as smart cities and smart health, as compared to precision agriculture that is still under development. To cater for a better performance in terms of reliability of information, scalability, interoperability and flexibility of the proposed architecture, multiple open-source IoT platforms such as CityPulse [13, 14], ThingSpeak™ [15], Freeboard [16], OpenIoT [17, 18], Kaa [19] that could be adopted were evaluated. Furthermore, the proposed architecture can be enhanced by adding a knowledge-based layer that can process farmer's requests using query languages and evaluate the queries to obtain results. For example, the amount of fertilizer to be applied to a certain region in the greenhouse is a request in real-time, it is then evaluated into queries and the result can lead to an automated action through the use of actuators in the greenhouse.

3.3 Heterogeneous Environmental Sensors

A commercial greenhouse already comes with a partially automated system that injects water and fertilizer (termed as fertigation) at timely intervals. However, the system does not adjust to continuous changes happening in the environment and causes either insufficient or excessive fertigation. The excess fertilizer and water is drained out of the greenhouse and causes pollution as well as wastage of valuable agricultural resources.

Environmental parameters directly influence the growth and yield of crops [20] and fertigation should be adjusted according to the crops' environment, for example, on a sunny day there will be higher photosynthesis requiring more fertigation as compared to a rainy day. In the proposed architecture, multiple heterogeneous sensors will be used to measure different environmental parameters. A few examples are given in Table 1.

3.4 Protocols and Topologies

In the proposed WSN architecture, the heterogeneous environmental sensors collect and send data to the microcontroller gateway node using the existing communication protocol IEEE802.11/WIFI. The microcontroller aggregates the sensed data and transfers it to the access point by using the same communication protocol. Sensor nodes will be spread across the greenhouse, will be unattended and should work independently [21]. This poses network challenges that traditional routing protocols and topologies cannot solve. Among the different existing topologies, an energy-efficient topology with clustering algorithms [22] will be the most suitable to achieve better utilization of sensor nodes as well as to make the WSN energy-efficient thereby extending the network's lifetime. Moreover, faulty readings from the sensors can affect decisions. Outliers should be spotted and removed with the use of fault detection algorithms so that the abnormal readings do not pollute decisions. Data duplicates should also be fused using efficient data fusion algorithms to optimize the performance of the transmission stream.

Table 1 Examples of sensors that monitor various environment parameters in a greenhouse

Sensors	Environmental parameters
Photosensor LM358 and LDR Light Sensor DHT22, ALS-PTl9 Light Sensor	Light irradiance
dfRobot pH meter kit Soil Test Kit—pH	Soil pH
YL-38 + YL-69 soil moisture sensor	Soil moisture
EC Meter, where TDS = EC reading (in microSiemens/cm) by 1000 and divide by 2	Soil Nutrients Level (Total Dissolved Solids, TDS and Electrical Conductivity, EC)
EC-probe for soil salinity	Soil salinity
NPK Micro-sensors PhotoHI83225: Nutrient Analysis Photometer Fiber Optic Sensor Soil Test Kit—colorimetry for NPK	NPK soil nutrients (Nitrogen, Phosphorus, Potassium)
237-L Leaf Wetness Sensor	Leaf moisture/wetness
Li-Cor, 6400XT, Lincoln, NE, USA	Leaf conductance and transpiration rate
Parallax TCS3200 color sensor Parallax ColorPAL color sensor	Leaf color
AgriHouse Smart Leaf Sensor (SG-1000)	Leaf thickness
HR202 Humidity Sensor DHT22 humidity/temperature sensor SY-HS-220 humidity sensor	Air humidity
DHT22 humidity/temperature sensor LM35 Temperature sensor	Air temperature
NDIR Carbon Dioxide Sensor IR for CO_2 Agriculture Greenhouse 0–1% COZIR-A CO_2 sensor, EE-89 series Figaro's TGS4161 CO_2 sensor	Air carbon dioxide content
Cl-340 handheld photosynthesis sensor	Plant hydrogen content
TPS-2 portable photosynthesis sensor Cl-340 handheld photosynthesis sensor PTM-48A photosynthesis monitor	Plant photosynthesis

3.5 Data Design

Heterogeneous environmental sensors have different data range and depend on the crop being monitored. In the proposed architecture, tomatoes are considered as the perfect crop for experimentation since they have a relatively short time to reach maturity and harvest of rarely exceeding 100 days in a greenhouse. Moreover, the Food and Agricultural Research and Extension Institute, FAREI in Mauritius [23] reported a marked decrease in Tomato yield due to unfavorable climatic conditions, and in the National Strategic Plan for Food crops, Livestock and Forest sectors 2016–2020, FAREI mentioned that developing protected culture systems such as a greenhouse is one of the main and key objectives to mitigate climate change effects on food crop

Table 2 Environmental parameters range for optimal growth of tomatoes in a greenhouse [4–7]

Environmental parameters	New transplants	At maturity	Fully grown plants
Light irradiance	DLI 6–8 mol m^{-2} d^{-1}	Daily Light Integral, DLI 22–30 mol m^{-2} d^{-1}	
Soil pH	Between 5.6 and 5.8 (acidic)		
Soil moisture (water for irrigation)	About 2 oz (50 ml) of water per crop per day	Up to 3 quarts (2.7 L) of water per crop per day	2 quarts per crop per day
Total dissolved solids, TDS	550–600 ppm TDS	800–1100 ppm TDS	1100–1600 ppm TDS
Electrical conductivity, EC	0.6–0.7	0.9–1.8	1.8–2.2
NPK soil nutrients (Nitrogen, Phosphorus, Potassium) in fertilizers	50–75 ppm Nitrogen NPK ratio as 5-10-10	100–125 ppm Nitrogen NPK ratio as 10-10-10	125–200 ppm Nitrogen NPK ratio as 10-10-10
Leaf moisture/wetness Leaf color and thickness	Used to detect deficiencies and greenhouse diseases		
Relative air humidity	Daytime ideal humidity between 80 and 90% Night-time ideal humidity between 65 and 75%		
	Pollination N/A	Pollination ideal humidity between 60 and 70% (3 times per week, usually mid-day)	
Air temperature	Daytime ideal temperature between 70 to 82 °F Night-time ideal temperature between 62 and 64 °F		
Air carbon dioxide content	800–1000 ppm CO^2		

production. Additionally, at the Massachusetts Institute of Technology, MIT, IoT-based agricultural experiments are carried out to enhance the growth and improve the taste of tomatoes as mentioned in the article IoT: The Internet of Tomatoes [24]. Table 2 shows the range of sensor values for the optimal growth of tomatoes in a greenhouse.

3.6 Application Design on the Remote Cloud Server

The proposed WSN architecture for precision agriculture as shown in Fig. 1 including the sensor design, protocols and topologies, and data design contributes in the reliability of data transferred through the first and second tiers. The remote cloud server further processes the data streams from the microcontroller (second tier) for analytics thereby increasing reliability of the proposed system. Finally, the farmer will be able to view greenhouse crops' conditions in real-time and take appropriate decisions whilst visualizing the greenhouse's analytics report on any device connected to the Internet.

4 Evaluation of the Proposed WSN Architecture for Precision Agriculture

Advancements in sensor technology are currently happening at an incredible rate, therefore sensors will continue to evolve and new sensors will continue to come on the market in the near future. The proposed WSN architecture should thus be flexible to accommodate more sensors without disrupting the current infrastructure. The WSN should also apt for non-reachable places in the greenhouse, such as near the roots of the growing crops. The implementation cost is inexpensive as it will use cheap sensors and free open-source cloud servers for analytics. The WSN avoids plenty of wiring as the sensors are wirelessly connected to the microcontroller. The transfer of sensed data is preprocessed at the microcontroller tier ensuring reliability of information before transferring to the cloud server for further processing, visualization, and analytics. The proposed WSN architecture for Precision Agriculture is evaluated at the three tiers that are (1) sensors tier, (2) gateway microcontroller tier, and (3) remote cloud server tier as shown in Table 3.

5 Challenges and Future Works

The proposed WSN architecture for precision agriculture has the potential to improve the crop yield by sensing almost all environmental parameters that can influence crop growth in a greenhouse. This system improves the overall farmer's experience where the latter no longer need to perform frequent manual checkups in the greenhouse. Reliability and flexibility are ensured throughout all the tiers in terms of data acquisition, processing, transferring, and visualization. The next challenge relates to the decision-making process. Instead of the farmer taking the decisions, actuators could be connected to the system that can automatically take appropriate actions whenever required in the greenhouse such as adjusting the fertigation with respect to the crops' conditions in real-time, spraying mist, opening of window outlets among others. Even though greenhouses are a protected environment, bugs can still penetrate and infest crops. Therefore, pest management could be considered where sensors are used to detect different pests and the system predicts the application of pesticides only on targeted areas in the greenhouse, which will definitely be beneficial for food safety. Furthermore, the second tier can further be enhanced by introducing a knowledge-based layer that can learn from trends in historical data using machine learning algorithms such as computational intelligence [25, 26] to ultimately predict

Table 3 Three tier levels evaluation of WSN architecture for precision agriculture

First Tier: Data Acquisition Tier (Sensors level)	
Sensors	All the sensors used in the greenhouse should send the raw sensory data to the gateway node. Sensitive sensors are water resistant and energy efficient. Moreover, the sensors should not obstruct the normal growth of crops in the greenhouse
Flexibility	The system should accommodate new sensors without disrupting the architecture in place
Data capture	Various environmental sensors are used to capture data efficiently using sensor technology. The data are sent to the microcontroller for preprocessing before sending to the cloud server
Data representation and transfer	At the sensor level, preprocessing, filtering and compression techniques are used to ensure efficient transfer of data to the microcontroller over the network using WIFI
Second Tier: Data Processing and Transfer Tier (Microcontroller level)	
Data fusion and aggregation	At the microcontroller's level, heterogeneous data are collected. To reduce the volume of transmitted data to the remote cloud server, data fusion techniques are used to improve both transmission time and energy efficiency. Furthermore, data aggregation techniques are also used for preliminary statistical analysis. This tier is also fault tolerant as it eliminates out-of-bounds erroneous data from sensors before transmission to the 3rd tier
Reliability	The sensed data are preprocessed at the microcontroller level ensuring the reliability of the data
Data transfer	The preprocessed data is transferred to the remote cloud server through the Internet at timely intervals
3rd Tier: Data Visualization and Analytics Tier (Remote cloud server level)	
Data analysis/interpretation/decision-making	The remote cloud server validates the incoming data integrity and saves it to the cloud's database. The data is then made available to the farmer in the forms of analytics (graphical charts as well as individual sensed values) for further analysis, interpretation and decision-making
Fault tolerance	The remote cloud server should continue to operate despite any failure at the first and second tiers
Security and privacy	The security and integrity of the transmitted data should be preserved

growth, pests, and fertigation requirements. Further challenges that will be addressed as future works of this research are as follows (1) quality of sensory data including completeness, timeliness, accuracy, reliability, usability, and relevancy issues, (2) flexibility, interoperability, and infrastructure issues that can be resolved by making use of existing open-source IoT frameworks as a middleware layer, (3) privacy and security issues by implementing algorithms such as encryption, cryptography, authentication protocols, and privacy mechanisms, (4) data management issues where data mining techniques will be explored for large-scale data sets, and (5) easy-to-use dashboard and easy-to-understand analytics for a better user acceptance.

6   Conclusion

The emerging technological advancements in WSNs has increased its use in precision agriculture. Farmers have started to accept sensor technology for better yield and efficiency of crops. However, there are further challenges to tackle for the effective implementation of the WSN system for precision agriculture. In this paper, the design of a reliable and flexible WSN using heterogeneous environmental sensor data streams have been described. The proposed system is used to monitor the ever-changing greenhouse environmental conditions that will allow the farmer to have access to real-time as well as historical data of the greenhouse. The farmer will also be able to remotely take decisions instead of frequently having to manually inspect the greenhouse. The next phase of this research includes the implementation of the proposed WSN architecture in a tomato greenhouse followed by an evaluation of the system. The challenges discussed and proposed future works will be addressed during the implementation of the WSN system for precision agriculture.

References

1. Alexandratos, N., Bruinsma, J.: World agriculture towards 2030/2050. Land Use Policy **20**(4), 375 (2012)
2. CityPulse. http://www.ict-citypulse.eu/page
3. EU FP7 CityPulse Project- Open Source Tools and Components. https://github.com/CityPulse
4. Flores, K.O., Butaslac, I.M., Gonzales, J.E.M., Dumlao, S.M.G., Reyes, R.S.: Precision agriculture monitoring system using wireless sensor network and Raspberry Pi local server. In: 10th IEEE International Conference, Proceedings/TENCON, pp. 3018–3021. IEEE (2017)
5. Freeboard. https://freeboard.io
6. Hamouda, Y.E.M., Elhabil, B.H.Y.: Precision agriculture for greenhouses using a wireless sensor network. In: Palestinian International Conference on Information and Communication Technology, pp. 78–83. IEEE (2017)
7. IoT Open Patforms. http://open-platforms.eu
8. Satapathy, S.C., Bhateja, V., Raju, K.S., Janakiramaiah, B. (eds.): Data engineering and intelligent computing. In: Proceedings of IC3T 2016, vol. 542. Springer, Heidelberg (2017)
9. Kaa. https://www.kaaproject.org

10. Kassim, M., Mat, I., Harun, A.: Wireless sensor network in precision agriculture application. In: Computer, Information and Telecommunication Systems, International Conference, pp. 1–5. IEEE (2014)

11. Khedo, K.K., Hosseny, M.R., Toonah, M.Z.: PotatoSense: a wireless sensor network system for precision agriculture. In: IST-Africa Conference Proceedings, pp. 1–11. IEEE (2014)

12. Mat, I., Kassim, M.R.M., Harun, A.N.: Precision agriculture applications using wireless moisture sensor network. In: Communications (MICC), IEEE 12th Malaysia International Conference, pp. 18–23. IEEE (2015)

13. Math, R.K., Dharwadkar, N.V.: A wireless sensor network based low cost and energy efficient frame work for precision agriculture. In: Nascent Technologies in Engineering, International Conference, pp. 1–6. IEEE (2017)

14. Mississipi State University: Greenhouse Tomato Handbook. Human Mutation, **35**(7) (2014)

15. MIT Technology Review Insights: IoT: The Internet of Tomatoes. https://www.technologyreview.com/s/601793/iot-the-internet-of-tomatoes

16. Ojha, T., Misra, S., Raghuwanshi, N.S.: Wireless sensor networks for agriculture: the state-of-the-art in practice and future challenges. Comput. Electron. Agric. **118**, 66–84. (Elsevier) (2015)

17. Open-IoT. https://github.com/OpenIotOrg/openiot

18. Rawat, P., Singh, K.D., Chaouchi, H., Bonnin, J.M.: Wireless sensor networks: a survey on recent developments and potential synergies. J Supercomputing **68**(1), 1–48 (2014)

19. Satapathy, S.C., Bhateja, V., Raju, K.S., Janakiramaiah, B.: Computer communication, networking and internet security. In: Proceedings of IC3T, vol. 5 (2016)

20. Satapathy, S.C., Bhateja, V., Das, S.: Smart computing and informatics. In: Proceedings of the First International Conference on SCI, vol. 1 (2016)

21. Singh, D.P., Bhateja, V., Soni, S.K.: Energy optimization in WSNs employing rolling grey model. In: Signal Processing and Integrated Networks (SPIN), International Conference, pp. 801–808. IEEE (2014)

22. Snyder, R.G.: Greenhouse Tomatoes Higher Quality & Value, pp. 1–18 (2017)

23. Statistics Mauritius: Digest of Agricultural Statistics (2014)

24. Suazo-López, F., Zepeda-Bautista, R., Sánchez-Del Castillo, F., Martínez-Hernández, J.J., Virgen-Vargas, J., Tijerina-Chávez, L.: Growth and yield of tomato (*Solanum lycopersicum* L.) as affected by hydroponics, greenhouse and irrigation regimes. Ann. Res. Rev. Biol. **4**(24), 4246 (2014)

25. ThingSpeak. https://thingspeak.com

26. Tubaishat, M., Madria, S.: Sensor networks: an overview. IEEE Potentials **22**(2), 20–23 (2003)

Terminating CU Processing in HEVC Intra-Prediction Adaptively Based on Residual Statistics

Kanayah Saurty, Pierre C. Catherine and Krishnaraj M. S. Soyjaudah

Abstract The current standard in video compression, High-Efficiency Video Coding (HEVC/H.265), provides superior compression performances compared to its H.264 predecessor. However, considerable increase in processing time is brought about with the large Coding Tree Unit (CTU) in H.265. In this paper, a method of terminating the Coding Unit (CU) earlier is proposed based on the luma residual statistics gathered during the encoding of the initial frames of the sequence. The gathered statistics are then formulated into thresholds adaptively and are used to overcome the unnecessary processing of potential CUs during subsequent frames. Experimental results obtained indicate that the encoding time can be reduced by 36.1% on average compared to HM16 along with a BD-Rate of only 0.29%.

Keywords HEVC · Intra-prediction · Encoder optimization · Residual statistics
Early CU size determination

1 Introduction

The new HEVC standard which was released in 2013 is a video project of the Joint Collaborative Team on Video Coding (JCT-VC), formed by ISO/IEC/MPEG and ITU-T/VCEG [6]. HEVC produces almost twice the coding efficiency of the Advanced Video Coding (AVC/H.264) [16] for equal perceptual video quality [14]. This superiority of HEVC over the AVC standard is clearly illustrated in the high

K. Saurty (✉)
Université des Mascareignes, Pamplemousses, Mauritius
e-mail: ksaurty@udm.ac.mu

P. C. Catherine
University of Technology, La Tour Koenig, Mauritius
e-mail: ccatherine@umail.utm.ac.mu

K. M. S. Soyjaudah
University of Mauritius, Réduit, Mauritius
e-mail: ssoyjaudah@uom.ac.mu

© Springer Nature Singapore Pte Ltd. 2019
S. C. Satapathy et al. (eds.), *Information Systems Design and Intelligent Applications*, Advances in Intelligent Systems and Computing 863,
https://doi.org/10.1007/978-981-13-3338-5_13

compression ratio delivered by the latest standard [2]. However, HEVC also comes along with a major shortcoming—its high complexity makes the encoding process very time-consuming.

Compression is achieved by predicting part of a picture and storing only the prediction information along with the residuals. Intra and inter-prediction are the two common techniques inherent in codecs. When the prediction of the blocks in the picture is based only on the information within that picture it is called intra-prediction. Inter-prediction is termed when prediction information is obtained by referring to locations in another frame of the same sequence.

In this paper, only the early termination of CU splitting for the HEVC intra-prediction will be considered. Many approaches with acceptable drops in quality have been proposed to alleviate the complexity in the HEVC standard. In [3], a progressive Bayesian classification is proposed that makes use a two-stage refinement method. In the first stage, classifications are made to separate CUs into split, non-split and an intermediate region. For the latter case, further refinement is conducted in a second stage. Unnecessary CU depth is skipped in [5] by comparing the variance of pixels within the original block and those of the four children.

In [7] a method of identifying the non-split CUs is proposed by analysing the computed gradients (directional and global) in order to determine the smoothness of the texture or the angular inclination. The Hadamard Absolute Difference (HAD) cost is the basis of terminating CUs earlier in [11]. CUs with HAD cost lower than the threshold value set are terminated. Moreover, for 8 8 CUs, the difference between the best mode HAD cost and the DC coefficient is further evaluated to terminate the CU.

Classification of the homogeneity of video content is made in [8] using an adaptive double threshold. The homogeneity of the CU is based on the energy distribution property of the Discrete Cosine Transform (DCT) coefficients. Partitioning data related to the neighbouring CUs along with co-located ones are used to avoid processing of smaller CU sizes. Fast intra-prediction is formulated around coding bits in [17]. Thresholds are formed using the transformed residual coefficients coding (RCC) along with intra-prediction mode coding (IPMC) bits.

The Grey Value Range (GVR) thresholds distribution is studied in [4] to formulate appropriate CU termination thresholds for different Quantisation Parameters (QPs) and depth levels. In [9], the Coefficient of Variance (CV) of the source block along with its four children is computed. Smooth regions are those with small CV values and are therefore terminated as non-split CUs. Furthermore, a high CV value of a child CU indicates a local complex texture leading to most probably a split CU. A fast CU size decision algorithm is proposed in [15] by using texture characteristics to reduce the intra-prediction complexity. Indeed the Mean Absolute Differences (MADs) along with neighbouring trends in CU splitting are optimised in this approach. While most works in this area try to identify smooth regions within the picture, the recognition of objects within a picture based on histogram of the pixel intensities and applying Otsus thresholding [10] may also be used to detect complex textures for which the CU need to be split.

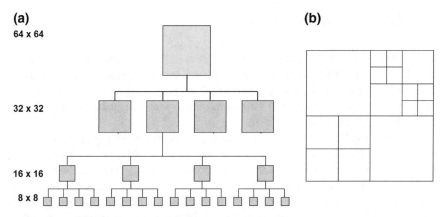

Fig. 1 **a** Subdivision of CTU in HEVC, **b** Resulting CTU structure

The residuals obtained following the Rough Mode Decision (RMD) operation are exploited in this paper in order to identify those CUs that can be terminated earlier as well as those CUs that will most probably be split. While residual data are already used in [12, 13] to optimise the inter-prediction process, the residuals are utilised in this paper to reduce the intra-prediction complexity. The rest of the paper is structured as follows. An overview of the intra-prediction process in HEVC is provided in Sect. 2. Section 3 elaborates on the proposed technique along with the formulation of the threshold values. The results of subsequent experiments are presented and discussed in Sect. 4. Finally, a brief conclusion is provided in Sect. 5.

2 Overview of HEVC Intra-Prediction

2.1 Coding Unit Structure

While the previous AVC standard divides a frame into 16×16 macroblocks [16], HEVC partitions allow up to 64×64 Coding Tree Units (CTUs). Each CTU is stored as luma and chroma Coding Tree Blocks (CTBs) together with related syntax information. A CTU can be coded as a single unit denoting a large square area of homogeneous region or it can be further split into multiple partitions of decreasing size down to the smallest block of 8×8 pixels. Figure 1a illustrates the subdivisions of a CTU.

The prediction data of the CU is stored within a Prediction Unit (PU). For intra-prediction, permitted PU configurations are $2N \times 2N$ and $N \times N$ for a $2N \times 2N$ CU. The PU size therefore either retains the same size as the CU or it can be split into four blocks, each one of them with separate prediction data [14].

2.2   Intra-Prediction Process

The selected intra-prediction mode of a PU is used to generate a prediction image block. The difference between the original block and the prediction image, which is termed as the residuals, is then transformed prior to being coded. The aim of the selection process is therefore to choose the mode that will guarantee an optimal output in terms of quality and bitrate.

Choosing the best prediction mode is done among the 35 possible modes (33 angular, DC and Planar). For the angular prediction modes, extrapolation of the reference pixels from the top and left neighbouring CUs are performed. Given the high number of arithmetic operations required for each pixel, this operation is very time-consuming. Once the residual block for each mode is obtained, it is again partitioned into multiple Transform Blocks (TBs) that are further processed by transform coding to remove spatial correlation within the transform block before it is quantised.

In order to finally choose the best mode, the full Rate–Distortion (RD) cost of these modes should ideally be computed. This approach would, however, consume enormous computing resources and therefore a simplified model is adopted by HEVC as follows:

1. A smaller set of modes is preselected based on the RMD cost. For PUs of size 4×4 and 8×8, a set of 8 modes is chosen while only 3 modes are selected for larger size PUs.
2. Up to 3 Most Probable Modes (MPMs) are then added to the set if these modes were not selected earlier. A simplified RD cost is then computed to identify the best mode from the set.
3. The full RD cost is eventually performed only on the cheapest mode obtained from the simplified RD operation.

    Nevertheless, complexity still remains high for HEVC intra-prediction.

2.3   Final Structure of the CTU

Once the CTU has been completely split and the corresponding RD cost computed for each subdivision, children CUs are merged to form bigger ones as far as possible. The splitting decision for a CU is based on the computed Rate–Distortion (RD) cost. The distortion from the original picture and the number of bits required to encode the CU are the elements considered in the cost function as given in Eq. 1.

$$J = D + \lambda \cdot B \tag{1}$$

where J is the RD cost, D is the distortion, λ is the Lagrangian constant and B represents the number of bits required.

A cost comparison is made between the parent node and the total cost of its four children for the leaf nodes at each depth. The large parent node is retained when its cost is smaller, otherwise, the four children are maintained in the final structure. A high compression rate is characterised by the selection of large CU sizes. Complex textures often result in smaller CUs sizes to reflect the different prediction modes of the individual CUs.

To reach this final structure, all CU sizes have been tried even though a subsequent portion of the computations are discarded and therefore not forwarded to the decoder. By anticipating the final structure, the unnecessary computation can thus be avoided in order to reduce the encoder complexity.

3 Early CU Termination Decision

While the brute force processing in determining the CU size yields a high performance, a substantial number of CUs with relatively low cost could already be identified at a higher depth. These CUs, if terminated, would relieve the encoder from the unnecessary processing of the smaller size CUs and thereby reducing the encoding time.

In this paper, a method of identifying these non-split CUs is proposed by analysing the luma residuals of the Rough Mode Decision (RMD) prediction. Once the RMD process is terminated, a list of fewer modes is preselected for the intensive Rate–Distortion operation (RDO). The list is ordered by increasing RD cost. The cost function considers the distortion and the number of bits required for the mode. The distortion is given by the Sum of Absolute Differences (SAD) for each mode. The luma residuals of the cheapest RMD mode is considered in the proposed approach.

The Mean Absolute Difference (MAD) of each 2×2 block forming the residuals are computed. The peak value, ρ, of the list of MADs is then found. Figure 2 illustrates the percentage occurrence of split and non-split CUs of size 16×16 during the initial 5 frames for the *Basketball* sequence (Class D). It clearly shows a high proportion of non-split CUs for low values of ρ while high peak values are normally split. It is also noted that the peak value for optimum non-split CU selection differs from sequence to sequence and also varies with different QP values.

An *adaptive* method is therefore proposed to identify the threshold values. For each sequence of 50 frames, the first two initial frames are used to collect the peak values for the split and non-split CU with CU sizes ranging from 8×8 to 64×64. The data is then analysed to formulate thresholds for early processing termination. The graph formed by split and the non-split CUs are denoted as functions f_{split} and $f_{\text{non-split}}$. Two threshold values are identified—the first one, $T_{\text{non-split}}$, to terminate the CU earlier and the second one, T_{split}, to force the splitting of the CU and are given below.

$$T_{\text{non-split}} = f_{\text{split}}(5\%) \qquad (2)$$

$$T_{\text{split}} = f_{\text{non-split}}(98\%) \qquad (3)$$

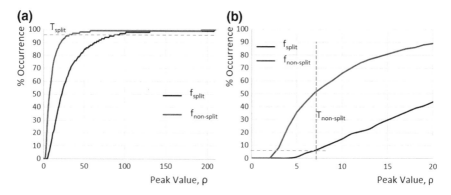

Fig. 2 Percentage variation of split and non-split CUs for different values of ρ for 16×16 CUs during the first five frames of the *BasketballDrive* sequence, **a** the full range **b** with peak values up to 20

where $T_{\text{non-split}}$ is set as the value of ρ to reach 5% on the f_{split} function. This denotes that a maximum of 5% of the split CUs not further divided. In fact, those incorrectly selected non-split CUs will induce an acceptable deterioration in the output given that they have quite small values of ρ. A value of ρ above 5% does produce a higher time savings but is also accompanied by unacceptable deterioration of the output. On the other hand, a high proportion (more than 50%) of terminating CUs will be correctly identified.

The T_{split} threshold is set at 98% of the $f_{\text{non-split}}$ function. Thus, 2% of the non-split CUs will actually be split. While this forced splitting will produce a higher bitrate, it will however improve the quality slightly. In general, it is assumed that CUs with large ρ values will normally be split as is the case during our analysis. CUs with ρ values falling between the two thresholds follow the normal HEVC conventional processing. These 2 these two thresholds are illustrated graphically in Fig. 2.

4 Experimental Results and Discussions

Experiments were conducted using the HM16 encoder with the *all intra high-efficiency* encoding configuration. The first 50 frames of 18 sequences (class A to class E) were encoded using QP values of 22, 27, 32 and 37. BD-Rate and BD-PSNR are computed using the Bjøntegaard metrics [1]. The performances are also reported based on the following formula:

$$\Delta\text{TET}(\%)\frac{\text{TET(proposed)} - \text{TET(HM}_{16})}{\text{Time(HM}_{16})} \times 100 \qquad (4)$$

$$\Delta\text{PSNR(dB)} = \text{PSNR(proposed)} - \text{PSNR(HM}_{16}) \qquad (5)$$

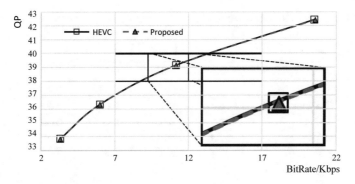

Fig. 3 Performance comparison of *BasketballDrill* sequence

$$\Delta \text{BitRate}(\%) = \frac{\text{BitRate(proposed)} - \text{BitRate(HM}_{16})}{\text{BitRate(HM}_{16})} \times 100 \qquad (6)$$

where TET stands for the total encoding time, PSNR for the peak signal-to-noise ratio and BitRate for the bit rate per second.

The two thresholds given in Eqs. 2 and 3 are formulated *adaptively* for each sequence and QP value after analysing the HM16 encoding pattern during the initial 2 frames. These thresholds are thereafter applied to the remaining frames. When the ρ value of a CU is below $T_{\text{non-split}}$, the CU is terminated without the need to process the subdivisions. CUs with ρ values above the T_{split} threshold are automatically split and since these CUs will be eventually partitioned there is no need to correctly identify the prediction mode. In such case, only the best candidate mode from the RMD process is used during the RDO operation.

Table 1 shows the experimental results for terminating the CU processing earlier along with early decision to split the CU. Compared to HEVC, an average of 36.1% time savings is observed coupled with a negligible increase in bitrate (0.02%) and a marginal drop in PSNR (0.013 dB). This leads to a BD-Rate of only 0.29%. The complexity reduction ranges from 25.6% (*BQMall*, Class C sequence) to 54.7% (*Flowervase*, Class C sequence). The high time savings reached (*Flowervase, KristenAndSara, Vidyo1, Vidyo3 and Vidyo4*) are primarily due to the large extent of smooth texture present in these sequences. Thus, larger CU sizes are selected while avoiding the processing of the smaller CUs. The reduced bitrate observed for *Parkscene, Flowervase (C and D)* and *BasketballPass* further denotes a higher compression rate for these sequences compared to the standard HEVC encoder.

Figure 3 illustrates the performance for the *BasketballDrill* sequence (worst case in the conducted experiment with a BD-Rate of 0.46%). It is observed that the proposed approach shows practically no deviation from the standard HEVC (HM16) over the full range of QP values.

Comparison of the proposed algorithm is made with that of Lu [8]. While [8] came up with an overall average reduction of 30.17% in encoding time, the proposed method yields a superior time savings of 36.1% together with a lower bitrate (0.29%

Table 1 Experimental Results of terminating CUs adaptively based on residual statistics

Class sequence		ΔTET	ΔPSNR	ΔBitRate	BD-PSNR	BD-Rate
		%	dB	%	dB	%
A	PeopleOnStreet	−30.1	−0.014	0.05	−0.013	0.37
	Traffic	−27.1	−0.010	0.06	−0.014	0.30
B	BasketballDrive	−38.3	−0.002	0.08	−0.008	0.15
	Tennis	−38.6	−0.004	0.06	−0.009	0.27
	BQTerrace	−32.3	−0.012	0.00	−0.011	0.07
	ParkScene	−28.3	−0.007	−0.06	−0.004	0.12
C	Flowervase	−54.7	−0.019	−0.15	−0.010	0.17
	Keiba	−29.2	−0.013	0.03	−0.017	0.30
	BasketballDrill	−26.2	−0.011	0.13	−0.029	0.46
	BQMall	−25.6	−0.013	0.11	−0.022	0.34
D	Flowervase	−49.6	−0.027	−0.14	−0.012	0.33
	Mobisode2	−40.6	−0.012	0.14	−0.023	0.45
	BasketballPass	−35.5	−0.016	−0.05	−0.016	0.27
	Keiba	−20.5	−0.013	0.00	−0.013	0.22
E	KristenAndSara	−45.5	−0.013	0.02	−0.024	0.37
	Vidyo1	−44.1	−0.011	0.03	−0.016	0.31
	Vidyo3	−42.7	−0.015	0.05	−0.021	0.35
	Vidyo4	−40.4	−0.011	0.07	−0.019	0.41
Average		−36.1	−0.013	0.02	−0.015	0.29

against 0.31%). Sequences common to these two works are compared and are shown in Table 2. It is to be noted that among the 9 sequences, the proposed approach illustrates a higher time savings with an average of 32.1% compared to 28.9% for the method in [8]. In addition, both the average BD-Rate and BD-PSNR are much lower in this work. The *BasketballDrill* sequence, for example, shows more or less the same encoding time reduction in both methods but the BitRate is lower in the proposed one by almost 0.73%, indicating very little deterioration in quality compared to [8].

5   Conclusion

A novel approach is proposed in this paper to accelerate the HEVC encoder processing by analysing the resulting luma residuals of the Rough Mode Decision (RMD) process. The residuals are partitioned into 2×2 blocks and the average value is computed for each one of them. The peak average value for each CU is then compared with the threshold values to determine whether to terminate the splitting of the CU. These thresholds are formulated adaptively by analysing the HEVC splitting decision pattern of the first two frames. This approach yields a time savings of 36.1% on

Table 2 Comparison of proposed method with that of Lu [8]

Rate		Lu [8]			Proposed method		
		ΔTET	ΔPSNR	ΔBitRate	ΔTET	ΔPSNR	ΔBit
Class	sequence	%	dB	%	%	dB	%
A	PeopleOnStreet	22.90	−0.02	0.35	−30.1	−0.013	037
	Traffic	27.20	−0.02	0.29	−27.1	−0.014	0.30
B	BasketballDrill	38.53	−0.02	0.88	−38.3	−0.008	0.15
	BQTerrace	28.03	−0.01	0.24	−32.3	−0.011	0.07
	ParkScene	28.82	−0.02	0.42	−28.3	−0.004	0.12
C	BasketballDrill	22.73	−0.02	0.34	−26.2	−0.029	0.46
	BQMall	24.19	−0.01	0.26	−25.6	−0.022	0.34
D	BasketballPass	28.01	−0.02	0.27	−35.5	−0.016	0.27
E	KristenAndSara	39.59	−0.02	0.36	−45.5	−0.024	0.37
Average		28.9	−0.018	0.38	−32.1	−0.015	0.27

average while the increase in bitrate and the drop in PSNR are negligible. In future, it is expected to further enhance the speed of the encoder by combining this approach with an appropriate reduced-mode algorithm for the RDO process.

References

1. Bjøntegaard, G.: Calculation of Average PSNR Differences between RDCurves. Doc. VCEG-M33, VCEG (April 2001)
2. Bossen, F., Bross, B., Suhring, K., Flynn, D.: HEVC complexity and implementation analysis. IEEE Trans. Circuits Syst. Video Technol. **22**(12), 1685–1696 (2012). https://doi.org/10.1109/TCSVT.2012.2221255
3. Chen, J., Yu, L.: Effective HEVC intra coding unit size decision based on online progressive Bayesian classification. In: 2016 IEEE International Conference on Multimedia and Expo (ICME), pp. 1–6 (July 2016). https://doi.org/10.1109/ICME.2016.7552970
4. Ding, H., Huang, X., Zhang, Q.: The fast intra CU size decision algorithm using gray value range in HEVC. Optik – Int. J. Light Electron Opt. **127**(18), 7155–7161 (2016). https://doi.org/10.1016/j.ijleo.2016.05.061, http://www.sciencedirect.com/science/article/pii/S0030402616305010
5. Ha, J.M., Bae, J.H., Sunwoo, M.H.: Texture-based fast CU size decision algorithm for HEVC intra coding. In: 2016 IEEE Asia Pacific Conference on Circuits and Systems (APCCAS), pp. 702–705 (Oct 2016). https://doi.org/10.1109/APCCAS.2016.7804070
6. ISO/IEC JTC1/SC29 WG11: Joint Call for Proposals on Video Compression Technology. Doc. VCEG-AM91, ITU-T (January 2010)
7. Jamali, M., Coulombe, S.: Coding unit splitting early termination for fast HEVC intra coding based on global and directional gradients. In: 2016 IEEE 18th International Workshop on Multimedia Signal Processing (MMSP), pp. 1–5 (Sept 2016). https://doi.org/10.1109/MMSP.2016.7813356
8. Lu, X., Xiao, N., Hu, Y., Martin, G., Jin, X., Wu, Z.: A hierarchical fast coding unit depth decision algorithm for HEVC intra coding. In: 2016 Visual Communications and Image Processing (VCIP), pp. 1–4 (Nov 2016). https://doi.org/10.1109/VCIP.2016.7805517

9. Öztekin, A., Erçelebi, E.: An early split and skip algorithm for fast intra CU selection in HEVC. J. Real-Time Image Process. **12**(2), 273–283 (Aug 2016). https://doi.org/10.1007/s11554-015-0534-2

10. Satapathy, S.C., Sri Madhava Raja, N., Rajinikanth, V., Ashour, A.S., Dey, N.: Multi-level image thresholding using Otsu and chaotic bat algorithm. Neural Comput. Appl. **29**(12), 1285–1307 (Jun 2018). https://doi.org/10.1007/s00521-016-2645-5

11. Saurty, K., Catherine, P.C., Soyjaudah, K.M.S.: Terminating CU splitting in HEVC intra prediction using the Hadamard Absolute Difference (HAD) cost. In: 2015 SAI Intelligent Systems Conference (IntelliSys), pp. 836–841 (Nov 2015). https://doi.org/10.1109/IntelliSys.2015.7361239

12. Saurty, K., Catherine, P.C., Soyjaudah, K.M.S.: Fast adaptive inter-splitting decisions for HEVC based on luma residuals. In: 2017 1st International Conference on Next Generation Computing Applications (NextComp), pp. 75–80 (July 2017). https://doi.org/10.1109/NEXTCOMP.2017.8016179

13. Saurty, K., Catherine, P., Soyjaudah, K.M.S.: Inter prediction complexity reduction for HEVC based on residuals characteristics. Int. J. Adv. Comput. Sci. Appl. (IJACSA) **7**(10) (2016). https://doi.org/10.14569/IJACSA.2016.071042

14. Sullivan, G., Ohm, J., Han, W.J., Wiegand, T.: Overview of the High Efficiency Video Coding (HEVC) Standard. IEEE Trans. Circuits Syst. Video Technol. **22**(12), 1649–1668 (Dec 2012)

15. Trang, D.L.D., Kim, K., Chang, I.J., Kim, J.: Texture characteristic based fast algorithm for CU size decision in HEVC intra coding. In: 2017 7th International Conference on Integrated Circuits, Design, and Verification (ICDV), pp. 88–93 (Oct 2017). https://doi.org/10.1109/ICDV.2017.8188645

16. Wiegand, T., Sullivan, G., Bjontegaard, G., Luthra, A.: Overview of the H.264/AVC video coding standard. IEEE Trans. Circuits Syst. Video Technol. **13**(7), 560–576 (July 2003). https://doi.org/10.1109/TCSVT.2003.815165

17. Yao, F., Zhang, X., Gao, Z., Yang, B.: Fast mode and depth decision algorithm for HEVC intra coding based on characteristics of coding bits. In: 2016 IEEE International Symposium on Broadband Multimedia Systems and Broadcasting (BMSB), pp. 1–4 (June 2016). https://doi.org/10.1109/BMSB.2016.7521942

Artificial Neural Networks Based Fusion and Classification of EEG/EOG Signals

Vikrant Bhateja, Aparna Gupta, Apoorva Mishra and Ayushi Mishra

Abstract Electroencephalogram (EEG) denotes to the brain waves whereas Electrooculogram (EOG) denotes the eye blinking signals. Both the signals are accompanied by various artifacts when they are recorded. Preprocessing becomes an important task in order to get rid of artifacts and use these signals in various biometric and clinical applications. Stationary Wavelet transform (SWT) with the combination of Independent Component Analysis (SWT + ICA) is used to preprocess EEG signal and Empirical Mode Decomposition (EMD) is used to preprocess EOG data. After the processing/filtering of both the signals, feature extraction is done. For the purpose of feature extraction, time delineation in the case of EOG and Auto-Regressive Modeling (AR) technique in the case of EEG signal is implemented. In order to minimize the number of features, fusion of extracted features is performed using Canonical Correlation Analysis (CCA). In order to perform dimensionality reduction, classification is performed which classifies the features into sets. Artificial Neural Network (ANN) is used to form suitable feature arrays and evaluate the classifier's performance. The chief goal is to develop a multimodal system which possesses high classification and recognition accuracy so that biometric authentication can be performed using the combination of EEG and EOG signals.

Keywords AR · ANN · CCA · EEG · EOG · EMD · SWT-ICA

V. Bhateja (✉) · A. Gupta · A. Mishra · A. Mishra
Department of Electronics and Communication Engineering,
Shri Ramswaroop Memorial Group of Professional Colleges (SRMGPC),
Lucknow 226028, UP, India
e-mail: bhateja.vikrant@gmail.com

A. Gupta
e-mail: aparnag2430@gmail.com

A. Mishra
e-mail: apoorvamishra3103@gmail.com

A. Mishra
e-mail: ayushimishra960@gmail.com

© Springer Nature Singapore Pte Ltd. 2019
S. C. Satapathy et al. (eds.), *Information Systems Design and Intelligent Applications*, Advances in Intelligent Systems and Computing 863,
https://doi.org/10.1007/978-981-13-3338-5_14

141

1 Introduction

EEG signals denote the electrical actions of the brain which is recorded by placing several electrodes on the scalp. In this case the single channel database is considered as the recordings of Fp1 electrode is considered. They have frequency range from 0.01 to 100 Hz and may vary from a few $\mu V–100\mu$. Due to the complexity in their nature, they possess various artifacts such as ocular, muscular, cardiac, glossokinetic and environmental [1–3]. EOG signals denote the electrical actions of eyeball and eye blinking motions which possess a frequency range from 0.5 to 15 Hz [4]. Usually, the EOG signals are a form of an artifact present in EEG signal. However, in this paper, EOG is considered as a source of features so that the recognition and classification accuracy [4] is enhanced. In order to suppress the various artifacts present in the signal, preprocessing is performed. Suitable features are extracted from the preprocessed signal and this stage is known as feature extraction. Various methodologies have been proposed for the processing of signals. For EEG signal filtering combination of SWT-ICA approach [5–8] is used as discussed in Mishra et al. [6]. Firstly, the un-processed EEG [9–11] signal is decomposed using Stationary Wavelet Transform (SWT). Once the decomposition is performed, Fast ICA methodology is applied to process the unfiltered signal. The EOG signals are acquired from EEG signals using Empirical Mode Decomposition (EMD). After the filtering of the two signals, AR modeling is used to extract EEG features and time delineation of eye blinks is used for EOG feature extraction as discussed in Gupta et al. [12]. In order to eliminate redundant features, feature fusion and classification is an important task. Various methodologies have been developed to fuse and classify the extracted features. Golz et al. [13] used Support Vector Machine (SVM) for fusing EEG/EOG features extracted during sleep but the error rate increased because fusion task was affected by high amplitudes observed in EOG signal. Kaur et al. [14] used a combination of Principal Component Analysis (PCA) and Genetic Algorithm for image fusion. The drawback of this approach yielded higher bit error rate. Zheng et al. [15] used CCA to examine the multichannel EEG database. The approach explored the coordinate system and yielded higher Signal to Noise Ratio (SNR) values. Muller et al. [16] used linear and nonlinear classifiers on EEG database but due to improper knowledge of database, it proved ineffective. Basheer et al. [17] used Artificial Neural Networks (ANN) for modeling curves of microbial growth and studied their dependence on temperature and pH. This approach was easier to address the real-time applications and was computationally efficient. The feature sets classified by ANN help in effective examination of the multilevel human authentication. The multilevel authentication here indicates the joint usage of EEG and EOG signals. The remaining part of this paper is organized as follows. Section 2 describes the proposed EEG/EOG Feature Fusion and Classification approach. The approaches used at each stage are explained in detail in the Sects. 2.1 and 2.2, respectively. The experiments performed and the achieved results for the fused and classified features from EEG/EOG signals are discussed in Sect. 3. Finally, the summation of concluded work is discussed in Section.

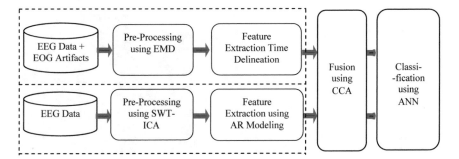

Fig. 1 Block diagram of EEG/EOG fusion and classification

2 Proposed EEG/EOG Fusion and Classification Approach

Based on the projected classification technique, firstly the EEG signal preprocessing is done using the SWT-ICA combinational approach [6] so as to suppress the noisy contents present in the signal. From the noise suppressed EEG, the EOG signals are mined using the EMD technique [18, 19]. After the preprocessing stage, feature extraction is performed. This is fulfilled using AR Modeling in the case of EEG, whereas time delineation in case of EOGs, respectively [12]. After the two feature matrices of respective signals are constructed, they are merged/fused using an approach known as CCA [15]. Then the data is processed for dimensionality reduction and classification is performed. This is done using ANN [20, 21]. The complete methodology has been pictured as per the block diagram in Fig. 1.

Various sub-modules of the above block diagram are explained in the following subsection.

2.1 Canonical Correlation Analysis (CCA)

CCA is implemented to maximize correlations amongst data arrays. The factors so obtained are denoted as the correlation coefficients and can be evaluated straight from the two data arrays or from depictions known as the covariance matrices. If the correlation coefficient comes out to be zero, the arrays are supposed to be uncorrelated and there subsists no direct connection amid them [15]. CCA is an efficient approach in resolving complications like dimensionality reduction, understanding dependency arrangements, etc. [22].

Fig. 2 Multi-layered
artificial neural network
system [23]

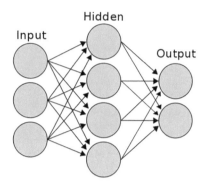

Table 1 Algorithm of ANN

BEGIN
Step 1: Allocate casual weights to the links
Step 2: Calculate the activation rate of hidden nodes using links between inputs and hidden nodes
Step 3: Then, find the activation rate of output nodes
Step 4: Find error rates and update the links between input and hidden nodes
Step 5: Cascade down the error to hidden nodes through the weights and inaccuracies found at the output nodes
Step 6: Update errors and weights between input and hidden nodes
Step 7: Repeat steps 2–6 until the convergence criteria is not reached
Step 8: Using the final link weights, score the activation rate of the output nodes
END

2.2 *Artificial Neural Networks (ANN)*

Artificial neural networks (ANNs) [17] are referred to as the calculation methods prepared by using huge amount of inflexibly interrelated adaptive features or neurons. These are capable of carrying out vastly comparable calculations for data handling and information depiction. Erudition in ANNs is achieved through distinct processes grounded on knowledgeable procedures supposed to imitate the learning methods of organic schemes. ANNs can also be skilled to identify outlines and nonlinear prototypes established in the course of training. These permit neural systems to simplify their suppositions and make a request to patterns not formerly faced [20, 23]. Figure 2 shows the structure of ANN with three inputs, four hidden states and two outputs (Table 1).

3   Results and Discussions

The EEG and EOG records during visual relief and eye flutters were taken from PhysioBank ATM [24]. EEG/EOG is customarily documented by placing numerous

Table 2 Various results acquired from classification

Performance parameters	ANN classifier
Training set	15
Testing set	14
True Positives (TP)	10
True Negatives (TN)	7
False Positives (FP)	3
False Negatives (FN)	2
Accuracy (%)	80.942
Sensitivity	0.833
Specificity	0.7

probes like Cp1, Fp1, etc. on the skull. But in this work, only a lone electrode location is measured, i.e., the Fp1 electrode. The Fp1 recordings contain 15 subjects with period of 10 s and 160 samples each frame. AWGN noise considering different SNRs (5, 10, 15, and 20 dBs) is mixed with the signal taken from databank. Figure 3 depicts noisy and reconstructed EEG waveforms for two respective signals with varying SNR values. The encircled parts in the reconstructed image shows clipping of peaks. This clipping signifies effective suppression of AWGN noise. The SNR (fidelity assessment parameter) is computed and it comes out to be -7.35 dB for the noisy signal. Preprocessing is then carried out by means of SWT-ICA technique in case of EEG whereas EMD in case of EOG. The After signal reconstruction the recorded SNR comes out to be as 25.20 dB. After the noise suppression stage, appropriate features are taken out. This is realized using AR modeling and time delineation methods in case of EEG/EOG, respectively. A group of 20 features is achieved from EEG and that of seven features from EOG signal feature extraction. These features form a matrix of 86×1 and 13×1 correspondingly. After attaining the two feature sets, their fusion is carried out. This is done so as to eradicate redundant features. CCA approach is applied to merge the EEG/EOG feature arrays. The subsequent feature matrix acquired has size of 1×86 which shows the abolition of undesirable features. Fused feature array is then handled to execute classification. ANN is used to classify the features and performance of the classifier is assessed. The grouping skill of a feature matrix can be examined from the classification accuracy. The greater value of correctness can be achieved by attaining more values of true positives, i.e., large number of coordinated groups. The classifier should own little sensitivity so that it does not produce variable outcomes. The classified parameters from ANN are tabulated in Table 2.

(a)   **(b)**

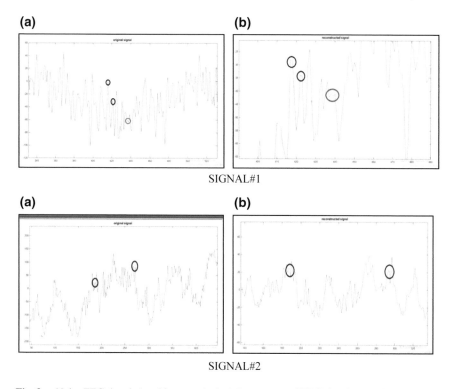

SIGNAL#1

SIGNAL#2

Fig. 3 **a** Noisy EEG signals 1 and 2, respectively. **b** Reconstructed EEG signals 1 and 2, respectively

4   Conclusion

In this paper, an approach is presented for fusing and classifying the extracted EEG/EOG features. To get rid of redundant features, feature fusion is performed and this helps to accomplish the final objective of human biometric recognition with enhanced recognition accuracy. Furthermore, the ANN classifier is used to perform the classification of fused signals. The features classified by ANN can be helpful to develop a multimodal system using which human biometric authentication can be performed. As per future perspective, the task of classification can be tested and compared with other classifiers. The proposed methodology states that the ANN classifier gives high accuracy which is an indication of no loss of useful information and error-free classification. This can be useful to perform the human authentication task precisely. For future application, higher recognition accuracy can be achieved by increasing the number of features.

References

1. Klass, D.W.: The continuing challenge of artifacts in the EEG. J. EEG Technol. **35**(4), 239–269 (1995)
2. Tatum, W.O., Dworetzky, B.A., Schomer, D.L.: Artifact and recording concepts in EEG. J. Clin. Neurophysiol. **28**(3), 252–263 (2011)
3. Lay-Ekuakille, A., Vergallo, P., Griffo, G., Urooj, S., Bhateja, V., Conversano, F., Casciaro, S., Trabacca, A.: Multidimensional analysis of EEG features using advanced spectral estimates for diagnosis accuracy. In: Proceedings of Medical Measurements and Applications, pp. 237–240, IEEE (2013)
4. Abbas, S.N., Zahhad, M.A.: Eye blinking EOG signals as biometrics. In: Biometric Security and Privacy in Signal Processing, pp. 121–140. Springer, Heidelberg (24 December, 2016)
5. Lay-Ekuakille, A., Vergallo, P., Griffo, G., Conversano, F., Casciaro, S., Urooj, S., Bhateja, V., Trabacca, A.: Entropy index in quantitative EEG measurement for diagnosis accuracy. IEEE Trans. Instrum. Meas. **63**(6), 1440–1450 (2014)
6. Mishra, A., Bhateja, V., Gupta, A., Mishra, A.: Noise removal in EEG signal using SWT-ICA combinational approach. In: 2nd International Conference on Smart Computing and Informatics (SCI-2018)
7. Zahhad, M.A., Ahmed, S. M., Abbas, S.N.: A new multi-level approach to EEG based human authentication using eye blinking. J. Pattern Recogn. Lett. **82**, 216–225 (Elsevier) (11 August, 2015)
8. Bhateja, V., Verma, R., Mehrotra, R., Urooj, S.: A non-linear approach to ECG signal processing using morphological filters. Int. J. Meas. Technol. Instrum. Eng. **3**(3), 46–59 (2013)
9. Verma, R., Mehrotra, R., Bhateja, V.: An improved algorithm for noise suppression and baseline correction of ecg signals, In: Proceedings of the International Conference on Frontiers of Intelligent Computing: Theory and Applications (FICTA), pp. 733–739, Springer, Heidelberg (2013)
10. Verma, R., Mehrotra, R., Bhateja, V.: An integration of improved median and morphological filtering techniques for electrocardiogram signal processing. In: 3rd International Advance Computing Conference (IACC), pp. 1223–1228, IEEE (February 2013)
11. Anand, D., Bhateja, V., Srivastava, A., Tiwari, D.K.: An approach for the preprocessing of EMG signals using canonical correlation analysis. In: Smart Computing and Informatics, pp. 201–208, Springer, Singapore (2018)
12. Gupta, A., Bhateja, V., Mishra, A., Mishra, A.: Auto regressive modeling based feature extraction of EEG/EOG signals. In: International Conference Information and Communication Technology for Intelligent Systems (2018)
13. Gold, M., Sommer, D., Chen, M., Trustschel, U., Mandic, P.D.: Feature fusion for detection of microsleep events. In: J. VLSI Signal Process. **49**(10), 329–342 (October 2007)
14. Kaur, R., Kaur, S.: An approach for image fusion using PCA and genetic algorithm. Int. J. Comput. Appl. **145**(6), 54–59 (2016)
15. Zheng, Y., Wan, X., Ling, C.: An estimation method for multi-channel EEG data based on canonical correlation analysis. J. Electron. **24**(3), 569–572 (2015)
16. Muller, K.R., Anderson, C.W., Birch, G.E.: Linear and nonlinear methods for brain computer interfaces. IEEE Trans. Neural Syst. Rehabil. Eng. **11**(2), 165–169 (2003)
17. Basheer, I.A., Hajmeer, M.: Artificial neural networks: fundamentals, computing, design and application. J. Microbiol. Methods **43**(1), 3–31 (2000)
18. Tiwari, D.K., Bhateja, V., Anand, D., Srivastava, A., Omar, Z.: Combination of EEMD and morphological filtering for baseline wander correction in EMG signals. In: Proceedings of 2nd International Conference on Micro-Electronics, Electromagnetics & Telecommunications, vol. 434, pp. 365–373, Springer, Heidelberg (2018)
19. Lay-Ekuakille, A., Griffo, G., Conversano, F., Casciaro, S., Massaro, A., Bhateja, V., Spano, F.: EEG signal processing and acquisition for detecting abnormalities via bio-implantable devices. In: International Symposium on Medical Measurements and Applications (MeMeA), pp. 1–5, IEEE (May, 2016)

20. Guler, I., Ubeyli, E.D.: A mixture of experts network structure for modeling doppler ultrasound blood flow signals. Comput. Biol. Med. **35**(7), 565–582 (2005)
21. Gautam, A., Bhateja, V., Tiwari, A., Satapathy, S.C.: An improved mammogram classification approach using back propagation neural network. In: Data Engineering and Intelligent Computing, vol. 542, pp. 369–376. Springer, Singapore (2018)
22. Mishra, A., Bhateja, V., Gupta, A., Mishra, A.: Feature fusion and classification of EEG/EOG signals. In: International Conference on Soft Computing and Signal Processing (2018)
23. Abraham, A.: Artificial neural networks. In: Handbook of Measuring System Design, 1st edn. Wiley publications, USA (2005)
24. PhysioBank ATM. https://physionet.org/cgi-bin/atm/ATM

Dolphin Swarm Algorithm for Cryptanalysis

Seeven Amic, K. M. Sunjiv Soyjaudah and Gianeshwar Ramsawock

Abstract This work focuses on the application of Dolphin Swarm Algorithm (DSA) to cryptanalysis. It is a relatively new optimization algorithm that is based on the collective behavior of dolphins to hunt for food in their natural habitat. Experimental results show that DSA is more efficient than well-known algorithms like genetic algorithm and ant colony optimization when utilized for the cryptanalysis of the Data Encryption Standard. It also produces solutions with higher fitness in a shorter period of time and with fewer individuals. The DSA can also be used to attack other block ciphers like Advanced Encryption Standard or Elliptic Curve Cryptography due to the independence of the fitness function with respect to the cipher.

Keywords Nature-inspired algorithms · DES · Genetic algorithm
Ant colony optimization · Dolphin swarm algorithm · Optimization
Cryptanalysis

1 Introduction

Cryptography is a fundamental component of modern digital information and communication security. Cryptanalysis is a scientific study of tools and techniques to decrypt a message without knowledge of the key or to indirectly identify the key which has been used for message encryption. One of the ways to establish the strength of an encryption algorithm is through ethical cryptanalysis. Cryptanalysis of modern ciphers falls in the category of hard problems as the key space may

S. Amic (✉)
Université des Mascareignes, Rose-Hill, Mauritius
e-mail: samic@udm.ac.mu

K. M. S. Soyjaudah · G. Ramsawock
University of Mauritius, Reduit, Mauritius
e-mail: ssoyjaudah@uom.ac.mu

G. Ramsawock
e-mail: gramsawock@uom.ac.mu

© Springer Nature Singapore Pte Ltd. 2019
S. C. Satapathy et al. (eds.), *Information Systems Design and Intelligent Applications*, Advances in Intelligent Systems and Computing 863,
https://doi.org/10.1007/978-981-13-3338-5_15

prove to be astronomically gigantic. Of late, researchers in the field of cryptanalysis have been investigating various techniques in breaking cryptographic systems. Precisely, nature-inspired algorithms have taken the research community by storm due to their ability to solve hard problems which cannot be tackled using traditional approaches [1]. It has been observed that nature performs inbuilt, distributed, and self-regulated information processing activities efficiently without any centralized control [2]. There is indeed an increasingly growing bunch of nature-inspired algorithms which have been effectively applied to problems in numerous fields. Three main disadvantages have been identified with metaheuristics: (a) they may get stuck in a local solution; (b) the time taken to reach a solution may be very long; and (c) they depend largely on specific parameters which are tedious to calibrate for optimal results. Though optimal solutions may not be guaranteed, they are still reached in paradoxically reasonable amount of time whenever successful. This work focuses on the experimental probe of a novel stochastic nature-inspired optimization method, namely, Dolphin Swarm Algorithm (DSA), on the problem of cryptanalysis of modern ciphers like the Data Encryption Standard (DES). The structure of this paper further dismantles as follows: Section 2 covers the related work pertaining to cryptanalysis using nature-inspired techniques. A brief description of DES is included in Sect. 3. The concepts of GA and ACO are explained in Sects. 4 and 5, respectively. A comprehensive discussion of DSA is given in Sect. 6 followed by a short description of the methodology and experiments conducted in Sect. 7. The analysis and discussion of experimental results are deliberated in Sect. 8. Section 9 summarizes the paper followed by a description of probable extensions of this work in future.

2   Related Work

In recent years, population-oriented nature-inspired as well as evolutionary algorithms such as Particle Swarm Optimization (PSO), Genetic Algorithm (GA), and Ant Colony Optimization (ACO) have attracted many researchers around the world in the scientific discipline of cryptanalysis. In 2007, Song et al. [3] demonstrated that GA can be efficiently used to cryptanalyze four-rounded DES and Feistel-like ciphers. They also proposed a sound and useful cipher-independent fitness function. Shahzad et al. [4] showed the higher success of PSO for cryptanalysis of four-rounded DES in comparison with GA. An improved cryptanalysis of elliptic curve cryptosystems has been proposed by Ribaric and Houghten in [5]. Khan et al. [6] used ACO for the systematic attack of four-rounded DES. Recently, the innovative firefly algorithm has been successfully implemented for cryptanalysis of knapsack cryptosystem and monoalphabetic substitution ciphers which usefully overcome the problem of local minima thereby addressing the time needed to find a solution. Binary firefly algorithm was proposed by authors in [7] for the cryptanalysis of 16-rounded DES. Garg compares the efficiency of GA, simulated annealing, and tabu search for the cryptanalysis of transposition ciphers [8]. Of late, a novel metaheuristic population-based algorithm simulating the preying behavior of cats has been demonstrated to successfully cryptanalyze DES with high efficiency [9]. The use of cuckoo search combined with Levy flight for cryptanalysis of Vigenère cipher has been demonstrated in [10]. The extensive use of metaheuristic search optimization algorithms applied to cryptanalysis, demonstrated by the large number of published papers in recent years, shows that nature-inspired algorithms are an improved strategy and a favorable technique to efficiently solve the problem of cryptanalysis.

3 Data Encryption Standard

In this work, the Data Encryption Standard (DES) algorithm is used as the cipher to be cryptanalyzed and is considered sufficiently challenging. DES was developed at IBM in the early 70s based on an initial design by Horst Feistel. The Federal Information Processing Standard (FIPS) published DES as an official cryptographic algorithm for the United States, the FIPS PUB 46 [11]. DES performs encryption of a 64-bit data block using a 56-bit key in 16 rounds. DES ensures high diffusion by complex permutation of data bits and confusion by using ingenious substitution tables. The relatively small key size of DES renders the latter particularly vulnerable to brute-force attack on modern computers. DES is severely predisposed to linear and differential cryptanalytic attacks. Although DES is considered as insecure for practical uses, it is still being employed in legacy systems and in the form of triple DES (3DES). Despite its various deficiencies, DES provides a noteworthy basis for experimental cryptanalysis.

4 Genetic Algorithm

In the 1970s, John H. Holland conceived a powerful heuristic optimization technique, namely, the Genetic Algorithm (GA) [12]. It is based upon the fundamental biological process of evolution pertaining to genetics and natural selection in a given population of chromosomes. Each chromosome or individual consists of a sequence of genes and represents a prospective solution to the problem. GA simulates the evolutionary process of the population of chromosomes from one generation to the next, each time attempting to improve the quality (or fitness) of the resulting population using natural selection and genetic operations. The outline genetic algorithm based on [3] is shown in Fig. 1.

Natural selection involves electing chromosomes for mating to mimic the survival of the fittest individuals best adapted to its environment. The selection process may be performed using the tournament or roulette wheel approach. Each selected pair of chromosomes is then coupled to produce new generation chromosomes using genetic operators: crossover, inversion, and mutation. The crossover operator mixes portions of parent chromosomes producing offspring chromosomes. Different types of crossover are as follows: uniform, single-point, double-point, and k-point. Inver-

Genetic Algorithm
Establish λ, the desired fitness
Let t_{max} be the maximum number of generations
Generate N keys and determine the fitness of each
Determine the fittest key, *Best*
$t = 0$
While *Fitness(Best)* < λ **and** $t < t_{max}$
If *elitism* = **True**
Save *Best* in new population
End if
For $i = 1$ **to** N
Select two keys using a selection rule
Crossover the selected keys producing two new keys
Mutate the two keys
Save the two resulting keys
End For
Evaluate fitness of all keys
Find *Best*
$t = t + 1$
End While
Return *Best*

Fig. 1 Genetic algorithm

sion involves reversing a sequential section of a chromosome. Mutation is a random change at some positions (genes) in a chromosome. This can be carried out by sequentially traversing through every gene and then bit-flipping the allele in case a generated random value is greater than a predefined limit known as the mutation rate. Elitism, if set to true, may be introduced in the algorithm by including the best chromosome in the current generation into the next generation automatically. This guarantees that the best chromosomes are always passed on to the next generation of chromosomes.

5   Ant Colony Optimization

The Ant Colony Optimization (ACO) is a metaheuristic algorithm that is inspired by the intelligent behavior of ants. ACO was initially proposed by Marco Dorigo in his Ph.D. thesis [13]. Ants, in nature, live in nests and have to forage their environment for food. In the quest for food, they take a route to the food source and deposit pheromone leaving a trail as they wander. Other ants may follow the same path depositing further pheromone which guides them toward food sources discovered by previous ants. With time, the pheromone may evaporate at a given rate, ρ, and decrease in concentration. So the different paths between the ants nest and the food source may be represented as a connected undirected graph $G = (V, E)$, where V are the vertices (or nodes) and E the edges. When an ant reaches a particular vertex, the probability of the ant heading to another particular vertex depends upon the distance and the pheromone concentration between the vertices. Suppose we have S nodes and an ant arrives at node i, and $d_{i,j}$ is the distance separating node i and node j, where $i, j \in S$, the probability of the ant choosing to move to node j is given by the formula [6]:

$$p_{i,j} = \frac{[\tau_{i,j}]^{\alpha}[\eta_{i,j}]^{\beta}}{\sum_{k \in S}[\tau_{i,k}]^{\alpha}[\eta_{i,k}]^{\beta}} \tag{1}$$

where $\tau_{i,j}$ is the pheromone level on the edge between nodes i and j, and $\eta_{i,j}$ is a heuristic value calculated as follows:

$$\eta_{i,j} = \frac{1}{d_{i,j}} \tag{2}$$

$\propto$ and β are parameters that influence the pheromone value and heuristic value, respectively. Initially, the edges are initialized with small random levels of pheromone and can be stored in a two-dimensional array.

The pheromone evaporation on edge $e(i, j)$ can be performed as

$$\tau_{i,j} = \tau_{i,j} \cdot \rho \tag{3}$$

The ACO algorithm is depicted in Fig. 2.

Ant Colony Optimization Algorithm

Establish λ, the desired fitness
Let t_{max} be the maximum number of iterations
Generate N ants and determine the fitness of each
Initialize edges with pheromone randomly
Identify the best tour, *Best*
$t = 0$
While $Fitness_{Best} < \lambda$ and $t < t_{max}$
    For $i = 1$ **to** N
        i^{th} ant completes its tour, deposit pheromone
        Evaluate its fitness
    End For
    Find *Best*
    Perform pheromone evaporation
    $t = t + 1$
End While
Return *Best*

Fig. 2 Ant colony optimization algorithm

6 Dolphin Swarm Optimization

The Dolphin Swarm Algorithm (DSA) was developed by Wu et al. [14] in 2016 based on the characteristics and behavior of dolphins while hunting for food. Dolphins possess remarkable communication features and are considered as being outstandingly intelligent within the animal kingdom. Dolphins cooperate in predation through the use of echolocation, sounds, teamwork, and division of labor. In DSA, the smart behavior of dolphins is simulated to achieve their goal. The DSA exhibits characteristics such as first-slow-then-fast and periodic convergence and, does not get stuck with local minimum as evidenced in [14]. Furthermore, DSA is more appropriate for low-dimensional unimodal problems with fewer individuals in the swarm [14].

6.1 Characteristics of Dolphins in Swarms

Echolocation
Echolocation is the ability to determine the position of objects by analyzing reflected sound. Although dolphins have good eyesight, they emit sound and use echolocation (also called biosonar) to determine the position of their prey and estimate their distance away from the latter.

Division of labor and cooperation
Dolphins rarely hunt in solo; they rather cooperate among themselves to achieve the goal especially when the prey is large [15]. They also organize themselves through role specialization (division of labor) wherein some dolphins in the pod play the role of driver (they head-on chase the prey), and others that of barrier (they barricade the prey on the flanks).

Communication
Dolphins exchange information with each other by making different sounds such as squeals, squeaks, whistles, screams, clicks, and crunches. As such, when preying, they can summon other dolphins and update them with the location of the prey.

6.2 Terminology

Dolphin
The main agent in the DSA is the dolphin and each dolphin represents a feasible solution to the problem. The number of dolphins in the pod or swarm is called the population size, N. Each dolphin is defined as a vector $\boldsymbol{Dol}$

$$\boldsymbol{Dol}_i = [x_1, x_2, \ldots, x_D] \tag{4}$$

where $i = 1, 2, \ldots, N$ and $x_j (j = 1, 2, \ldots, D)$ is the component of the D-dimensional space of the problem to be solved. For cryptanalysis, $x_j = \{0|1\}$, a binary digit.

Individual optimal solution and neighborhood optimal solution

Each dolphin in the pod has two variables: its individual optimal solution (represented by L) and the neighborhood optimal solution (represented by K). The individual optimal solution L is the best solution that a dolphin finds in a single time. The neighborhood optimal solution K is the best solution found by a dolphin or its surrounding mates.

Fitness

The fitness of a solution is a measure of its proximity to the actual solution. In this work, we aim at maximizing this value. The fitness of a dolphin $\mathbf{Dol}$ is calculated using the formula:

$$fitness(\mathbf{Dol}) = \frac{\partial}{N} \tag{5}$$

where ∂ is the number of matching bits between the expected ciphertext and the obtained ciphertext from the Dolphin $\mathbf{Dol}$. N is the number of dimensions of the search space. The value of the fitness of a solution lies in the range [0..1] inclusive.

Distance

The distance that separates two dolphins influences the communication between the two and can be expressed as the distance between two vectors. The distance, d, between two vectors $\mathbf{u}$ and $\mathbf{v}$ in Euclidean N-space is calculated using the formula:

$$d = \|\mathbf{u} - \mathbf{v}\| = \sqrt{(u_1 - v_1)^2 + (u_2 - v_2)^2 \ldots (u_N - v_N)^2}, \quad i = 1, 2, \ldots, N \tag{6}$$

In DSA, three particular distances are important:

1. The distance between two dolphins $\mathbf{Dol}_i$ and $\mathbf{Dol}_j$ is denoted as $DD_{i,j}$

$$DD_{i,j} = \|\mathbf{Dol}_i - \mathbf{Dol}_j\|, \quad i, j = 1, 2, \ldots, N \quad i \neq j \tag{7}$$

2. The distance between dolphins $\mathbf{Dol}_i$ and K_i is denoted as DK_i

$$DK_i = \|\mathbf{Dol}_i - K_i\|, \quad i = 1, 2, \ldots, N \tag{8}$$

3. The distance between dolphins L_i and K_i is denoted as DKL_i

$$DKL_i = \|L_i - K_i\|, \quad i = 1, 2, \ldots, N \tag{9}$$

Transmission Time Matrix

The transmission Time Matrix, TS, is a square matrix of order N, which stores inter-dolphin information exchange. $TS_{i,j}$ represents the remaining time for sound to move from $\textbf{\textit{Dol}}_j$ to $\textbf{\textit{Dol}}_i$.

6.3 Dolphin Swarm Algorithm

DSA consists of six important pivotal phases, namely, initialization phase, search phase, call phase, reception phase, predation phase, and termination phase. While the first and last phases are trivial, the intermediate phases are disseminated below.

6.3.1 Search Phase

In this phase, each dolphin examines its vicinity by uttering M sounds in random directions. Each sound denoted by vector $\textbf{\textit{V}}$ is composed of D dimensions:

$$V = [v_1, v_2, \ldots v_D] \tag{10}$$

At time t, a dolphin $\textbf{\textit{Dol}}$ makes a sound $\textbf{\textit{V}}$, and it searches for a new solution $\textbf{\textit{X}}$, as follows:

$$X = Dol + V \cdot t \tag{11}$$

In order for a dolphin not to get stuck in the search phase, it is allowed to search for only $t \leq T_1$ times, where T_1 is a fixed maximum search time. So for t times, a dolphin might have t solutions. The fitness value of the solutions is calculated, and $\textbf{\textit{L}}$ is set to the best solution. Further,

$$\textbf{if } fitness(\textbf{\textit{L}}) > fitness(\textbf{\textit{K}}), \textbf{ then } \textbf{\textit{K}} \text{ is set to } \textbf{\textit{L}}, \textbf{ otherwise } \textbf{\textit{K}} \text{ does not change} \tag{12}$$

6.3.2 Call Phase

In this phase, each dolphin informs other dolphins of the outcome of the search phase if an improved solution has been located. The transmission time matrix TS is modified accordingly. For each pair of dolphins $\textbf{\textit{Dol}}_i$ and $\textbf{\textit{Dol}}_j$,

$$\textbf{if } fitness(K_i) < fitness(K_j) \textbf{ and } TS_{i,j} > \frac{DD_{i,j}}{A \cdot speed}$$

$$\textbf{then } \quad TS_{i,j} = \left[\frac{DD_{i,j}}{A \cdot speed} \right] \tag{13}$$

where $speed$ represents the speed attribute of sound and A is an acceleration constant.

6.3.3 Reception Phase

In the reception phase, all the elements of the transmission time matrix, **TS**, are decreased by one to designate sound flow in unit time. Then all the elements of **TS** are checked.

$$\text{if } TS_{i,j} = 0 \text{ then } TS_{i,j} = T_2 \tag{14}$$

where T_2 is the maximum transmission time.

For a given pair of dolphins **Dol$_i$** and **Dol$_j$**,

$$\text{if } fitness(K_i) < fitness(K_j) \text{ then } K_i = K_j \tag{15}$$

6.3.4 Predation Phase

In the predation phase, each dolphin is required to determine the encircling radius R_2 and update itself as **new Dol**. R_2 represents the distance between the dolphin's neighborhood optimal solution **K** and its position after the predation phase. The outline pseudocode based on [14] for a given dolphin **Dol$_i$** to calculate its new position is as follows.

Calculate DK_i and DKL_i using (7) and (8), respectively.
Calculate the search radius, R_1, using (16).

$$R_1 = T_1 \times speed \tag{16}$$

if $DK_i \leq R_1$ **then**

$$R_2 = \left(1 - \frac{2}{e}\right)DK_i, e > 2 \tag{17}$$

$$\textbf{new Dol}_i = K_i + \frac{\textbf{Dol}_i - K_i}{DK_i}R_2 \tag{18}$$

else if $DK_i \geq DKL_i$ **then**

$$R_2 = \left(1 - \frac{\frac{DK_i}{fitness(K_i)} + \frac{DK_i - DKL_i}{fitness(L_i)}}{\frac{e \cdot DK_i}{fitness(K_i)}}\right)DK_i, \quad e > 2 \tag{19}$$

else

$$R_2 = \left(1 - \frac{\frac{DK_i}{fitness(K_i)} - \frac{DKL_i - DK_i}{fitness(L_i)}}{\frac{e \cdot DK_i}{fitness(K_i)}}\right)DK_i, \quad e > 2 \tag{20}$$

endif

$$newDol_i = K_i + \frac{Random}{\|Random\|} R_2 \qquad (21)$$

endif

The skeleton pseudocode for DSA is shown in Fig. 3.

7 Methodology and Experiments

The proposed algorithms were implemented in Java programing language on an Intel CORE i7 computer at 2.2 GHz and 8 GB of active memory. Java, being an object-oriented programming language, provides a good user-friendly platform for plugging-in different cryptographic and optimization algorithms for easier comparison. However, the same can be implemented in MATLAB as well. The proposed framework to implement the designed experiments is shown in Fig. 4. As shown in the diagram, different cryptographic algorithms may be plugged-in as the fitness calculator, which evaluates the fitness of a given key, does not depend upon the internal design of the cipher at hand. The software was executed with Simplified DES (SDES) to check if it was working, then later with DES.

This works focuses on the justification of efficiency with the implementation of DSA for cryptanalysis in comparison with GA and ACO. A key is represented as an array of bytes. For the sake of stochasticity, random keys and random plaintext were

---

Dolphin Swarm Algorithm

Establish λ, the desired fitness
Let t_{max} be the maximum number of iterations
Generate N dolphins with random position
Determine each dolphin's fitness, L and K
Find the dolphin with highest value of K, $KBest$
$t = 0$
While $Fitness_{KBest} < \lambda$ **and** $t < t_{max}$
   For $i = 1$ **to** N
      Do Search phase
      Do Call phase
      Do Reception phase
      Do Predation phase
      Calculate the position of $newDol_i$
      Update L and K
   End For
   Find $KBest$
   $t = t + 1$
End While
Return $KBest$

Fig. 3 Dolphin swarm algorithm

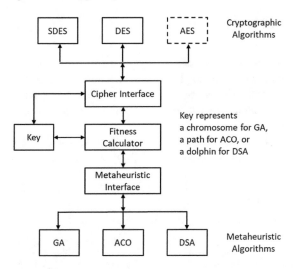

Fig. 4 Software framework

used. In order to generate solutions of higher fitness, the initial population of keys of an experiment is generated using

$$Key_i = P_i \oplus C_i \tag{22}$$

where P_i is the plain text, and C_i is the corresponding ciphertext obtained when P_i is encrypted with the real key. $\oplus$ is the bitwise XOR operator.

Three experiments were designed in which each of the algorithms was executed for $R = 100$ times. For each experiment, the solution with highest fitness was logged at generations $t = 100, 200, \ldots, 1200$ for a given population size, N. Then the average fitness value of the best solutions obtained was calculated for generations $t = 100, 200, \ldots, 1200$ for each algorithm. The population size varied in the range $N = 10, 20, \ldots, 60$. The parameters used for each algorithm are summarized in Table 1.

Table 1 Parameters used for experiments

Algorithm	Parameters
GA	*elitism*=true, *uniformRate*=0.5, *mutationRate*=0.05
ACO	$\alpha = 1.5, \beta = 1, \rho = 0.3$
DSA	$T_1 = 3, T_2 = 1000$, speed=1, $A=5, M=3, e=4$

8   Results and Discussion

The results of the experimentations are displayed in Figs. 5 and 6. There are two noticeable observations from the plots: (a) the average fitness of optimum solutions increases with the population size for all algorithms GA, ACO, and DSA; and (b) the optimum solutions obtained with DSA are significantly higher than those obtained from GA and ACO.

The comparison of DSA versus ACO and DSA versus GA is shown in Fig. 7 using the Root Mean Square Difference (*RMSD*) method. *RMSD* is a measure of the

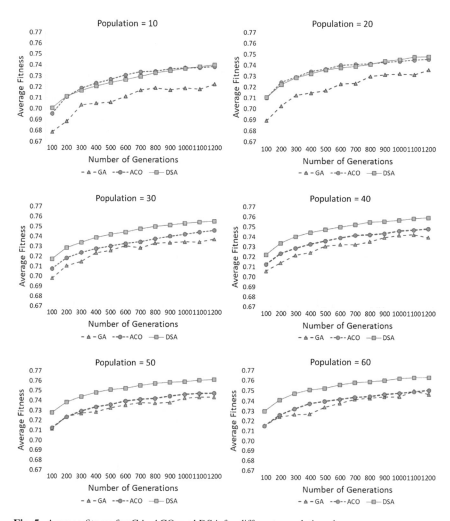

Fig. 5 Average fitness for GA, ACO, and DSA for different population sizes

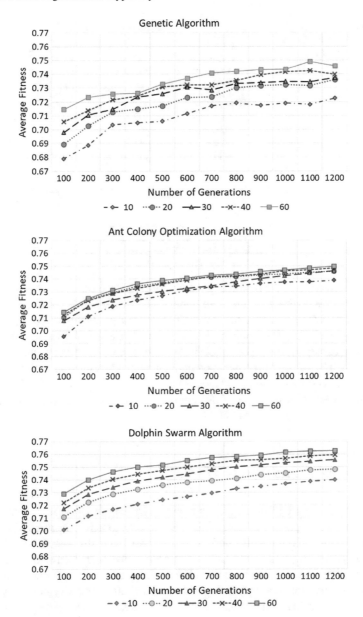

Fig. 6 Average fitness for different algorithms: GA, ACO, and DSA

difference between two sets of data x_1 and x_2 and is calculated as formulated in (23). The higher the *RMSD* value, the larger the gap between the data sets.

Fig. 7 RMSD in average fitness for ACO versus DSA and GA versus DSA

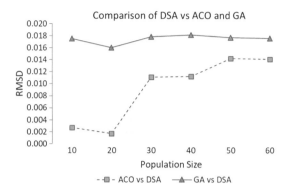

$$RMSD = \sqrt{\frac{\sum_1^N \left(x_{1,i} - x_{2,i}\right)^2}{N}} \qquad (23)$$

It can be seen from the plot in Fig. 7 that the difference between DSA and GA is high and almost constant for different population sizes, whereas the average difference of ACO compared with DSA is initially small and gradually increases, but does not reach that of GA. This confirms that DSA is more efficient than both ACO and GA for cryptanalysis. However, a precaution should be taken before generalizing the conclusion. The values of the parameters of ACO and GA utilized in this work are those that are commonly used, and those of DSA have been proposed by Wu in [14]. We cannot predict the behavior of the algorithms under other parameter values and for population sizes greater than 60.

9   Conclusion and Future Works

In the current research work, we have made an attempt to carry out a comparison of efficiency of cryptanalysis using DSA with respect to nature-inspired algorithms like GA and ACO. The results obtained in this work evidently lead to the inference that DSA yields in optimum keys of higher fitness on average when compared to GA and ACO algorithms under similar conditions. DSA may be used with other block ciphers like Advanced Encryption Algorithm and Elliptic Curve Cryptography as the fitness calculator does not depend on the cryptographic algorithm used, but rather on the plaintext and related ciphertext pairs only. It is essential to note that due to the high diffusion and confusion inbuilt in modern product ciphers such as DES and AES, it is still difficult to suggest an efficient and reliable fitness (objective or cost) function. The one used in this work being only an approximate measure, there is potential room in defining an improved fitness function for cryptanalysis. This work may be further extended by comparing DSA with other prominent nature-inspired optimization algorithms like firefly, cuckoo search, and bat algorithms. Moreover, the results obtained from DSA may be enhanced by tuning different parameters of the algorithm or combination of the latter with other algorithms like GA.

References

1. Yang, X.: Nature-Inspired Metaheuristic Algorithms, 2nd edn (2010)
2. Siddique, N., Adeli, H.: Nature inspired computing: an overview and some future directions. Cognit. Comput. **7**, 706–714 (2015)
3. Song, J., Zhang, H., Meng, Q., Wang, Z.: Cryptanalysis of four-round des based on genetic algorithm. In: 2007 International Conference on Wireless Communications, Networking and Mobile Computing, WiCOM 2007, pp. 2326–2329 (2007)
4. Shahzad, W., Siddiqui, A.B., Khan, F.A.: Cryptanalysis of four-rounded DES using binary particle swarm optimization. Simulation, 1757–1758 (2009)
5. Ribaric, T., Houghten, S.: Genetic programming for improved cryptanalysis of elliptic curve cryptosystems. IEEE Congr. Evol. Comput. CEC 2017 - Proc. 419–426 (2017)
6. Khan, S., Shahzad, W., Khan, F.A.: Cryptanalysis of four-rounded DES using ant colony optimization. In: 2010 Int. Conf. Inf. Sci. Appl. ICISA (2010)
7. Amic, S., Soyjaudah, K.M.S., Mohabeer, H., Ramsawock, G.: Cryptanalysis of DES-16 using binary firefly algorithm. In: IEEE International Conference on Emerging Technologies and Innovative Business Practices for the Transformation of Societies, EmergiTech (2016)
8. Garg, P.: Genetic algorithms, Tabu search and Simulated Annealing: a comparison between three approaches for the cryptanalysis of transposition cipher. Inst. Manag. Technol. India (2005)
9. Amic, S., Sunjiv Soyjaudah, K.M., Ramsawock, G.: Binary cat swarm optimization for cryptanalysis. 2017 IEEE Int. Conf. Adv. Networks Telecommun. Syst., pp. 1–6 (2017)
10. Bhateja, A.K., Bhateja, A., Chaudhury, S., Saxena, P.K.: Cryptanalysis of Vigenere cipher using Cuckoo Search. Appl. Soft Comput. J. **26**, 315–324 (2014)
11. Barker, W.C., Barker, E.B.: Data encryption algorithm. Recomm. Triple Data Encryption Algorithm Block Cipher (2012)
12. Holland, J.H.: Adaptation in Natural and Artificial Systems : An Introductory Analysis With Applications to Biology, Control, and Artificial Intelligence (1975)
13. Dorigo, M.: Optimization, learning and natural algorithms. PhD Thesis, Politecnico di Milano, Milano (1992)
14. Wu, T., Yao, M., Yang, J.: Dolphin swarm algorithm. Front. Inf. Technol. Electron. Eng. **17**, 717–729 (2016)
15. Gazda, S.K., Connor, R.C., Edgar, R.K., Cox, F.: A division of labour with role specialization in group-hunting bottlenose dolphins (*Tursiops truncatus*) off Cedar Key, Florida. Proc. R. Soc. B Biol. Sci. **272**, 135–140 (2005)

A Cloud-Based Energy Monitoring System Using IoT and Machine Learning

Zoya Nasroollah, Iraiven Moonsamy and Yasser Chuttur

Abstract Finding means to encourage consumers to monitor their energy use is an important step toward optimizing on depleting natural resources used for energy production. The current proposal employs cloud computing and machine learning to analyze energy data collected from a nonintrusive IoT system to display energy consumption from several appliances connected to the same power line. For scalability purpose, all collected data from sensors are processed on the cloud and useful information such as appliance monitored and energy consumed can easily be accessed on a mobile app. Preliminary results indicate that the proposed system promises to be a suitable alternative for traditional monitoring systems to deliver instant and historical energy consumption data to consumers, who can, in turn, adopt efficient and smarter ways to use energy.

Keywords IoT · Machine learning · Energy monitoring · Cloud computing

1 Introduction

To meet increasing global energy demand, power plants are expanding their exploitation of depleting natural resources. Consequently, efforts to find alternative energy production mechanisms are underway. Finding alternative sources of energy is an important direction to explore but until a viable solution is found, strategies to efficiently use energy could be a means to optimize on the current resources used in energy production [1]. For that purpose, it is essential to understand consumer energy use behavior and find means to optimize the way energy is utilized. By understanding

Z. Nasroollah · I. Moonsamy · Y. Chuttur (✉)
University of Mauritius, 80837 Reduit, Mauritius
e-mail: y.chuttur@uom.ac.mu

Z. Nasroollah
e-mail: zoya.nasroollah@umail.uom.ac.mu

I. Moonsamy
e-mail: iraiven.moonsamy1@umail.uom.ac.mu

© Springer Nature Singapore Pte Ltd. 2019
S. C. Satapathy et al. (eds.), *Information Systems Design and Intelligent Applications*, Advances in Intelligent Systems and Computing 863,
https://doi.org/10.1007/978-981-13-3338-5_16

how energy is being consumed, energy optimization strategies could be set up and consumers may be sensitized, through awareness programs, to reduce their energy consumption and/or avoid energy wastage [2]. Through a self-learning process, consumers could also be encouraged to manage their energy use and adopt smarter ways in consuming energy. A suitable method to accurately monitor energy consumption is therefore required [3].

Traditionally, utility meters are used to monitor energy consumption on a given premise. Energy consumption data supplied by traditional meters, however, is aggregated across all appliances making it difficult to obtain useful information like energy consumption, in duration and cost, for any given device. Furthermore, readings available on meters do not instantly inform consumers of their energy consumption and the relevant costs incurred. Instead, bills are supplied only after a certain period of use such that it is already too late to adopt energy-saving strategies. As there is a lack of accurate and timely energy consumption information, therefore, traditional meters are not very helpful in assisting consumers in understanding their energy use or to adopt energy-saving strategies.

In contrast, modern energy monitoring systems allow consumers to monitor their energy use at equipment level and in real time. Several energy monitoring devices are already available on the market with different functionalities. A review of all the features provided by energy monitoring systems is beyond scope but all monitoring systems have typically the same architecture: sensors connected along various points on a power line, a central processing unit to collect and process all the data received from the sensors, and a control panel or application on the user side to provide information on energy consumption. Users are also given the possibility to control the power supplied to a given segment or appliance on the monitored premise.

The current proposed monitoring system is similar to the existing energy monitoring system in many ways. However, it adopts a different architecture in that only one IoT sensor is required and all data are processed on the cloud using machine learning algorithm. The goal is to monitor a site from a single point without the need to install sensors on each and every energy-consuming device. The added benefit is that equipment can be added and removed from a power line without any concern for additional sensors. By having quick and accurate energy consumption data, it is expected that consumers would be in a better position to optimize their energy use [4].

2 Related Works

Past studies reveal interesting facts about the effect of energy use feedback on consumer behavior. In parallel, several researchers have developed energy monitoring systems, which make use of IoT devices or applied machine learning to energy management. In the coming sections, relevant literatures are presented and discussed in the context of the proposed system.

2.1 Energy Consumption Feedback on Consumer Use Behavior

Faruquia et al. [5] investigated the degree of consumers' responsiveness to direct feedback on their electricity use. Their research showed that the data provided to consumers on their electricity use motivates them in reducing their electricity consumption. Furthermore, consumers described as active users of monitoring systems were found to lower their energetic use on average by about 7%. In addition, Meyers et al. [6] reviewed the effect of electronic monitoring and control systems on the reduction of energy consumption in households. Observations revealed that significant amounts of energy are wasted through inefficient energy distribution in households. For examples, spaces that were unused were often heated or cooled, whole premises would be overheated or undercooled and use of inefficient appliances would be common practice. In general, it was found that more than 39% of the domestic main energy is lost through wastage. The authors also highlighted the need for monitoring systems that would reduce the installation process to the minimum.

Another study by Bonino et al. [7] surveyed participants to determine the requirements for an efficient energy monitoring system of smart homes. Results obtained indicated that most users would like to have a home energy monitor installed in a central location in their residence, which can provide direct feedback to assist them in reducing their electricity use. Sundramoorthy et al. [8] further investigated the success of energy consumption feedback information provided by their Digital Environment Home Energy Management System (DEHEMS) on energy consumption behavior. Following preliminary installations in households, survey and focus group studies, participants were found to modify the way they use their appliances. In some cases, consumers even replaced appliances, which were considered to be energy inefficient. The outcome of the introduction of the feedback mechanism in the DEHEMS also showed a reduction of 8% in the users' daily electricity use within the first week of feedback delivery. Further support to the effect of providing instant feedback on energy consumption was found by Darby [9], who determined that immediate direct feedback could be very important in order to realize savings in daily energy use. The author claims that energy savings up to 15% were noted following the delivery of feedback to consumers.

It follows that when timely energy consumption data is provided in real time and in a user-friendly manner to consumers, it is very likely that consumers will react positively and will modify their energy consumption habit to adopt energy savings measures.

2.2 Energy Monitoring Using IoT

Several researchers have already developed energy monitoring applications using IoT technology. In general, most IoT implementations make use of sensors, a microcontroller, and a display but some variations do exist in terms of the number of sensors required, type of, connectivity used, data storage, and the method used to deploy sensors on a power line. Some examples follow.

Noman et al. [10], for instance, implemented an IoT system to measure voltage, current, and other energy parameters on a power line using two sensors, one for the current and the other one for the line voltage. Data collected were fed into a NodeMCU microcontroller where other useful energy parameters like active power, reactive power, apparent power, and power were computed using different mathematical equations. Processed data was then displayed on a small LCD screen or an Android device using a Bluetooth module. Mashuque et al. [11] developed an energy monitoring system using the NodeMCU Mega 1280 microcontroller along with low-pass filters and additional circuits to obtain and process voltage and current data from a power line. Power calculated using measured current and voltage values were then displayed to users on an LCD screen. Historical data about energy consumption was also stored on an SD card for later reference. The home energy management system developed by Fatima [12] involved the use of smart plugs, which were made up of a NodeMCU microcontroller, a set of sensors, an Xbee radio and a relay box to enable close monitoring and controlling of appliances on a given site. Appliance to be monitored had to be connected to the smart plugs.

The systems described above reported accurate monitoring results demonstrating the feasibility of using IoT devices in energy monitoring systems. In the present work, the IoT system proposed demarcates from previous implementations by (1) providing a scalable solution for energy monitoring using cloud computing (2) limiting the number of monitoring sensors to one device only for easy and nonintrusive installation (3) applying machine learning as a novel approach to automatically identify and detect energy-consuming devices on a power line and (4) allowing electric devices to be added or removed from a monitored site without the need for reconfiguration of the energy monitoring system.

2.3 Machine Learning for Energy Monitoring

Artificial Neural Network (ANN) is one of the many techniques used in machine learning. Through well-defined algorithms, ANN mimics the operation of the human brain by making use of interconnected graphs of artificial neurons to solve complex problems involving tasks such as classification, prediction, and generalization at very fast speed. By using historical data, ANN undergoes a learning process to find suitable correlations between the input and output data to the algorithm [13]. Figure 1 depicts a typical representation for neural networks.

Three main parts: the input layer, the hidden layer, and the output layer make up a neural network. Hidden layers can be present or absent in a neural network. Their main role is to process raw input data into a meaningful output based on different weights assigned to different inputs. The importance of a feature is decided by the weight assigned to an input. Thus, raw data fed into the input of the ANN algorithm

Fig. 1 Multi-layered neural network representation

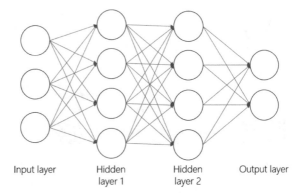

Input layer   Hidden layer 1   Hidden layer 2   Output layer

goes through at least one hidden layer of the algorithm, with each layer acquiring a new set of features. Upon reaching the last hidden layer, the algorithm gets to learn which group of features is observed to occur together, thus, identify an output class. In supervised mode, ANN must be trained using representative data to learn the fundamental features of input signal to observe prior to deciding on an output [14]. In case, the output does not match the expected outcome, a technique known as back propagation redistributes an error factor from the output to the input so as to modify the different weights assigned to the hidden layers and obtain better results.

Due to its accuracy, ANN has been used in various domains including that of energy management. In [15], for instance, ANNs are applied to forecast the energy generation of power plants. In [16], ANNs are used for demand-side management and accurate load forecasting. Similarly in [17], energy consumption was successfully forecasted for industrial buildings and transport networks using ANNs. Several other studies have applied ANN to energy management. For an in-depth review, see [18]. In general, past studies have used ANN mostly for the forecasting of energy consumption and demand planning or control. In the present study, ANN is used to analyze an aggregated load curve, L from a monitored site. The goal is to extract the constituent load curves of all devices that contributed to the curve L so as to be able to identify the energy consumption of a given equipment using energy data supplied by one sensor only.

3 Implementing the Energy Monitoring System

Figure 2 provides a general overview of the architecture for the implemented cloud-based energy monitoring system. A single current transformer (CT) sensor collects energy data from a power line. Collected data is sent to a NodeMCU ESP8266 board, which then transfers the data to an IoT Hub for further processing by the Azure Stream Analytics service. Data stored on the cloud database is easily accessible for further

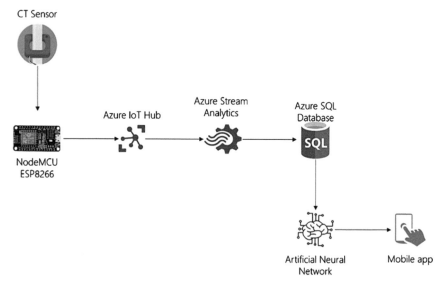

Fig. 2 General architecture for the implemented energy monitoring system

analysis and energy consumption data is made available to users on a mobile app. Further description for each part follows.

The CT sensor used in this project is the SCT-013-000 noninvasive current sensor, which can measure up to 100 A. The SCT-013-000 CT sensor has a split core, which allows the sensor to be placed around a power line without having to disconnect or disrupt any wiring. Such an arrangement makes it easy to clip the sensor to a power line and monitor the energy flow within a given site. Through magnetic induction, raw current data is read from a monitored line and fed into the NodeMCU board via a signal conditioning circuit. The NodeMCU ESP8266 microcontroller is an open source IoT platform specifically designed to read data very quickly from connected sensors. The microcontroller also provides the Arduino IDE for custom programming. With the benefit of having built-in support for wireless connectivity, the board can be easily programmed to stream monitored energy data from the CT sensor via an Internet connection to the Azure cloud service.

Azure cloud service[1] is a Microsoft product that provides several benefits to the development of cloud-based apps. In addition to providing suitable APIs to read data from IoT systems, Azure cloud provides the necessary support for data storage and analysis. Furthermore, by using a cloud service, the current proposal ensures that the energy monitoring system is scalable and can accommodate a network of IoT systems with energy data accessible to consumers anywhere anytime. The following Azure cloud services are used for implementing the current energy monitoring system: Azure IoT hub, Azure Stream Analytics, and Azure SQL database.

---

[1] https://azure.microsoft.com/.

In the current proposal, the Azure IoT hub acts as a bridge between the various services that the Azure cloud offers and the NodeMCU ESP8266 microcontroller. Energy data collected from a CT sensor is fed into the NodeMCU ESP8266 device, which then transfers the data over the Internet to the Azure IoT hub for further processing by the Azure Stream Analytics service. The latter easily captures streaming data from multiple IoT devices to distribute the data to other cloud services. Near real-time analysis of the data for prediction and forecasting purposes can also be performed. The service is scalable and can easily accommodate a large amount of sensor data without the need to cater for additional infrastructure. For the purpose of the present system, Azure stream analytics was used (1) to convert the data received by the IoT hub into a format suitable for storage in the Azure SQL database and (2) to store monitored energy data in the Azure SQL database for further analysis. Data collected in the database was further processed by a mobile app service and a mobile app for instant energy data monitoring at the user's end. Developed using Android Studio, the mobile app allowed users to create an account to connect to their IoT device and the power line being monitored on the cloud.

As demonstrated by [19, 20] electric equipment exhibits unique electric signatures when consuming power. By extracting specific features of a load curve consisting of power aggregated by multiple loads, it is possible to identify the duration and amount of energy consumed by a single device using ANN. In the present work, the ANN algorithm supplied by WEKA[2] was used to analyze the apparent and reactive power data calculated from energy data collected from the CT sensor. The ANN algorithms on WEKA are coded in JAVA and could thus easily be called in Android Studio for inclusion in the mobile app developed for the system. A dataset consisting of apparent power, reactive power and appliance name collected for several loads was used as input to train the ANN algorithm from WEKA. Weight values for the network, number of hidden layers, training speed of the algorithm were then fine-tuned to recognize the presence or absence of a device on a monitored power line. The in-built classifier provided by WEKA made it easy to generate the name of the device detected based on the input provided. For testing purposes, three loads were used: a hairdryer, a toaster and an electric iron. Those devices were chosen because of their well-known significant power demand. Figure 3 shows a sequence of sample screenshots of the mobile app interface when (1) connected to the line being monitored with no load turned on, (2) when a load is turned on and detected, and (3) when the detected equipment is in operation.

4 Discussions

The simple interface developed for the mobile app, as shown in Fig. 3 instantly presented power consumption data to users. The operation of appliances was also automatically detected in real time by relying on the electric signatures identified

---

[2]https://www.cs.waikato.ac.nz/ml/weka/.

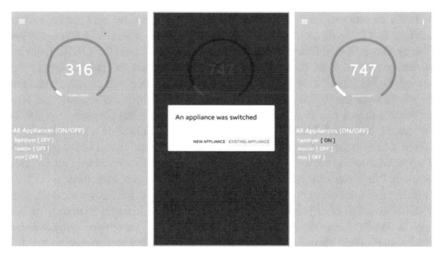

Fig. 3 Mobile app interface for the implemented energy monitoring system

by the ANN algorithm used for the study. With IoT and cloud-based applications becoming more interoperable and affordable, the present system represents an easy to install, reliable and better alternative to the commercially available energy monitoring systems in several ways. The system presented here makes use of one sensor only and therefore can easily be installed on any site without the need to disrupt preexisting electric installations. At the same time, using its unique approach to automatically detect appliances on a power line, there is no need to reconfigure the monitoring system to accommodate new appliances. The cloud further provides a reliable and secure platform for storing energy data. Consumers can, therefore, obtain useful insights of both current and past consumption. As energy information can be made available at appliance level, consumers are given the possibility to understand their energy consumption better and are thus able to take remedial actions to use energy in a smarter and more efficient manner.

5   Conclusions and Future Works

The combination of machine learning, IoT, and cloud computing offers a scalable nonintrusive real-time energy monitoring solution that can accommodate a large number of users. The prototype developed for this study showed promising results that would warrant further research. Potential directions to explore would be to test a similar system on actual users over a period of time and to evaluate the eventual effect on energy use behavior. Other machine learning algorithms can also be used to evaluate appliance detection accuracy. Further research can also be conducted on the interface presented to users such as the kind of information to be displayed to users and the kind of features desired by users to assist them in saving energy. Data collected on the cloud can further be analyzed to understand user energy consumption behavior on a larger scale. It would be interesting to visualize how energy consumption varies by regions, communities, seasons, etc.

Acknowledgements Special thanks goes to Mr. Y. Beeharry from the University of Mauritius FoICDT computer lab for his valuable advice during the implementation phase of the prototype.

References

1. Cominola, A., Giuliani, M., Piga, D., Castelletti, A., Rizzoli, A.E.: A hybrid signature-based iterative disaggregation algorithm for non-intrusive load monitoring. Appl. Energy **185**, 331–344 (2017)
2. Basu, K.: Classification techniques for non-intrusive load monitoring and prediction of residential loads. Doctoral dissertation, University of Grenoble (2014)
3. Mashuque, E., Shaikh, R.H., Nafis, A.K., Hafiz, A.R.: Smart energy monitoring using off the-shelf hardware and software tools. In: IASTED International Conference on Power and Energy Systems, pp. 160–167. Thailand (2013)
4. Wood, G., Newborough, M.: Influencing user behaviour with energy information display systems for intelligent homes. J. Energy Res. **39**(4), 495–503 (2007)
5. Faruquia, A., Sergici, S., Sharif, A.: The impact of informational feedback on energy consumption—a survey of the experimental evidence. Energy **35**(4), 1598–1608 (2010)
6. Meyers, R.J., Williams, E.D., Matthews, H.S.: Scoping the potential of monitoring and control technologies to reduce energy use in homes. Energy Build. **42**(4), 563–569 (2010)
7. Bonino, D., Corno, F., De Russis, L.: Home energy consumption feedback: a user survey. Energy Build. **47C**, 383–393 (2012)
8. Sundramoorthy, V., Liu, Q., Cooper, G., Linge, N., Cooper, J.: DEHEMS: a user-driven domestic energy monitoring system. In: Internet of Things (IOT), pp. 1–8. Tokyo (2010)
9. Darby, S.: The effectiveness of feedback on energy consumption. Technical Report. Environmental Change Institute, University of Oxford (2006)
10. Noman, A.A., Rahaman, M.F., Ullah, H., Das, R.K.: Android based smart energy meter. In: 4th National Conference on Natural Science and Technology, pp. 1–3. Bangladesh (2017)
11. Mashuque, E., Shaikh, R.H., Nafis, A.K., Hafiz, A.R.: Smart energy monitoring using off the-shelf hardware and software tools. In: IASTED International Conference on Power and Energy Systems, pp. 160–167. Thailand (2013)
12. Fatima, E.B.: Smart home energy management system monitoring and control of appliances using an arduino based network in the context of a micro-grid. Dissertation, Al Akhawayn University, Morocco (2015)
13. Sholahudin, S., Han, H.: Simplified dynamic neural network model to predict heating load of a building using Taguchi method. Energy **115**, 1672–1678 (2016)
14. Tamizharasi, G., Kathiresan, S., Sreenivasan, K.S.: Energy forecasting using artificial neural networks. Energy **3**(3), 7568–7576 (2014)
15. Olivencia, P.F., Ferrero, B.J., Gomez, F.J.F., Crespo, M.A.: Failure mode prediction and energy forecasting of PV plants to assist dynamic maintenance tasks by ANN based models. Renew. Energy **81**, 227–238 (2015)
16. Chaturvedi, D.K., Sinha, A.P., Malik, O.P.: Short term load forecast using fuzzy logic and wavelet transform integrated generalized neural network. Electr. Power Energy Syst. **67**, 230–237 (2015)
17. Kumar, R., Aggarwal, R.K., Sharma, J.D.: Energy analysis of a building using artificial neural network: a review. Energy Build. **65**, 352–358 (2013)

18. Kalogirou, S.A.: Applications of artificial neural networks in energy systems. Energy Convers. Manage. **40**, 1073–1087 (1999)
19. Kushiro, N., Ide, T., Tomonaga, K., Ogawa, Y., Higuma, T.: Can electric devices be identified from their signatures of waveform? In: 12th Annual IEEE Consumer Communications and Networking Conference (CCNC), pp. 531–536. Las Vegas (2015)
20. Fogarty C., Carolyn A., Hudson, S.E.: Sensing from the basement: a feasibility study of unobtrusive and low-cost home activity recognition. In: Proceedings of the 19th Annual ACM Symposium on User Interface Software and Technology, pp. 91–100. Switzerland (2006)

A Prototype Mobile Augmented Reality Systems for Cultural Heritage Sites

Arvind Ramtohul and Kavi Kumar Khedo

Abstract Augmented Reality (AR) is an emerging technology that is currently rev-olutionising the worldly activities. The evolution of mobile devices has pioneered AR as a state-of-the-art technology in the last decade giving rise to more and more Location-Based Mobile AR (LBMAR) systems. This technology is widely been used by heritage industry to create the missing sparks from their static environment thus making it livelier and entertaining. Yet, current systems have not exploited the emerging sensory technologies in smartphones which led to low-quality AR experi-ences. In this work, we will model a real-time Mobile AR system for a cultural site in Mauritius that will both promote the cultural tourism and enhance the visiting expe-rience. Moreover, the proposed prototype will also attract historians and students so that they can carefully study those heritage aspects that were not accessible before.

Keywords Augmented reality · Mobile augmented reality
Location-based mobile augmented reality · Real-time · Cultural tourism
Heritage

1 Introduction

The recent advances in new technologies have undoubtedly created various break-throughs in many spheres of the worldly activities. One such invention is Augmented Reality (AR) technology; AR is a technique that combines a live view in real-time with virtual computer-generated images creating a real-time augmented experience of reality [1]. The increasing demands of mobile devices have propelled industries

A. Ramtohul (✉) · K. K. Khedo
Faculty of Information, Communication and Digital Technologies,
University of Mauritius,
Moka, Mauritius
e-mail: arvind.ramtohul@outlook.com

K. K. Khedo
e-mail: k.khedo@uom.ac.mu

© Springer Nature Singapore Pte Ltd. 2019
S. C. Satapathy et al. (eds.), *Information Systems Design and Intelligent Applications*, Advances in Intelligent Systems and Computing 863,
https://doi.org/10.1007/978-981-13-3338-5_17

to continuously focus on their research and development programmes to bring new innovations. This has accelerated the developments of mobile AR since they now have more battery, processing, memory and bandwidth resources. Due to the lightness and portability, various application areas have adopted mobile AR; not only it enriches user experiences, it also offers a sense of uniqueness making the moments more memorable.

On another perspective, industries are more capable of engaging users and identifying users' tastes to continuously improve its day to day activities. The widespread use of mobile AR has contributed to a huge growth in various industries, including heritage industry. In most of the cases, Cultural Heritage sites do not have useful information or they lack user guides which affect the visiting experience. Although Cultural Heritage (CH) sites have employed traditional mediums such as dashboards, booklets and maps for sharing of information, visitors find it uninteresting and not motivating enough [2]. In this context, CH sites can adopt the high technologies to bring more liveliness to their static environment and also taking full advantages to showcase all the detailed heritage aspects to visitors.

Mauritius has a vivid history and consequently, there are a number of CH sites as a reference to visualise the timelines of different historical aspects. This study will showcase a prototype of a mobile AR system for enhancing the visiting experiences and to engage visitors towards a better CH experience. The design is modelled for Aapravasi Ghat site, a UNESCO World Heritage Site [3]. This paper is structured as follows: Sect. 2 introduces the background study and related works, the detailed prototype design is discussed in Sect. 3 followed by the evaluation of the proposed AR system in Sect. 4. Section 5 highlights the research challenges and directions. Lastly, Sect. 6 is the conclusion of the study.

2   Brief Literature Review

There are several works that have focussed on the development of AR applications, and here our focus is mainly on Mobile AR systems at CH sites. The evolution of smartphones and the introduction of new sensors have increased the popularity of such systems. In the early 1990s, a paradigm shift happened and the term 'Augmented Reality' was associated with a head mounted system to assist the users for their human-involved operations [4]. However, a true mobile AR application was later developed by Hollerer et al. [5], called as MARS. The application provided location-based information to tourist in a city.

Alleto et al. [6] designed an indoor location-aware system for an Internet of Things (IoT)-based smart museum. It is based on a client–server architecture, whereby the clients rely on a wearable devices and the server hosts the central processing logic. The wearable device has an image recognition and positioning capabilities that works jointly with the server to provide real-time cultural information to users. The authors used Bluetooth infrastructure for localisation and cloud storage for multimedia contents. In addition, the system also checks the users' movement to activate multimedia cultural contents. The authors claim their system could be expanded to other IoT features such as monitoring temperature and light, maintenance scheduling, and 4D animations. On the contrary, Bluetooth is not a common existing infrastructure found in buildings thus requiring more resources in terms of workforce and cost. Moreover, the coverage is also limited to few metres thus impacting the scalability of the project.

Banterle et al. [7] developed LeeceAR, an AR application at the MUST museum in Leece, Italy. The application has two main functionalities: the matching and tracking module and the 3D rendering module. The matching and tracking module processes the images or the videos from the camera of mobile devices, and the 3D rendering module overlays historical and architectural information on top of it. This marker-less system might work efficiently with small dataset, but it remains uncertain if the performance would be the same with huge amount of data thus it can impact on the system's responsiveness.

Julier et al. [8] presented VisAge, an AR model based on a set of spatially distributed Point of Interests (POI). The prototype is a web-based application with a map view of the POIs attached with an online database. Contents of POIs are stored in the database, and it may include text, images and audio. Moreover, this system provides an interesting feature whereby authors can add POIs or modify existing POIs and subsequently the new POIs can be viewed by other users or they can incorporate them within their own experiences. On the other end, this feature can be irrational if the added contents are not related to that respective POI.

Recently, Vera et al. [9] proposed an approach to enhance user experience in POIs with AR. As per previous works studied, their work also focussed on personalisation to value the human factors but classified it into Content based, Collaborative and Hybrid—combination of Content and Collaborative. In addition, the authors introduced the gamification aspect to motivate users in finding new ways of interaction with objects and improve their engagements towards this technology. Gamification is indeed a bonus to an AR environment, but it also requires a good localisation accuracy to make it work flawlessly. Therefore, this can still be considered a research problem that can be further investigated and improved.

Vainstein et al. [10] presented a study on Head-Worn Display (HWD) in Cultural heritage. The authors listed the user requirements on enhancing the museum visiting experience with HWD devices. Amongst the requirements, it also includes device characteristics, personalisation and content presentation to enrich the interaction of users. A prototype was thus developed as illustrated in below Fig. 1, the design did not cover all the requirements but it could be further extended in the future. On this study, the authors primarily demonstrated the use of HWD so that users relish the real environment instead of forcing user to look at a screen. However, this kind of implementation requires an extensive know-how of HWD and it can be costly to implement it on a large scale.

Table 1 illustrates a comprehensive analysis of the previously discussed related works and additional AR systems. The systems have been evaluated in terms of the mobile sensors, techniques, its lightweight function and its flexibility.

Table 1 Comparison of the existing system

Location-based mobile AR systems	Sensors						Techniques						Lightweight	Flexibility	Limitations
	Compass	Gyroscope	Accelerometer	Bluetooth	Wi-Fi	Camera	GPS	POI	Movement pattern	Orientation	Tracks			
AREAv1 [11]	Y	–	Y	–	–	Y	Y	Y	–	Y	–	Moderate	Y	Pattern of movement not taken into account
AREAv2 [11]	Y	Y	Y	–	–	Y	Y	Y	–	Y	Y	Moderate	Y	Same as AREAv1
VisAge [8]	–	–	–	–	–	Y	–	Y	–	–	–	Low	–	Customised POI content could be irrelevant
Capece et al. [12]	Y	–	–	–	–	Y	Y	Y	–	Y	–	Low	Y	Degrading user experience
Paucher and Turk [13]	Y	Y	Y	–	–	Y	Y	–	–	Y	–	Low	–	Works only in uniform environment
Alletto et al. [6]	–	–	–	Y	–	Y	–	Y	–	–	–	Moderate	–	Bluetooth is limited to few metres

(continued)

Table 1 (continued)

Location-based mobile AR systems	Sensors						Techniques					Lightweight	Flexibility	Limitations
	Compass	Gyroscope	Accelerometer	Bluetooth	Wi-Fi	Camera	GPS	POI	Movement pattern	Orientation	Tracks			
Vainstein et al. [10]	–	–	–	–	–	Y	–	–	–	–	–	Low	–	Costly to implement in large scale
Banterle et al. [7]	–	–	–	–	–	Y	–	Y	–	–	–	Low	–	Performance degrades with large amount of data
Tan and Chang [14]	Y	Y	Y	–	–	Y	Y	–	Y	Y	–	Low	–	2D coordinates not sufficient for real-time interaction

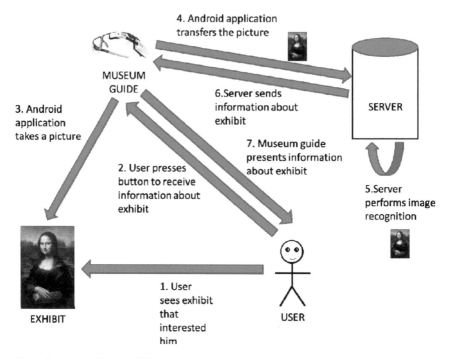

Fig. 1 System architecture [10]

3 Mobile AR Prototype @ Aapravasi Ghat

The design of the mobile AR is divided into several subsystems: (1) Positioning Infrastructure, (2) Point of Interests, (3) Positioning Service, (4) Processing Centre, (5) Recommender Centre and (6) IoT services. It is supported by a multitier system as illustrated in Fig. 2.

- **Positioning Infrastructure**

    Aapravasi Ghat is already covered with Wi-Fi infrastructure networks, the system will use the existing infrastructure to provide positioning services and augmented cultural contents to visitors. The system will be made available on Android and iPhone devices, visitors will be able to download the app at any time in Play Store and App Store, respectively. The Wi-Fi networks will act as a gateway between the system and the mobile devices, all the requests and responses will be passed through the Wi-Fi networks and will be transferred to the related subsystems.

- **Point of Interests (POI)**

    The composition of Aapravasi Ghat is surrounded with several Object of Interests (OOI), and in turn the OOIs can be grouped into a single or multiple POIs. Initially, a brief survey will be carried out to identify the POIs in terms of its historical

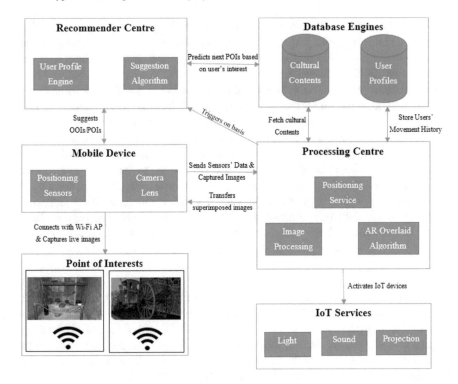

Fig. 2 Multi-tier architecture of mobile AR prototype

importance, order of time and particular civilisation. At each identified POI, a Wi-Fi Access Point (AP) will be setup and configured to work incoordination with existing Wi-Fi nodes. In this perspective, the OOIs will be indexed with its particular POI and these datasets will be clustered in the Cultural Contents database.

- **Processing Centre**

The processing centre is the core nucleus of the system, it comprises all the logical flows including the positioning service, the image processing algorithm, the AR object identification algorithm and the cultural contents database. In addition, it acts as the middleware system for IoT devices and the recommender centre.

Positioning Service: As discussed in POI section, the Cultural Contents database engine will also have the received signal strength indicator (RSSI) from Wi-Fi AP, the associated OOIs and POIs. Identically, the positioning mechanism will be able to locate the visitors autonomously by only detecting the closest Wi-Fi AP. However, the system will activate the smartphone camera if the location information is not sufficient for the processing centre to translate the request. For example, the system might require the assistance of camera at a POI comprising of several OOIs. The location ID, the POI ID and OOI ID will be sent to the processing centre for further processing.

Image Processing Algorithm: This segment works inline with the positioning service, and it is triggered when required by the processing centre. It is divided into two sub-process: (1) Matching the live images and (2) Fetching the cultural contents information. Initially, it will compare the live images against the pre-stored data on the database engines. As the database is already indexed with the POI field, the system will not carry out a whole database search. The output will be sent to retrieve the best-fit cultural content of the same exhibit. A weight may be assigned to the fetched cultural contents to declassify the unimportant ones. Overall, this section of the system is critical to the proper functioning of the system. The fetching and retrieval of data should be within a reasonable time delay, so that visitors can have a real-time interaction and they can enjoy their informal learning to the maximum.

AR Object Overlaid Algorithm: Once the AR information is available, this subsystem will calculate the inclination degree of the camera's field view from readings on the mobile sensors. The system superimposes the live images with the AR contents, and adjusts it to the focal point on the smartphone screen.

- **Recommender Centre**:

This add-on feature monitors the movement of the visitors to derive a mobility pattern. Substantial information about users' interests can be retrieved, and the system further uses it to propose the next interesting OOIs or POIs. It has a separate database engine that hosts users' profiles data that gets populated from the processing centre at an evenly mode. A suitable algorithm should be carefully chosen to derive the mobility pattern and predict the interesting exhibits to the visitors.

- **IoT Services**:

The integration of IoT is another add-on feature to bring more liveliness to the environment. It is connected with the simplest IoT devices like light, sound and projection systems at Aapravasi Ghat. At regular time intervals, it verifies with the processing centre if there is a need to activate those services. The processing centre checks the number of visitors present in respective POIs, and decides to activate those services.

Figure 3 demonstrates the sequential procedure calls for each subsystem.

4 Qualitative Evaluation

This prototype has been designed taking into consideration the physical dynamics of Aapravasi Ghat. Though the design has not been validated, it can be evaluated through a methodical approach to find its strength and weakness. Overall, this system has been designed to work with smartphones as it is more convenient and cost-effective. The visitors will have augmented information at a real-time basis, it will not only enrich the user experiences but it will also offer a sense of uniqueness making the moments more memorable. Yet, too much usage of smartphone can distract the visitors from seeing the real objects and this can negate the cultural experience. The

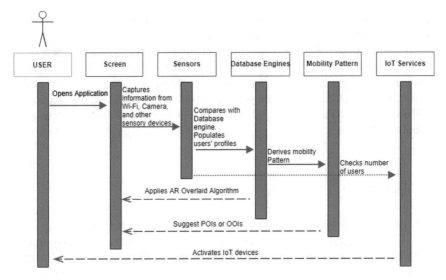

Fig. 3 Sequential diagram

usage of AR should be limited only when there is a need for visualising complex objects. Moreover, two-dimensional (2D) coordinate is not sufficient to provide real-time interaction as it challenges the human capacity to understand complex objects. The system is modelled by using a simplified database search technique taking into consideration the limited capacities of mobile devices. This prototype has been conceived using Wi-Fi network infrastructure since it is commonly available and the same design can be modelled at different sites demonstrating its robust flexibility.

5 Research Challenges and Directions

From the above related work comparison, it is noticed that all existing systems have used camera lens for object' recognition. Given the wider applicability of camera, an LBMAR system should rather concentrate on providing the right location at the right time. Most of the existing systems have focussed on enriching the visual contents and did not cater on their positioning approaches. Only the systems AREAv2 and Tan and Chang have incorporated a series of mobile sensors for users' and objects' tracking; demonstrating a robust localisation scheme using mobile AR, however, they have lacked in strength to provide a lightweight solution and a flexible approach. In our research, the term 'lightweight' is defined as low-consuming energy technique that is practicable on resource-constraint mobile devices and the term 'flexibility' caters for the physical characteristics of other CH sites. To the best of our knowledge, these two terms are new to mobile AR and it can be further expanded to improve the visiting experience in a CH site. In addition, the pattern of movement is a potential subject that can be exploited in LBMAR systems, it derives a pattern based on the users' past location histories and suggests interesting POIs to them. Tan and Chang is among the few studies that have modelled it, but they did not mention how they could implement it in a different context such as in a CH site. In this work, we integrated both parameters on the design to increase the engagement level of visitors and to promote cultural places among the general public. However, further works will be carried out to implement the proposed mobile AR model and an evaluation of the system will be done.

6 Conclusion

In this study, we modelled an LBMAR system for the Aapravasi Ghat site. The proposed design comprises of many sub-systems which will work jointly to provide a real-time augmented information to users. Multi-sensing capabilities of smartphones are employed to capture live images, calculating the users' field of view and the rotation view. A lightweight technique has been proposed for the image processing segment whereby the system does not carry out a full database search. The recommender system tracks all the movement of the visitors suggesting them new interesting POI that they can visit. In this way, the users are more engaged and they have a real-time interacting along the way keeping them focus on their cultural learning. The adoption of AR to CH sites can be very beneficial, it could create the missing link between culture and tourism thus contributing towards more sustainable growth.

References

1. Van Kleef, N., Noltes, J., van der Spoel, S.: Success factors for augmented reality business models. Study Tour Pixel, 1–36 (2010)
2. Pendit, U.C., Zaibon, S.B., Bakar, J.A.A.: Mobile augmented reality for enjoyable informal learning in cultural heritage site. Int. J. Comput. Appl. **92**(14), 19–26 (2014)
3. UNESCO: http://whc.unesco.org/en/list/1259
4. Caudell, T.P., Mizell, D.W.: Augmented reality: an application of heads up display technology to manual manufacturing processes. In: Proceedings of the Twenty-Fifth Hawaii International Conference on System Sciences, vol. 2, pp. 659–669. IEEE, Kauai, HI, USA (1992)
5. Höllerer, T., Feiner, S., Terauchi, T., Rashid, G., Hallaway, D.: Exploring MARS: developing indoor and outdoor user interfaces to a mobile augmented reality system. Comput. Graphics **23**(6), 779–785 (1999)
6. Alletto, S., Cucchiara, R., Del Fiore, G., Mainetti, L., Mighali, V., Patrono, L., Serra, G.: An indoor location-aware system for an IoT-based smart museum. IEEE Internet Things J. **3**(2), 244–253 (2016)
7. Banterle, F., Cardillo, F.A., Malomo, L., Pingi, P., Gabellone, F., Amato, G., Scopigno, R.: LecceAR: an augmented reality app. In: Fifth International Conference on Digital Presentation and Preservation of Cultural and Scientific Heritage (DiPP), pp. 99–108, Veliko Tarnovo, Bulgaria (2015)

8. Julier, S.J., Blume, P., Moutinho, A., Koutsolampros, P., Javornik, A., Rovira, A., Kostopoulou, E.: VisAge: augmented reality for heritage. In: Proceedings of the 5th ACM International Symposium on Pervasive Displays, pp. 257–258. ACM, Oulu, Finland (2016)

9. Vera, F., Sánchez, J.A., Cervantes, O.: Enhancing user experience in points of interest with augmented reality. Int. J. Comput. Theory Eng. **8**(6), 450–457 (2016)

10. Vainstein, N., Kuflik, T., Lanir, J.: Towards using mobile, head-worn displays in cultural heritage: user requirements and a research agenda. In: Proceedings of the 21st International Conference on Intelligent User Interfaces, pp. 327–331. ACM, Sonoma, CA, USA (2016)

11. Pryss, R., Geiger, P., Schickler, M., Schobel, J., Reichert, M.: Advanced algorithms for location-based smart mobile augmented reality applications. Procedia Comput. Sci. **94**, 97–104 (2016)

12. Capece, N., Agatiello, R., Erra, U.: A client-server framework for the design of geo-location based augmented reality applications. In: 20th International Conference Information Visualisation (IV), pp. 130–135. IEEE, Lisbon, Portugal (2016)

13. Paucher, R., Turk, M.: Location-based augmented reality on mobile phones. In: Computer Vision and Pattern Recognition Workshops (CVPRW), Computer Society Conference on Computer Vision and Pattern Recognition—Workshops, pp. 9–16. IEEE, San Francisco, USA (2010)

14. Tan, Q., Chang, W.: Location-based augmented reality for mobile learning: algorithm, system, and implementation. Electron. J. e-Learn. **13**(2), 138–148 (2015)

A Secured Communication Model for IoT

Reza Hosenkhan and Binod Kumar Pattanayak

Abstract In the current scenario, global Internet is oriented around human-to-human communication only. With the emergence of Internet of Things (IoT), this communication is supposed to be transformed to a communication between everything using the services of global Internet where human-to-human, human-to-machine, and machine-to-machine communications are made viable. With billions of smart devices referred to as "things" being connected to the global Internet, the major concern arises around the security aspects of IoT. In this paper, we address a publish/subscribe security model for IoT which can usefully contribute to secured communications on IoT.

Keywords IoT · Security · Smart devices · Communication

1 Introduction

Global Internet has practically captured all spheres of human life. As a medium of global communication, it has evolved in time and innovations are being incorporated into this communication technology almost every day. One of such innovations we came across in the recent years is Internet of Things (IoT) that has facilitated revolutionary changes in the methods of implementation of various aspects of global Internet. IoT can be considered to be an integrated component pertaining to the future Internet that is capable of integrating a wide range of physical as well as virtual components referred to as "things" that are self-configurable. More entities are supposed to form a completely dynamic global network infrastructure on the basis of standard as well as interoperable communication protocols. Here, every individ-

R. Hosenkhan
Université des Mascareignes, Beau Bassin-Rose Hill, Mauritius
e-mail: rhosenkhan@udm.ac.mu

B. K. Pattanayak (✉)
Siksha 'O' Anusandhan Deemed to be University, Bhubaneswar, India
e-mail: binodpattanayak@soa.ac.in

© Springer Nature Singapore Pte Ltd. 2019
S. C. Satapathy et al. (eds.), *Information Systems Design and Intelligent Applications*, Advances in Intelligent Systems and Computing 863,
https://doi.org/10.1007/978-981-13-3338-5_18

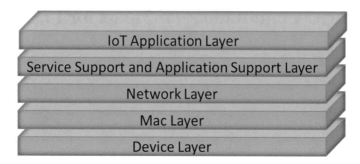

Fig. 1 IoT protocol architecture

ual entity ("things") must possess its identity, physical attributes along with virtual personality. "Things" in IoT are supposed to participate actively in various processes relating to business, social activities, information sharing where a need arises and they need to interact and communicate among themselves and at the same time with the environment too by virtue of exchanging data and information. They are supposed to sense the environment autonomously and react to it accordingly without human intervention. Intelligent interfaces in the form of services facilitate various interactions among the smart "things" over the global Internet [1].

2 IoT Architecture

IoT operates on a four-layer protocol architecture (Fig. 1). IoT application layer represents an interface for an IoT application. Service support and application support layer includes some generic service capabilities that are applicable to practically all IoT applications and at the same time, it incorporates some specific service capabilities too.

Network layer here performs the roles of the network as well as transport layer functionality. MAC layer is responsible for traditional data link layer functions such as framing, error and flow control, etc. Device layer plays the role of physical layer in the protocol stack [2].

3 Challenges of IoT

IoT as an integration of huge infrastructural resources across the globe face with a set of challenges that need to be addressed. These challenges are listed below:

(a) Provisioning of smart devices:

Although the methods of successful implementation of IoT with traditional IoT infrastructure are driven by a set of well-formulated guidelines, the actual challenge is expected to emerge when smart devices will be everywhere.

(b) Need for Open Standards:

IoT comprises of a wide range of heterogeneous devices with varying specifications. Growth of IoT technology solely depends upon the cooperation among the devices. The problem even further depends when more and more smart devices are included. This necessitates implementation of open standards for communication protocols of IoT.

Increased energy requirements

Gartner predicted that the number of smart devices in IoT would increase from 4.9 billion in 2015 to 25 billion in 2020. It refers to a 100% increase in number of smart devices every year. It consequently leads to a significantly increased demand in energy requirements too.

(c) Waste disposal:

Disposal of e-waste such as obsolete computers, phones, peripherals, etc., represent a huge challenge. Considering that smart devices are built in the way the computers are built and increase in smart devices is expected to be exponential, the problem of e-waste disposal can deepen further.

(d) Storage Bottleneck:

Although it is presumed that data generated by smart devices need not be stored longer before being communicated, still some relevant information like timers for devices need to be stored for some period of time. Other relevant control information too needs storage. Hence, it is crucial to leverage a policy about which data to be stored.

(e) Lack of Privacy:

Smart devices in IoT may very often tend to exchange confidential information. However, even smart devices can be tracked which means privacy of information cannot be secured which consequently may lead to legal issues too.

(f) Lack of Security:

Considering that basic devices such as routers, satellite receivers, TVs, etc., can be comfortably hacked, IoT is obviously not an exception to it. Thus security provisioning needs to be strengthened.

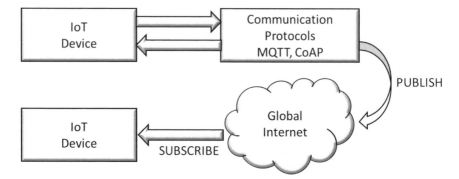

Fig. 2 Publish/subscribe communication model for IoT devices

4   Publish/Subscribe Model for IoT Security

For the reason that IoT devices are heterogeneous and in order for communication
with global Internet an IoT device needs to undergo bidirectional real-time connection. It is extremely challenging to secure such a communication. Traditional network
security protocols appear to be insufficient for IoT environment since it may often
involve communications among smart devices. It necessitates the design of a robust
security model that can facilitate secure communication across IoT devices in various
paradigms. In order for designing such a robust and reliable security model for IoT
environment the following five requirements must be taken care of.

(a) Devices with No Open Inbound Ports

In traditional network model, when a device sends data, a listening device keeps
open an inbound port infinitely until data are received. Such a strategy is prone to
potential risk from external attacks like malware, Denial of Service (DoS), etc. On
the other hand, IoT devices must support only outbound connections that are less
vulnerable to external attacks. Such a communication relies on publish/subscribe
communication design pattern that enables IoT devices securely send data bidirec-
tionally. Publish/subscribe strategy as a method of reliable communication among
IoT devices can be achieved using secure communication protocols like Message
Queuing Telemetry Transport ("MQTT"), Constrained Application Protocol (CoAP)
(Fig. 2).

(b) End-to-End Encryption:

Traditional data encryption standards like Transportation Layer Security ("TLS")
appear to be incapable of ensuring transmission security of data generated by IoT
devices. However, Advanced Encryption Standard ("AES") can usefully provide
end-to-end security for data transferred between IoT Devices.

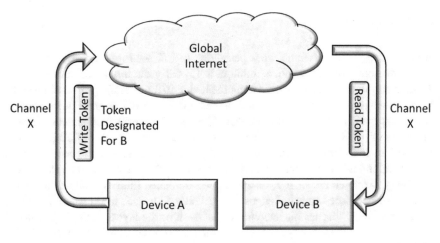

Fig. 3 Token-based access control for IoT communication

(c) Token-Based Access Control

Another challenge in IoT communication is to guarantee that the transmitted data are received by the designated recipient only while millions of IoT devices might be listening to the communication channel. In publish/subscribe communication paradigm, tokens can be distributed to IoT devices to grant access to specific data channels only (Fig. 3).

(d) Monitoring of Status of IoT Device:

For efficient implementation of IoT communication, it is essential to monitor "presence" of a device (online or offline) which may result from power failure in Internet link breakdown or even a tampering. In order to assure this, separate and secure data channels must be used to monitor the status of IoT devices especially. Online or offline status of a device can be determined by different thresholds of signal strength across the channel.

(e) User Friendliness

IoT devices need a user-friendly setup process and at the same time, getting the devices upgraded following software as well as firmware upgrades must be achieved at ease. A communication model supporting publish/subscribe paradigm using outbound ports 80 and 443 can be easily set up with provisioning for IoT devices. In addition to this, manufacturers of IoT devices need to make use of the devices secure publish/subscribe channel to notify the devices to download and install software as well as firmware upgrades as and when these upgrades become available.

5   Related Work

A wide spectrum of research work pertaining to IoT security has been conducted by researchers around the globe. Authors in [3] carry out a detailed survey on the existing communication protocols implemented in IoT, mechanisms for secure communications along with open research issues relating to IoT as the future Internet. A new approach to design and implementation of IoT security mechanism has been proposed in [4] where authors attempted to explore explicit roles of all the actors involved in IoT communication taking into account the security aspects such as identification/authentication, trust, privacy, responsibility, autoimmunity, safety, and reliability. Architecture of IoT, its application domains, future development trends along with key issues and challenges are explicitly addressed in [5] wherein authors comprehensively discuss the specific role of each of the layers in the IoT protocol stack. An embedded security framework for IoT is proposed in [6] wherein the design of the framework mostly relies upon three aspects such as environmental factor, security objectives, and basic requirements. Security threats pertaining to IoT along with possible countermeasures are addressed by authors in [7].

Here authors focus on all possible attacks relating to each layer of IoT protocol stack proposing the respective countermeasures for it. The authors in [8] also gives detailed layer-wise security issues relating to IoT protocol architecture thereby focussing on specific IoT security goals like data confidentiality, data integrity, and data availability. Suchitra and Vandana [9] describe the security issues relating to IoT wherein principal focus is made on security issues that are applicable to Wireless Sensor Networks ("WSN") and Radio Frequency Identification RFID technology. The authors Qiang et al. [10] also address the security issues pertaining to RFID technology and wireless communication information security along with privacy protection in IoT.

6   Conclusion and Future Work

IoT being regarded as the future Internet is expected to enrich the present human-to-human communication over the internet by enabling human-to-machine as well as machine-to-machine communications too. However, it carries with itself a lot of issues and challenges related to information security as well as communication security. In this paper, we address a publish/subscribe security model for IoT that would hopefully be capable of resolving many of the security issues. Nevertheless, this model can further be enhanced by virtue of adding an intelligent component to cater to the security needs of IoT.

References

1. Vermesco, O., Harrison, M., Vogt, H., Kalabourkas, K., Tomasellar, M., Wouters, K., Glusmessli, S., Haller, S.: Internet of Things: Strategic Research Roadmap (2009)
2. Burel, B., Barker, L., Divitini, M., Peret, F.A.F, Russell, I., Siever, B., Tudor, L.: Courses, content and tools for internet of things in computer science edcuation. In: Proceedings of ITICSE Conference on Working Group Reports, pp. 125–139, Italy (2017)
3. Granjal, J., Monterio, E., Silva, J.S.: Security for the internet of things. A survey of existing protocols and open research issues. IEEE Commun. Surv. Tutorials **17**(3), 1294–1312 (2015)
4. Riachi, A., Challal, Y., Natalizio, E., Chtourou, Z., Bouabadallah, A.: A systemic approach for IoT security. In: IEEE Dcoss, pp. 351–355 (2013)
5. Kahn, R., Khan, S.V., Zaheer, R., Khan, S.: Future internet: the internet of things architecture, possible applications and key challenges, possible applications and key challenges. In: Proceedings of 10th International Conference on Frontiers of Information Technology ("FIT"), pp. 257–260 (2012)
6. Babar, S., Stango, A., Prasad, N., Sen, J., Prasad R.: Proposed embedded security framework for internet of things ("IoT"). In: Proceedings of 2nd, IEEE Conference on Wireless Communication, Vehicular Technology, Information Theory and Aerospace & Electronic systems Technology ("VITAE"), pp. 1–6 (2011)
7. Ahmed, A.W., Ahmed, M.M., Khan, O.A., Shah, M.A.: A comprehensive analysis on the security threats and their countermeasures of IoT. Int. J. Adv. Comput. Sci. Appl. ("IJACSA") **8**(7), 489–501 (2017)
8. Pandey, E., Gupta, V.: An analysis of security issues of internet of things (IoT). Int. J. Adv. Res. Comput. Sci. Softw. Eng. ("IJARCSSE") **5**(11), 768–773 (2015)
9. Suchitra, C., Vandana, C.P.: Internet of things and security issues. Int. J. Comput. Sci. Mob. Comput ("IJCSMC") **5**(1), 133–139 (2016)
10. Qiang, C., Quan, G., Yu, B., Yang, L.: Research on security issues on the internet of things. Int. J. Future Gener. Commun. Networking (IJFGCN) **6**(6), 1–10

Alphabetic Cryptography: Securing Communication Over Cloud Platform

Sanjeev K. Cowlessur, B. Annappa, M. V. Manoj Kumar, Likewin Thomas, M. M. Sneha and B. H. Puneetha

Abstract This paper introduces alphabetic cryptography inspired by bidirectional DNA encryption algorithm. Alphabetic cryptography first offers higher randomization and secure communication over the cloud computing platform, and second supports the exchange of complete UNICODE character set. Alphabetic cryptography has been implemented on mobile and desktop platforms. Through experimental studies, it has been observed that randomness of encryption increases exponentially with the increase in the number of alphabets of the alphabetic encryption scheme.

Keywords Data security issues · Bidirectional DNA encryption algorithm
Alphabetic cryptography · Alphabetic digital coding · Socket programming

S. K. Cowlessur (✉)
Department of Software Engineering, Université des Mascareignes, Swami Dayanand Campus, Beau Plan, Pamplemousses, Mauritius
e-mail: scowlessur@udm.ac.mu

B. Annappa
Department of Computer Science and Engineering, National Institute of Technology Karnataka, Surathkal, Karnataka, India
e-mail: annappa@ieee.org

M. V. Manoj Kumar · M. M. Sneha
Department of Information Science and Engineering, NITTE Meenakshi Institute of Technology, Yelahanka, Bengaluru, India
e-mail: manojmv24@gmail.com

M. M. Sneha
e-mail: snehahr1990@gmail.com

L. Thomas · B. H. Puneetha
Department of Computer Science and Engineering, P.E.S. Institute of Technology and Management, Shivamogga, India
e-mail: likewinthomas@gmail.com

B. H. Puneetha
e-mail: puneeth.bh02@gmail.com

© Springer Nature Singapore Pte Ltd. 2019
S. C. Satapathy et al. (eds.), *Information Systems Design and Intelligent Applications*, Advances in Intelligent Systems and Computing 863,
https://doi.org/10.1007/978-981-13-3338-5_19

1   Introduction

Nowadays digital and electronic systems are generating data continuously. To store, process, and analyze the information does require a lot of computation and storage, so personal computers cannot accommodate such massive amount of storage and computation. The services offered by cloud computing can address this technological gap. Many organizations today are depending on cloud service providers for computer resources.

Cloud computing has three different service models: private, public, and hybrid.

- *Private cloud* runs for a single organization managed internally or by a third party.
- *Public cloud* provides services over a network and is open for public use (also known as "pay-per-use" model).
- *Hybrid cloud* is the combination of both private and public cloud services.

Cloud computing facilitates a range of on-demand services such as Software (SaaS), Infrastructure (IaaS), Network (NaaS), Platform (PaaS), etc. Many organizations are unenthusiastic to use cloud services due to the lack of data security for the data that resides on the cloud services provider's servers. So various approaches have been developed to address the issues of data security for strengthening the data stored and communicated over cloud platforms.

Data security remains the biggest challenge in cloud computing. Information stored on the cloud is sensitive. It is, therefore, essential to take care of this information so that it is not vulnerable to data leakages. Hence, various cryptographic algorithms have been adopted in cloud computing to enhance data security. Alphabetic cryptography is one such approach, which is used to strengthen the security over cloud platforms; it is based on inspiration and extension of bidirectional DNA encryption method. Alphabetic cryptography offers the following benefits:

1. Exponential randomization and secured communication during the encryption and decryption process.
2. It supports the exchange of complete UNICODE character set over cloud platforms.

Upcoming sections are organized as follows: Sect. 2 gives a brief overview of the related work. Section 3 explains the encryption and decryption procedures, Sect. 4 presents the results of experimental study, Sect. 5 details future directions and applications. Finally, the conclusion of the paper is given in Sect. 6.

2   Related Work

One of the main concerns during data transmission is data security. Data security is achieved by encryption, decryption, authentication, hash functions, and digital signature [1–3]. According to Diffie et al. [4], cryptography is a technique by which

Table 1 Alphabetic digital coding

Binary value	Alphabet coding
000	AA
001	AB
010	AC
011	AD
100	AE
101	BA
110	BB
111	BC

one can store and transmit data so that authenticated individuals can access it. The two important concepts in cryptography are encryption and decryption. Encrypted data is called ciphertext, while unencrypted data is referred to as plain text [5].

DNA cryptography [6] is one of the rapidly emerging technologies, which uses the concepts of DNA computing. Goyat et al. [7] proposed a DNA encryption technique on SaaS, which is less computational, more cost-effective and efficient. R. Pragaladan et al. introduced one of the initial DNA confidentiality structure in cloud infrastructure, which converts transactional data into binary data [8].

3 Methodology

In information science, the binary digital coding is encoded by two states 0 and 1. On similar lines, in this paper, we have used English alphabets to generate the encoding as shown in Table 1.

3.1 Key Combination

Here in this work, we are using ABCDE as a key to encode and decode a message. Key combinations are generated as shown in Table 2. The key ABCDE is exchanged with the receiver using the Diffie–Hellman key exchange algorithm [9]. Key combination values generated will be changed randomly every time.

3.2 Encryption Method

The method of encryption using alphabetic cryptography is illustrated in Fig. 1. Initially, the original message is converted to its binary equivalent. Binary values are

Table 2 Key generation
table for five alphabets

Key combination	Patterns
AA	01001
AB	10110
AC	01110
AD	01111
AE	00111
BA	01011
BB	00001
BC	01101
BD	10000
BE	00011
CA	00000
CB	10011
CC	11000
CD	01010
CE	00101
DA	10101
DB	10001
DC	10111
DD	10010
DE	10100
EA	00010
EB	00110
EC	00100
ED	01000
EE	00100

converted to alphabetic coding (shown in Table 1) and then key combination (shown
in Table 2) is applied to generate the encrypted message.

3.3 Decryption Method

Decryption is the exact reverse of the encryption process. It is the process of converting the ciphertext back to plain text (human-readable format). The procedure for
decrypting the encrypted message is shown in Fig. 2.

Once the encrypted message is received, key combination and alphabetic coding
are used to decrypt the message. The reverse of encryption method (binary to hex to
ASCII) is applied to obtain the plain text.

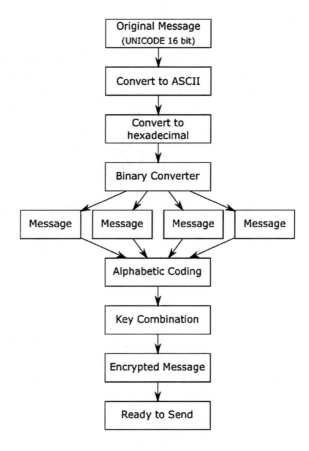

Fig. 1 Encryption procedure

3.4 Implementation Details

The proposed method for alphabetic cryptography (encryption and decryption) has been implemented on Android and Windows platform. At first, the cryptography process should enable users to compose the encrypted message and it send to the intended recipient. Once the message is received at the receiver, the decryption process takes place using the same key combination as that of the sender. Both encryption and decryption processes are illustrated in Figs. 1 and 2.

In our experiments, we have used the Java language for both Android and Windows platforms. Particularly, Android Studio IDE [10] is used for developing the mobile application development and NetBeans IDE is used for the desktop application. The evaluation of the randomness of alphabetic cryptography for increasing alphabetic combination has been done using the R programming language.

Fig. 2 Decryption
procedure

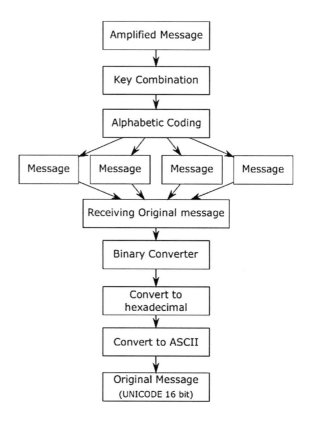

4 Results and Discussion

To understand the process of encryption, let us consider the following example. In this example, we will illustrate the process of encrypting Kannada text (one of the Indian languages) using alphabetic cryptography method detailed in Fig. 1. The detailed operation of the encryption and decryption implementation of alphabetic cryptography on the Android platform is given in Fig. 3.

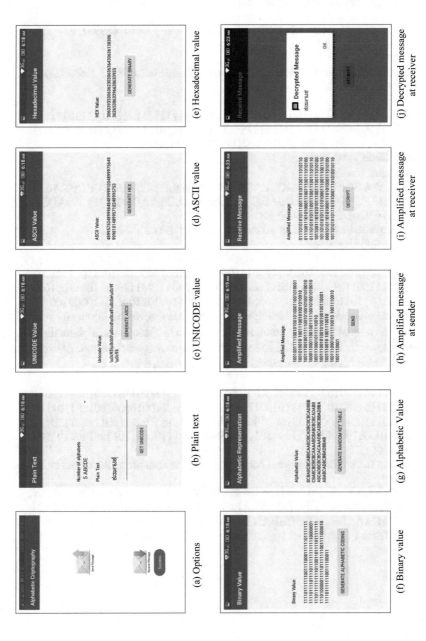

Fig. 3 Alphabetic cryptography: encrypting and decrypting the message

- Plain text: ಕರ್ನಾಟಕ

- Unicode: \nu0c95\nu0cb0\nu0ccd\nu0ca8\nu0cbe\nu0c9f\nu0c95

- ASCII:489957534899984848999910048999756489998101489957102489957
  53

- Hexadecimal:3063393530636230306363643063613830636265306339663066
  33935

- Binary:11110111111100111100011111110111111111110111101111011111
  11111110000001111011111111110110011011110111111111110100000111
  11011 111110011110000101111011111110011110001 1
From Table 1, we can write
- Alphabetic coding:
  BCBABCBCABBCAABCBCADBCBCBCADBBBCBABCBCBCBCAAA
  ABCBABCBCBCADABBABCADBCBCBCACAAADBCADBCBBADBB
  AABABCADBCBBADBBAB
The amplified message is generated by making use of Table 2:
- Amplified message:
  1001101110100111001100100100110010010011100110110010011100111 0
  0110110010010110110100111001100111001100100001001001101110100 1
  1100111001101100001000111010011011001001110011100111011100100 0
  1100100110110010011011000010001110100110110010010110110000100
At the receiver side, once the amplified message is received it is converted to plain
text with the help of the key generation table (shown in Table 2) and alphabetic digital
coding (shown in Table 1). The following sequence of operations will illustrate the
process of deciphering the amplified message (illustrated in Fig. 2).
- Amplified message:
  1001101110100111001100100100110010010011100110110010011100110 01
  1001101100100110111010011100111001110011001000010010011011100
  1001110011100110110000100011101001101100100111001110011101111
  0010001100100110110010010011011000010001110100110110010011011 00
  00100
By using the alphabetic digital coding and key combination, the original message
is retrieved:
- Alphabetic digital coding: BCBABCBCABBCAABCBCADBCBCBCA
  DBBBCBABCBCBCBCAAAABCBABCBCBCADABBABCADBCBCB
  C ACAAADBCADBCBBADBBAABABCADBCBBADBBA
From Table 1 of Alphabet Digital Coding, we can generate

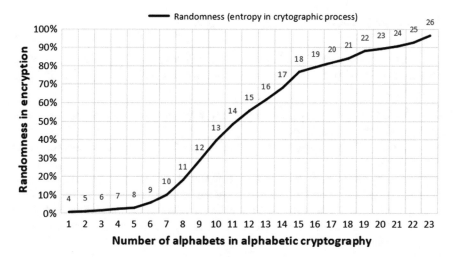

Fig. 4 Graph illustrating an increase in randomness in the cryptographic process with the increase in the number of alphabets in alphabetic cryptography

- Binary:111110111111100111100011111110111111111110111101111011111
  11111110000001111011111111111011001101111011111111111110100000111
  11011 11111001111000010111101111111001111100011

- Hexadecimal:
  30633935306362303063636430636138306362653063339663306 33935

- ASCII:
  48995753489998484899991004899975648999810148995710248995 3

- Unicode: \nu0c95\nu0cb0\nu0ccd\nu0ca8\nu0cbe\nu0c9f\nu0c95

- Plain text: ಕರ್ನಾಟಕ

The above-illustrated methods have considered five letters of the English alphabet for the encryption and decryption processes. As the number of alphabets in the encryption process increases, randomness and robustness of encryption exponentially increase as shown in Fig. 4.

The randomness in the encryption process is high for 26 letters and less for 4 letters. Randomness in the encryption process is measured with the help of information gain, i.e., bits [9]. This is the same measuring unit used to measure information gain in the message communication process (also known as entropy).

5   Future Direction and Applications

Alphabetic cryptography is a simple and effective way of securing communication over the cloud environment. There are various avenues for improving the alphabetic cryptography:

- Evaluating the performance of alphabetic cryptography with respect to video, photo, and other data formats.
- Analyzing alphabetic cryptography on a range of available mobile and computer platforms.
- Developing the method for selecting optimal number of encryption alphabets based on the length and format of the data (here we have illustrated for 5 alphabets).
- Proposed method works with 4–26 alphabetic letters. There is further evaluation needed if more than 26 letters are to be used (e.g., aa to zz, i.e., combination of more than one letter).

The use of alphabetic cryptography is appropriate in the following situations:

- Time-critical communication systems such as military, surveillance, health care, etc.
- Transmitting UNICODE characters.
- Devices having limited computation, memory, and battery power, such as mobile and laptops.
- Where the bandwidth of communication is limited and latency is not guaranteed (i.e., when the timely delivery of data is not guaranteed).

The use of alphabetic cryptography is not suitable particularly in situations where the size of data to be transferred is large (more than gigabytes of data).

6   Conclusion

This paper presented alphabetic encryption method for secured data transmission over cloud platforms. Alphabetic cryptography technique achieved exponential randomness in the cryptography process. In addition to this, it also supports the transfer of UNICODE characters over cloud platform. The proposed method has been tested on both desktop and mobile platforms.

References

1. Hojabri, M., Heidari, M.: Union of RSA algorithm, digital signature and KERBEROS in cloud security. In: International Conference on Software Technology and Computer Engineering, pp. 978–993 (2012)
2. Mao, W.: Modern Cryptography: Theory and Practice. Prentice Hall Professional Technical Reference (2003)
3. Rewagad, P., Pawar, Y.: Use of digital signature with Diffie-Hellman key exchange and AES encryption algorithm to enhance data security in cloud computing. In: IEEE International Conference on Communication Systems and Network Technologies, pp. 437–439 (2013)
4. Diffie, W., Hellman, M.E.: Privacy and authentication: an introduction to cryptography. Proc. IEEE **67**(3), 397–427 (1979)
5. Rescorla, E.: Diffie-Hellman Key Agreement Method, RFC 2631. IETF Network Working Group (1999)
6. Xiao, G., Mingxin, L., Lei, Q., Xuejia, L.: New field of cryptography: DNA cryptography. Chin. Sci. Bull. **51**(12), 1413–1420 (2006)
7. Goyat, S., Jain, S.: A secure cryptographic cloud communication using DNA cryptographic technique. In: IEEE International Conference on Inventive Computation Technologies (ICICT), vol. 3, pp. 1–8 (2016)
8. Pragaladan, R., Sathappan, S.: High confidential data storage using DNA structure for cloud environment. In: IEEE International Conference on Computation System and Information Technology for Sustainable Solutions, pp. 382–387 (2016)
9. Shannon, C.E.: Communication theory of secrecy systems. Bell Labs Tech. J. **28**(4), 656–715 (1949)
10. Zapata, B. C.: Android Studio Application Development. Packt Publishing Ltd, UK (2013)

An Exploratory Study of Evidence-Based Clinical Decision Support Systems

Sudha Cheerkoot-Jalim, Kavi Kumar Khedo and Abha Jodheea-Jutton

Abstract With the rapid expansion of medical research, patients worldwide are having higher expectations in terms of medical services and clinical care. Huge research efforts are invested with the objective to minimize medical errors, improve patient care efficiency and increase patient safety. One initiative towards this endeavour is the development of clinical decision support systems (CDSS), which give recommendations on patient diagnosis, treatment options and follow-up, to healthcare providers. Evidence-based CDSS has proved to boost up the use of traditional CDSS, since they have mechanisms to integrate new evidence from literature-based research findings into their knowledge base, for more informed decision-making. This work conducts an exploratory study of different evidence-based CDSS, which provides a comparative study and based on the findings, proposes future research challenges for such types of systems. This paper will provide both researchers and healthcare practitioners with the current state of affairs in the domain of evidence-based CDSS.

Keywords Evidence-based medicine · Clinical decision support systems
Clinical decision-making · Patient care

1 Introduction

In this era of globalization and technology outburst, huge efforts are invested in the promotion and adoption of Information and Communication Technologies (ICT) to improve patient care in healthcare institutions worldwide [1]. These institutions have to keep up with the expectations of patients, who are more than ever, seeking

S. Cheerkoot-Jalim (✉) · K. K. Khedo · A. Jodheea-Jutton
University of Mauritius, Reduit, Mauritius
e-mail: s.cheerkoot@uom.ac.mu

K. K. Khedo
e-mail: k.khedo@uom.ac.mu

A. Jodheea-Jutton
e-mail: a.jutton@uom.ac.mu

© Springer Nature Singapore Pte Ltd. 2019   207
S. C. Satapathy et al. (eds.), *Information Systems Design and Intelligent
Applications*, Advances in Intelligent Systems and Computing 863,
https://doi.org/10.1007/978-981-13-3338-5_20

enhanced healthcare treatment and services. According to Petterson et al. [2], the number of patient visits to primary care physicians in the US is expected to rise considerably by 2025. This places great demands on the healthcare industry to improve its information-handling abilities through electronic health records (EHR). Knowledge generated by EHRs can be applied to patient diagnosis, prognosis and treatment by healthcare providers [3].

The abundance of information available in EHRs to researchers and physicians has led to their integration with clinical decision support systems (CDSS). CDSS are computerized systems designed to assist healthcare providers in making decisions about individual patients in a timely manner. They have the potential to significantly improve the quality and efficiency of patient care, as well as alleviate the ever-increasing cases of medical errors, delayed clinical decisions and patient safety.

Along with CDSS, the use of evidence-based medicine (EBM) in improving clinical outcomes has also proved to be commendable. EBM is commonly used in decision-making for patient cases based on the physician's clinical expertise, patient values and the best available clinical evidence from systematic research [4]. This approach is useful to the physician in case of missing, incomplete or low-quality evidence. The authors of [5] have brought forward the concept of evidence-adaptive CDSS, which derives its knowledge not only from EHRs but also from the up to date evidence from literature-based research. The knowledge base in this case is automatically updated as new research becomes available.

This work conducts an exploratory study of the existing evidence-adaptive CDSS systems, performs a comparative study of the systems in terms of criteria such as the area of application, extent to which EBM is applied and sources of evidence considered and identifies the challenges which have to be considered in future research. This paper will provide both researchers and healthcare practitioners with the current state of affairs in the domain of evidence-based CDSS.

2   Literature Study

This section describes the different concepts, which are important in understanding the importance of evidence-based clinical decision-making.

2.1   Evidence-Based Medicine

A medical definition of EBM is provided in [6] as: 'the judicious use of the best current available scientific research in making decisions about the care of patients. EBM is intended to integrate clinical expertise with the research evidence and patient values'. EBM therefore greatly relies on findings from clinical research literature, which is usually available from a variety of sources including randomized trials, systematic reviews and guidelines literature [5]. Before the new evidence is used in clinical decision-making, it has to be verified for validity and applicability by domain experts. Healthcare providers highly commend the practice of EBM, in addition to their traditional practice guidelines. Decision-making is, therefore, not only influenced by the experiences and opinions of the healthcare provider but also by data extracted from clinical research and studies.

Main barrier of EBM. The practice of EBM is, however, greatly hindered by the exponential growth of clinical research literature, due to the large investment in medical research worldwide. With their very busy schedules, healthcare providers find it very difficult to keep updated with the unmanageable research literature. According to the study, a medical practitioner would need to spend around 20 h per day to keep up with new evidence represented by around 8000 relevant articles published monthly [7].

2.2 Clinical Decision Support Systems

Medical errors are quite common in healthcare institutions worldwide, mostly due to inadequate time for assessing patients. CDSS is one attempt to reduce medical errors. They are software designed to aid clinical decision-making by healthcare providers. The medical profile of the current patient is matched to a computerized knowledge base, usually derived from EHRs. Assessments and recommendations specific to the patient's profile are then presented for more informed decision-making by the healthcare provider. The main aim of CDSS is to decrease the incidence of medical errors and delayed clinical decisions, as well as improve patient safety and the quality of patient care [3, 8].

The most common type of CDSS is the knowledge-based CDSS, which consists of:

- a knowledge base
- the inference or reasoning engine and
- a mechanism to communicate with the user,

as shown in Fig. 1.

The knowledge base is usually represented as if-then rules. It usually includes probable symptoms-diagnoses associations, drug–drug or drug–food effects [9]. The reasoning or inference engine's function is to match current patient data with rules in the knowledge base. The communication mechanism deals with the input of data (mostly from EHRs and medical research) into the knowledge base and the presentation of the output recommendations to the user.

Application of CDSS. CDSS have been used for decision-making in various application areas like breast cancer [10], diabetes [11], neurosurgery [12], depressive disorders [13], hypertension [14], amongst others. The adoption of CDSS has shown to improve patient care in different ways. The varied uses of CDSS include, among others [9]:

Fig. 1 A general
knowledge-based CDSS [9]

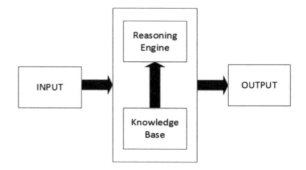

- indication of potentially harmful drug interactions
- more accurate diagnoses
- identification of optimal treatment strategies
- reminders about patient follow-up
- warnings for data submission
- formulation of health policies.

Previous studies have demonstrated the effectiveness of CDSS implementation in various settings. The authors of [15] conducted a meta-regression analysis of 70 studies using 15 factors relevant to CDSS success. As a result of their analysis, the authors concluded that in 68% of the trials, CDSS brought about a significant improvement. The systematic reviews of [16] and [17] showed that implementation of CDSS resulted in an improvement on practitioner performance and diagnosis in 57 and 64% of the studies, respectively. However, they also noticed a quite low impact on patient outcomes, more specifically, 30% in [16] and 13% in [17], indicating that CDSS still needs improvement.

2.3 Evidence-Based CDSS

One solution to improve the effectiveness of CDSS is to use evidence-based or evidence-adaptive CDSS. This subclass of CDSS continuously integrates new evidence from the research literature with the knowledge base of traditional CDSS [5]. Mechanisms should be formulated in order to incorporate findings from new, relevant studies available in the literature as soon as they become available, so as to provide updated recommendations. Figure 2 shows the integration of the knowledge base of evidence-adaptive CDSS with the evidence extracted from online research literature.

The aim of evidence-adaptive CDSS is not only to incorporate more clinical research evidence but also to ensure the quality, usefulness, accessibility, accuracy and reliability of the incorporated evidence, so as to positively impact on patient-specific assessment and recommendations. Cochrane Library, Best Evidence and Clinical Evidence are among the best electronic resources for EBM [5]. Their con-

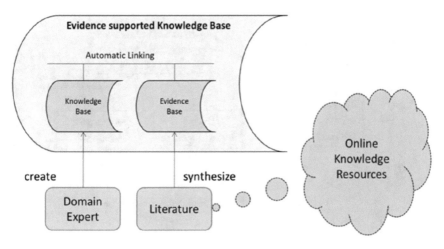

Fig. 2 Evidence-adaptive CDSS [18]

tents are mostly textual and therefore not machine-interpretable by traditional CDSS. A major challenge for evidence-adaptive CDSS is, therefore, the automated update of the knowledge base through machine-interpretable input from the research literature, therefore eliminating the main barrier posed by EBM, as pointed out in Sect. 2.1.

3 Critical Review of Evidence-Based Clinical Decision Support Systems

The contribution of EBM for clinical decision-making has been acknowledged by researchers, who have developed a myriad of CDSS, which rely not only on EHRs and clinical patient data but also on new evidence from clinical research, to provide recommendations. Seven such systems have been studied for the purpose of this work and are described below.

Intelligent Neurosurgical CDSS. Sakr et al. [12] propose an intelligent neurosurgical decision support system framework, whose main aim is to predict the ideal treatment method for lumbar disc patients and evaluate the treatment plan. The system is designed to make an integrated decision based on the reuse of historical data, benefits from latest published scientific research, usage of previous experience from clinical pathways and patient perspective, as shown in Fig. 3. The two main subsystems of the proposed framework are the Electronic Medical Record (EMR) subsystem, which is the core data source of the CDSS, and the data mining engine, which uses machine learning techniques to predict treatment and evaluate treatment planning. The authors claim that their framework will improve physician performance, improve accuracy and reduce the time for decision-making. The system does allow for the integration of new results and evidence from the medical domain, resulting in

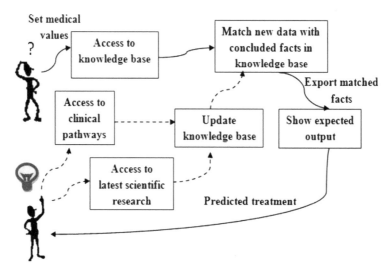

Fig. 3 Procedures for decision-making process [12]

a knowledge base which is up to date. However, the inputs from the latest scientific researches have to be integrated manually into the system. It does not include any automated mechanism to incorporate new medical research findings.

CompTMAP. The authors of [13] designed and developed a CDSS, CompTMAP, relying on evidence-based guidelines for decision-making in the treatment of major depressive disorder. The software was designed for both psychiatrists and primary care physicians. The knowledge base consists of patient information, medication information and dosages, appointment information and progress notes. The features of CompTMAP include diagnosis, treatment suggestions, recommendations for follow-up, access to physicians' order entry, adverse drug event alert systems and electronic documentation. CompTMAP has been extensively tested to ensure accuracy and reliability and had positive response from physicians following surveys conducted. However, the application does not cater for the integration of external sources of information. It contains a link to the American Psychiatric Association, which provides guidelines on how to perform a diagnostic evaluation, but these guidelines are not available in a comprehensive computerized format in the diagnostic system.

ATHENA DSS. The Assessment and Treatment of Hypertension: Evidence-based Automation (ATHENA) DSS [14] is among the earliest CDSS to cater for EBP. The system provides guideline-based management of hypertension. It encourages the control of blood pressure and recommends the choice of drug therapy, concerning co-morbid diseases. ATHENA DSS has two components: a knowledge base with hypertension data and a guideline interpreter that recommends treatment on an individual basis. So as to avoid conflicts between the data model of the system with that of integrated EMRs, ATHENA DSS has a flexible interface which allows integration with EMRs from different institutions. Since research on hypertension is continuously evolving, the knowledge base of the CDSS is designed to be modifiable so as to incorporate new evidence and guidelines. These changes to the knowledge base are done as soon as major new evidence becomes available by clinical domain experts and therefore is not automated.

ADDIS. In their work, the authors of [19] propose an open-source evidence-based drug-oriented strategy decision support system, Aggregate Data Drug Information System (ADDIS). The decision support tool is based on individual studies or evidence synthesis. The primary goal of the system was to enable the assessment of benefits and risks of different drugs, based on evidence extracted from clinical trials. ADDIS, however, has a few limitations. First, it is used only for strategic decision-making and not operational decision-making. Furthermore, ADDIS assumes a specific structure for the trials, therefore requiring appropriate mapping in many cases. Data was therefore mostly extracted manually with some assistance from trial registries and research publications. Data extraction from medical literatures in ADDIS is only semi-automated.

TADS. The decision support application, Triple Assessment Decision Support (TADS), developed by [20] focuses on a key part of breast care pathway, known as Triple Assessment. One main aim of developing TADS was to investigate how to construct the knowledge required for the application, based on sources like clinical practice guidelines and research literature, and how this can improve the compliance of clinicians with best practice. TADS deliver advice based on clinical guidelines, together with patient-specific decision support. The system also has hyperlinks to direct users to the clinical practice guideline and the research evidence which were used to make the decision. TADS, unfortunately, has quite limited capability to automatically extract information from evidence-based guidelines. Out of the seven clinical practice guidelines selected, only two provided recommendations in 'clear and easy-to-understand' form. The others have a mostly textual format and considerable effort is required to integrate the evidence into the knowledge base.

Decision support tool for glucose lowering therapy. The authors of [21] designed a patient-specific decision support tool to help healthcare providers set individual glycemic targets and select treatment options for type 2 diabetes mellitus patients (T2DM). The software, which is implemented using an online decision support system, Diascope, was developed, based on a systematic analysis of expert opinion and clinical trials. The authors commend the need for incorporating scientific knowledge into clinical practice, thus the requirement for a CDSS. The reasoning engine is based on a highly structured method, which provides recommendations based on the best available published evidence from clinical studies and practitioner experience. The authors wish to make their tool as 'evidence-adaptive' as possible. However, only annual updates are planned to incorporate new evidence, changing guidelines and increased expertise.

KnowledgeButton. KnowledgeButton is a 'comprehensive model for evidence adaptation from online credible knowledge sources in a well-defined and established manner' [22]. It is mainly applied in the domain of head and neck cancer. The main aim of the KnowledgeButton project is to continually allow the knowledge base of

CDSS to automatically incorporate new research evidence from journal publications in the PubMed database. To ensure relevancy, it uses automatic and semi-automatic query generation. Relevant information is extracted from the PubMed database using its API service support. The extracted articles are first ranked using some relevancy attributes and then presented to physicians. An article is considered as evidence, only if the physician approves it. The concepts from relevant research articles, with relationships are presented in visual format, thus helping the domain expert to create knowledge rules, which are validated, compiled and saved to the knowledge base of the CDSS. One limitation of KnowledgeButton is that it retrieves articles only from PubMed and does not consider other sources which may have relevant articles with new evidence. Furthermore, the integration of new evidence is only semi-automated, since, after being presented with the required information, physicians have to create the knowledge rules themselves.

4   Comparative Study

Table 1 shows the results of a comparative study conducted for the systems discussed in Sect. 3 using different criteria.

An overview of the different studies shows that a lot of effort is being put in incorporating the principles of EBM into modern CDSS in various application areas, for more efficient and informed decision-making in the clinical setting. The objectives of the decision support tool in the systems studied include diagnosis, treatment suggestions, patient follow-up, alerts and reminders, adverse drug effects, among others. These need to be continually adapted to new research in the medical field, thus the adoption of EBM principles.

All of the above systems, except CompTMAP, allow the integration of new evidence from varied sources (Table 1) in the knowledge base of the CDSS. In three of the systems, the modification to the knowledge base is only manual, that is, clinical experts are provided with an interface to add new rules, reflecting the results of new research. ADDIS, TADS and KnowledgeButton implements a semi-automated technique to incorporate new evidence, and therefore not yet fully evidence-adaptive. In ADDIS, an XML interface is used to import studies from ClinicalTrials.org. However, some fields have to be manually mapped to the entities in the database. Also, ADDIS assumes that a structured database of clinical trials is available. Out of the seven high-quality clinical practice guidelines considered in TADS, only two provided recommendations in clear and well-structured format. The others were mostly textual and it was, therefore, difficult to automatically extract relevant information. KnowledgeButton automates extraction of knowledge from textual sources and presents concepts with relationships to the user in visual format. But the creation of the knowledge rule is still done by the domain expert. No CDSS with fully automated update of the CDSS knowledge base has been encountered during the course of this work. It is also to be noted that the systems studied made use of various tools and techniques for the development of the CDSS.

Table 1 Comparative study of different evidence-based CDSS

CDSS	Area of application	Main purpose	Include EBM principles	Allow update of knowledge base with new evidence	Types of knowledge-based update	Sources of evidence/guidelines	Tools/techniques used
Intelligent Neurosurgical CDSS (2014)	Neurosurgery (lumbar disc patients)	Predict ideal treatment method for patients	Yes	Yes	Manual	Clinical pathways, varied sources	EMR, Rough Set Theory (AI technique)
CompTMAP (2004)	Depressive disorders	Diagnosis, treatment suggestions, follow-up, preventive care	Yes	No	None	American Psychiatric Association practice guidelines	Not applicable
ATHENA DSS (2000)	Hypertension management	Blood pressure control, recommendation for choice of drug therapy	Yes	Yes	Manual	Mostly clinical trials	EON architecture, Protege (for the knowledge base)
ADDIS (2013)	Drug-oriented strategy	Guideline formulation, marketing authorization, reimbursement	Yes	Yes	Semi-automated	Clinical trial registries (ClinicalTrials.gov), journal publications	External network meta-analysis library, JSMAA (benefit–risk models)
TADS (2008)	Breast care pathway	Improving compliance of clinicians with best practice	Yes	Yes	Semi-automated	Clinical practice guidelines, medical literature	CommonKADS methodology, PROforma decision support technology
Decision support tool for glucose lowering therapy (2015)	Type 2 diabetes mellitus	Setting of individual glycemic targets and selection of treatment options	Yes	Yes	Manual	Clinical studies	DiaScope (online decision support tool)
KnowledgeButton (2014)	Head and neck cancer	Decision-making in the clinical setting	Yes	Yes	Semi-automated	PubMed database	eUtils (Entrez programming utilities)

5 Research Challenges

All the works considered in this study have incorporated EBM principles in their decision support tools. However, the authors have identified some limitations, which can be catered for in future research initiatives. The following research avenues have been identified as a result of this work.

Integration of sources other than clinical trials. Most of the systems developed considered clinical trials as their main source of evidence. Researchers were successful in devising techniques for information extraction from trial data to some extent, mainly due to their somewhat established reporting format. However, some healthcare decisions may rely on some medical criteria, which are not reported in clinical trials [19]. Therefore, future research on evidence-based CDSS systems should pursue techniques for information extraction from diverse sources of medical research.

Automated information extraction. As outlined in Sect. 4, the update of the CDSS knowledge base with new evidence is mostly either manual or semi-automated, therefore, imposing some burden on the domain experts/physicians to create the knowledge rules. Systems which allow semi-automated information extraction do provide some assistance to the user, but the rule creation, through a rule editor, is still manual. Rigorous text/literature mining and information extraction techniques should be explored and developed, so as to allow the automated creation of knowledge rules from synthesized knowledge. These rules would be added to the knowledge base, only after verification by the domain experts.

Consideration of clinical context. For any evidence-based system to be successful and efficient, the context of the domain plays an important role [22]. When designing evidence-based CDSS, researchers and developers should devise methods to specify the context in which the knowledge is applicable and integrate local context into new evidence.

Human–Computer Interaction (HCI) issues. Clinician resistance has been found to be one of the major barriers in the adoption of evidence-based CDSS, despite their proven efficacy. One of the main reasons behind this is the inappropriate design of the user interface for knowledge rule creation and information presentation. Research initiatives should include efforts to improve the usability of such systems, so as to improve clinician acceptance. The system should also be feasible to use in the real-world clinical practice settings and be able to provide the right information, at the right time and in the appropriate format.

6 Conclusion

Research in the medical and clinical field is continually evolving, thus contributing to new evidence, which when incorporated in CDSS, result in the reduction of medical errors, improved healthcare quality and patient safety. The importance of EBM to improve the efficiency of clinical decision-making has been well recognized among the research community and well appreciated in the clinical setting by healthcare providers. The concept of evidence-adaptive CDSS has been coined to refer to the ability of the system to automatically keep up with new evidence as soon as it becomes available. This work included a study of different evidence-based CDSS which have been applied in various domains. This was followed by a comparative study, using various criteria, to investigate the level of EBM adoption in decision support tools. It was found that most of the systems studied had appropriate mechanisms to incorporate new evidence into their knowledge base, thus resulting in improved decision-making by healthcare providers. However, the systems were not completely evidence-adaptive, since the addition of new knowledge was not fully automated, posing a burden on the clinical experts. The comparative study also highlighted that the main source of new evidence was from clinical trials. More effective systems should be able to incorporate findings from other sources too, while considering local clinical context. Additionally, further research is required in order to improve the usability of such CDSS so as to increase its adoption among healthcare practitioners. Following this study, the research challenges identified are expected to give useful insights to future researchers and developers, so as to aspire for better adoption of evidence-based CDSS.

References

1. Zakaria, N., Affendi, S., Zakaria, N.: Managing ICT in healthcare organization: culture, challenges, and issues of technology adoption and implementation. In: Health Information Systems: Concepts, Methodologies, Tools, and Applications, pp. 1357–1372. IGI Global, Hershey (2010)
2. Petterson, S.M., Liaw, W.R., Phillips, R.L., Rabin, D.L., Meyers, D.S., Bazemore, A.W.: Projecting US primary care physician workforce needs: 2010–2025. Ann. Fam. Med. 10(6), 503–509 (2012)
3. Castaneda, C., Nalley, K., Mannion, C., Bhattacharyya, P., Blake, P., Pecora, A., Goy, A., Suh, K.S.: Clinical decision support systems for improving diagnostic accuracy and achieving precision medicine. J. Clin. Bioinform. 5(1), 4 (2015)
4. Sim, I., Gorman, P., Greenes, R.A., Haynes, R.B., Kaplan, B., Lehmann, H., Tang, P.C.: Clinical decision support systems for the practice of evidence-based medicine. J. Am. Med. Inform. Assoc. 8(6), 527–534 (2001)
5. The medical definition of evidence-based medicine. https://www.medicinenet.com/script/main/art.asp?articlekey=33300
6. Majid, S., Foo, S., Luyt, B., Zhang, X., Theng, Y.L., Chang, Y.K., Mokhtar, I.A.: Adopting evidence-based practice in clinical decision making: nurses' perceptions, knowledge, and barriers. J. Med. Libr. Assoc. JMLA 99(3), 229 (2011)
7. Berner, E.S., La Lande, T.J.: Overview of clinical decision support systems. In: Clinical Decision Support Systems, pp. 1–17. Springer, Cham (2016)
8. What is evidence-based practice. https://guides.mclibrary.duke.edu/c.php?g=158201&p=1036021
9. Alther, M., Reddy, C.K.: Clinical decision support systems. In: Reddy, C.K., Aggarwal, C.C. (eds.) Healthcare Data Analytics. Taylor and Francis, UK (2015)
10. Alaa, A.M., Moon, K.H., Hsu, W., Van der Schaar, M.: ConfidentCare: a clinical decision support system for personalized breast cancer screening. IEEE Trans. Multimedia 18(10), 1942–1955 (2016)

11. Zarkogianni, K., Litsa, E., Mitsis, K., Wu, P.Y., Kaddi, C.D., Cheng, C.W., Wang, M.D., Nikita, K.S.: A review of emerging technologies for the management of diabetes mellitus. IEEE Trans. Biomed. Eng. **62**(12), 2735–2749 (2015)
12. Sakr, A.A., Mosa, D.T., Shehabeldien, A.: Development of an intelligent approach for medical knowledge discovery and decision support. Development **99**(6) (2014)
13. Trivedi, M.H., Kern, J.K., Grannemann, B.D., Altshuler, K.Z., Sunderajan, P.: A computerized clinical decision support system as a means of implementing depression guidelines. Psychiatr. Serv. **55**(8), 879–885 (2004)
14. Goldstein, M.K., Hoffman, B.B., Coleman, R.W., Musen, M.A., Tu, S.W., Advani, A., Shankar, R., O'connor, M.: Implementing clinical practice guidelines while taking account of changing evidence: ATHENA DSS, an easily modifiable decision-support system for managing hypertension in primary care. In: AMIA Symposium p. 300. American Medical Informatics Association, USA (2000)
15. Kawamoto, K., Houlihan, C.A., Balas, E.A., Lobach, D.F.: Improving clinical practice using clinical decision support systems: a systematic review of trials to identify features critical to success. BMJ **330**(7494), 765 (2005)
16. Jaspers, M.W., Smeulers, M., Vermeulen, H., Peute, L.W.: Effects of clinical decision-support systems on practitioner performance and patient outcomes: a synthesis of high-quality systematic review findings. J. Am. Med. Inform. Assoc. **18**(3), 327–334 (2011)
17. Garg, A.X., Adhikari, N.K., McDonald, H., Rosas-Arellano, M.P., Devereaux, P.J., Beyene, J., Sam, J., Haynes, R.B.: Effects of computerized clinical decision support systems on practitioner performance and patient outcomes: a systematic review. JAMA **293**(10), 1223–1238 (2005)
18. Van Valkenhoef, G., Tervonen, T., Zwinkels, T., De Brock, B., Hillege, H.: ADDIS: a decision support system for evidence-based medicine. Decis. Support Syst. **55**(2), 459–475 (2013)
19. Patkar, V., Fox, J.: Clinical guidelines and care pathways: a case study applying PROforma decision support technology to the breast cancer care pathway. Stud. Health Technol. Inform. **139**, 233–242 (2008)
20. Ampudia-Blasco, F.J., Benhamou, P.Y., Charpentier, G., Consoli, A., Diamant, M., Gallwitz, B., Khunti, K., Mathieu, C., Ridderstråle, M., Seufert, J., Tack, C.: A decision support tool for appropriate glucose-lowering therapy in patients with type 2 diabetes. Diab. Technol. Ther. **17**(3), 194–202 (2015)
21. Afzal, M., Hussain, M., Khan, W.A., Ali, T., Lee, S., Kang, B.H.: KnowledgeButton: an evidence adaptive tool for CDSS and clinical research. In: Innovations in Intelligent Systems and Applications (INISTA) IEEE, pp. 273–280 (2014)
22. Towards evidence adaptive clinical decision support system. http://uclab.khu.ac.kr/resources/publication/C_336.pdf

Modelling the Effects of Wind Farming on the Local Weather Using Weather Research and Forecasting (WRF) Model

B. O. Jawaheer, A. Z. Dhunny, T. S. M. Cunden, N. Chandrasekaran and M. R. Lollchund

Abstract Exploitation of wind energy is rapidly growing around the world with large wind farms being set-up for the generation of electricity. It is reported in the literature that while converting the wind's kinetic energy into electrical energy, the wind turbines may modify the transfer of energy, momentum and moisture within the atmospheric layers in the surroundings of the farms. In this work, an attempt is made to study whether the wind farm situated at Roches Noires (operational since January 2016), in the north-east part of the island of Mauritius, creates such changes within a sufficiently large space around the farm. The Weather Research and Forecasting (WRF) numerical model is employed for this endeavour due to the unavailability of measured weather data in regions close to the farm. The WRF model results are first validated with recorded meteorological data from several meteorological stations around the island and then simulations are carried out for the years 2015, 2016 and 2017. Analysis of results for two selected locations (one upstream and one downstream) around the Roches Noires wind farm demonstrates a slight decrease both in wind speed and precipitation, one year after installation of the farm.

Keywords Wind farm · WRF model · Local weather · Wind speed
Precipitation

B. O. Jawaheer · A. Z. Dhunny · M. R. Lollchund
Department of Physics, University of Mauritius, Reduit, Mauritius

T. S. M. Cunden (✉)
Department of Electromechanical Engineering and Automation, Université
des Mascareignes, Rose Hill, Mauritius
e-mail: tcunden@udm.ac.mu

N. Chandrasekaran
Université des Mascareignes, Rose Hill, Mauritius

© Springer Nature Singapore Pte Ltd. 2019
S. C. Satapathy et al. (eds.), *Information Systems Design and Intelligent Applications*, Advances in Intelligent Systems and Computing 863,
https://doi.org/10.1007/978-981-13-3338-5_21

1   Introduction

A remarkable increase in wind farm installations was observed in the last decades around the world. The wind industry obtained a huge amount of investments in 2015 and is becoming one of the most flourishing industrial firms across the globe. Furthermore, the industry contributed in creating more than a million employments globally by the setting up of around 12 TW of offshore wind farms to generate more than 3.5% of the global electricity demand [12]. Many countries have also seen promising results, after installation of wind farms, from both environmental and economical perspectives [7]. As the most active approach, wind energy is believed to be an attractive renewable energy source because of its high efficiency and zero carbon emission.

It has been reported in the literature that while converting the wind's kinetic energy into electrical energy, a wind turbine may modify the transfer of energy, momentum and moisture within the atmospheric layers in its surroundings [18, 19, 22]. This effect is more significant in large wind farms. For instance, in Inner Mongolia, it has been observed that there has been an exceptional drought since 2015, and such a phenomenon has developed more rapidly in areas where wind farms are installed [7]. Zhou et al. [23] have studied the impacts of wind farms on the land surface temperature. They focused their research in the West Texas, where four of the world's largest wind farms are located. Their results showed a warming trend of up to 0.72 °C per decade. The authors argued that this change in climate can be due to the wind farm. In their recent study of the effects of wind farming on the local climate, Baidya et al. [2] showed that the near surface temperature and wind around a wind farm is indeed affected. Their study takes into account only localized processes with timescales of the order of days. It was found in another study [10] that precipitation is also affected by wind turbines. The authors argued that this change in climate can be due to the wind farm.

The work presented in this article aims to study the local weather conditions before and after setting up a recently installed wind farm in Mauritius, with an attempt to investigate whether the farm has an effect on the local climate. The paper is organized as follows: Sect. 2 describes the Roches Noires wind farm, in terms of its geographical location, topography and energy capacity. Section 3 presents the Weather Research Forecast (WRF) model which is used to generate the climate data for the region as well as the data analysis tools used to evaluate the performance of the model. Section 4 discusses the results, and finally Sect. 5 presents our conclusions.

2   Area Under Study—Roches Noires

Roches Noires is a village located on the north-eastern coast of the island of Mauritius at latitude 20.13°S and longitude 57.72°E and has an elevation of 34 m.a.s.l. On the west side of Roches Noires, there is a village of less than 6000 inhabitants, and on the

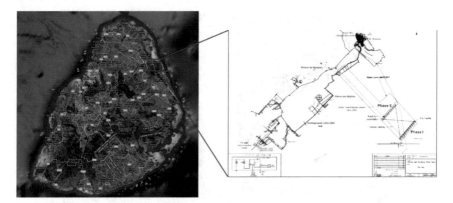

Fig. 1 Location of wind farm at Roches Noires (left: Mauritius map; right: the farm layout)

east side, there are secluded coves of white beaches. The topography is very rocky with a dry and bare land, which makes it inappropriate for agricultural purposes, animal grazing and construction purposes. Hence, Roches Noires is a suitable place for the setting up of a wind farm as with its rough topography and barren land, it will not hamper any future developments nor is a menace to the physical environment [23]. Recently in 2016, a wind farm comprising of 11 wind turbines, each of capacity 850 kW, has been established at Roches Noires with an estimated energy production of 81.9 GWh per year. The geographical location of the wind farm is shown in Fig. 1.

3 Methodology

3.1 The Weather Research and Forecasting (WRF) Model

The WRF model is a numerical weather prediction (NWP) and atmospheric simulation system designed for multiple applications [20]. It is an open source community model that can be used for research as well as operational weather forecasting. The flexible and modular nature of the software allows the continuous development of the code by a broad community of researchers worldwide.

The WRF model consists of two dynamical cores, namely, the Eulerian Mass (EM) core, also known as the Advanced Research WRF (ARW) and the Non-hydrostatic Mesoscale Model (NMM) core. Both are non-hydrostatic Eulerian dynamical cores with terrain-following pressure-based vertical coordinates. The ARW model solves the non-hydrostatic compressible Euler equations. The prognostic variables produced are the horizontal wind velocity components u and v in the Cartesian coordinate, the vertical wind velocity w, perturbation potential temperature, perturbation geopotential and perturbation surface pressure of dry air. WRF comes with a number of physical and dynamical options which fall into several categories, each containing

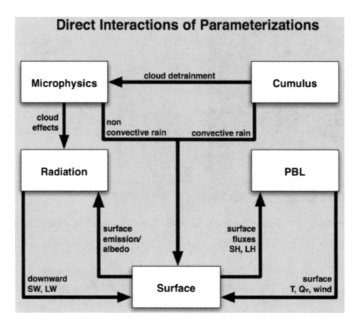

Fig. 2 The interactions between the physics options in WRF [20]

several choices. Physics packages compute tendencies for the velocity components, potential temperature, and moisture fields (Fig. 2). Carvalho et al. [5] have performed a sensitivity study of WRF. Good agreement was observed between modelling results and measured data.

While WRF was mainly intended for weather forecasting applications, it has been widely used by researchers in the field of wind resource assessment due to its simplicity, quality of downscaling and its ability to be coupled with many microscale models. For instance, WRF has been employed in wind-related studies such as wind resource assessment of Dragsha, Kosovo [13], wind resource mapping in Norway [4], mesoscale modelling for the wind atlas of South Africa [14], among others. It has also been used extensively to evaluate wind flow patterns in complex terrains in Portugal [6], Norway [3], Northern China [11]; for climate modelling in Egypt [8], in Australia [9] among many others. Other applications of WRF include the simulation of low-level jets in the central USA [21], wind climate over complex terrain [15], gravity waves [17], extreme winds [16].

3.2 Model Set-up for Mauritius

Figure 3 shows the configurations for the set-up of WPS: Outer domain, d01: 33 × 33 grid points, resolution 27 km × 27 km; First nested domain, d02: 37 × 37 grid

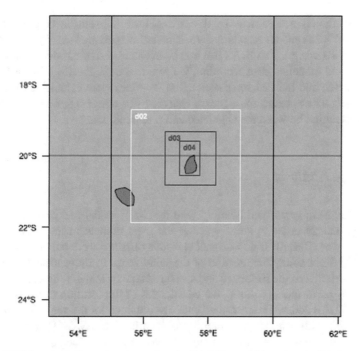

Fig. 3 WPS domain configuration. Outer domain d01: 27 km × 27 km; domain d02: 9 km × 9 km; domain d03: 3 km × 3 km; domain d04: 1 km × 1 km

points, resolution 9 km × 9 km; Second nested domain, d03: 52 × 52 grid points, resolution 3 km × 3 km; Third nested domain, d04: 61 × 91 grid points, resolution 1 km × 1 km. The vertical grid consisted of 41 eta levels with model top at 50 hPa. The lowest 12 of these levels are within 1000 m of the surface and the first level is located at approximately 14 m.a.g.l.

Initial and boundary conditions as well as fields for grid nudging come from NCEP FNL Operational Model Global Tropospheric Analyses, subset data of ds083.2, available at the Research Data Archive (RDA) of the National Centre for Atmospheric Research (NCAR), Boulder, Colorado, USA. The NCEP FNL (Final) Operational Global Analysis data have a resolution of one-degree by one-degree and are prepared operationally every six hours. The analyses are available on the surface, at 32 mandatory (and other pressure) levels from 1000 to 10 millibars, in the surface boundary layer and at some sigma layers, the tropopause and a few others. Parameters include surface pressure, sea-level pressure, geopotential height, temperature, sea surface temperature, soil values, ice cover, relative humidity, u and v winds, vertical motion, vorticity and ozone.

The model was first validated over the island of Mauritius so that it can further be used for modelling weather in Mauritius. The validation of the model was done by comparing the simulated data against the measured data obtained from two inland stations: Vacoas and Plaisance. Vacoas station is situated on the central plateau and

Plaisance station is located on the south-east of Mauritius. From the modelled outputs, data of wind speed and temperature were extracted at 10 m.a.g.l as the measured data was also at 10 m.a.g.l. The BIAS has first been calculated to correct and adjust the measured and modelled data accordingly. From that data, the Root Mean Squared Error (RMSE) and the Pearson's correlation (r-value) were calculated as given in Eqs. (1)–(3). The r-values for all cases analyzed was above 0.6, which indicated a strong correlation between the simulated and measured data.

3.3 Data Analysis

Simulations were performed monthly with the WRF model and evaluated against measured data in order to know whether the local climate in the vicinity of the wind farm was affected. The WRF model results obtained were only for the defined domains and not solely for location of the wind farm. In order to make a correct comparison between the simulated and measured data from wind stations, extraction of data was performed by setting the coordinates of the location of the wind farm using MATLAB codes. The measured data was obtained for the year 2016 and 2017 at 40 m.a.g.l with a temporal sampling of 10 min. The AWS [1] wind resource assessment handbook was used as a guide for data filtering. The data were checked thoroughly for homogeneity, outliers and missing records before being processed for the study. Errors and missing data as well as cyclonic winds (wind exceeding 40 m/s) were eliminated. We hence restricted our study for flow of wind on any normal day. The data was then averaged so that we have temporal sampling of 1 h to match with the WRF output which has temporal sampling of 1 h. The performance of the model is evaluated by using the BIAS, Root Mean Squared Error (RMSE) and the Pearson's correlation (r-value) as given in Eqs. (1) to (3).

$$\text{Bias} = \frac{1}{N}\sum_{i=1}^{N}(M_i - O_i) \tag{1}$$

$$\text{RMSE} = \sqrt{\frac{1}{N}\sum_{i=1}^{N}(M_i - O_i)^2} \tag{2}$$

$$r = \frac{\sum_{i=1}^{N}(M_i - \overline{M})(O_i - \overline{O})}{\sqrt{\sum_{i=1}^{N}(M_i - \overline{M})^2}\sqrt{\sum_{i=1}^{N}(O_i - \overline{O})^2}} \tag{3}$$

In the above equations, N is total number of data used for comparison, M_i is the modelled value and O_i is the observed value. $\overline{M}$ and $\overline{O}$ are the mean values of modelled and observed values, respectively. Bias provides information on the trend of the model to overestimate or underestimate the variable and quantifies the systematic error of the model. RMSE is similar to finding the absolute error in the data but it is more sensitive to occasional large errors due to its quadratic term. In

Eq. (3), a value of $|r| < 0.3$ indicates that there is a weak correlation between the two variables, a value of $|r|$ between 0.3 and 0.6 indicates moderate correlation and a value of $|r| > 0.6$ indicates strong correlation between the datasets.

4 Results, Analysis and Discussions

In this study, WRF simulations were performed for one year before the installation of the Roches Noires wind farm (i.e. for 2015), the installation year (i.e. for 2016) and one year after the installation (i.e. for 2017). Relevant data for the analysis presented in this section was extracted from the WRF output files at 40 m above ground level (m.a.g.l) with a temporal sampling of 1 h.

4.1 Comparison of Modelled and Measured Data

A full statistical analysis (Eqs. 1–3) has been performed to compare the extracted data at Roches Noires (wind farm location) from the model to the measured data for 2016 and 2017. Table 1 shows the results obtained. It should be noted that no measured data for January 2016 and October to December 2017 were available for comparison. It can be seen from Table 1 that the values of bias are all positive which means that the simulated values are an overestimation of the measured ones. The r-values show strong correlation between measured and simulated data of the wind speed for both the years 2017 and 2016. The RMSE values are all below 11% which indicate good accuracy of the model.

4.2 Wind Speed Before and After Installation of Wind Farm

Wind speed data from the simulations were extracted at two locations A and B (see Fig. 4) perpendicular to the wind turbine arrangement that is in the direction of the wind, for the year 2015 (before installation of wind farm) and 2017 (1 year after installation of wind farm) at 10 m.a.g.l. Figure 4 also shows a wind rose which illustrates the direction of wind blowing at Roches Noires. Due to the dimension of the grid size in domain d04, we chose point A to be 12 km downstream the wind farm and point B to be 3 km upstream the wind farm so that they are not in the same control volume of the WRF model.

Figure 5 shows the monthly variations of mean wind speed and direction for 2015 and 2017 at 10 m.a.g.l at point A. It can be observed that in 2015, the mean wind speed is around 7.3 m/s with highest speed in January at 11 m/s and lowest in March at 3 m/s while for 2017, the mean wind speed is 6.6 m/s with maximum speed in April at 9 m/s and minimum in September at 4.2 m/s. It is noted that in 2015, the

Table 1 Bias, RMSE and *r*-value of wind speed at Roches Noires wind farm for 2016 and 2017

Year	2016				2017			
Month	Mean wind speed (m/s)	Bias (m/s)	RMSE (m/s)	*r*-value	Mean wind speed (m/s)	Bias (m/s)	RMSE (m/s)	*r*-value
January	–	–	–	–	6.13	3.05	0.45	0.61
February	6.79	1.69	0.75	0.68	5.10	2.41	0.57	0.70
March	6.50	1.28	0.50	0.81	5.94	2.96	0.23	0.64
April	7.71	3.78	0.82	0.81	6.20	2.61	0.67	0.78
May	7.49	2.75	0.84	0.58	6.28	1.43	0.64	0.89
June	8.89	1.74	0.91	0.85	4.43	1.94	0.25	0.66
July	9.47	2.80	0.53	0.57	5.58	2.40	0.61	0.71
August	9.08	2.26	0.24	0.75	7.47	1.37	0.78	0.79
September	4.71	0.73	0.49	0.65	4.31	0.63	0.49	0.88
October	4.73	2.25	0.52	0.72	–	–	–	–
November	4.91	3.14	0.39	0.54	–	–	–	–
December	5.26	2.01	0.27	0.75	–	–	–	–

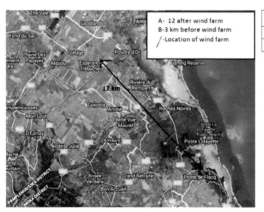

A- 12 after wind farm
B-3 km before wind farm
/-Location of wind farm

	Point A	Point B	W. Farm
Lat	20.05	20.15	20.13
Lon	57.65	57.74	57.72

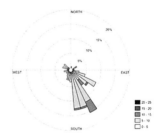

Fig. 4 Chosen locations for wind speed extraction (Points A and B) and wind rose which shows prevailing direction of wind

maximum wind speed was in January for all heights recorded and minimum was in February and March. But after the installation of the wind farm in 2017, the maximum wind speed was recorded in April and minimum in September for all the heights. This demonstrates that wind speed at 12 km away from the wind farm has somehow changed.

Figure 6 shows the monthly variations of mean wind speed and direction for 2015 and 2017 at 10 m.a.g.l at point B. In 2015, the mean wind speed is 7.2 m/s with maximum speed in January at 11 m/s and minimum in March at 3 m/s while for

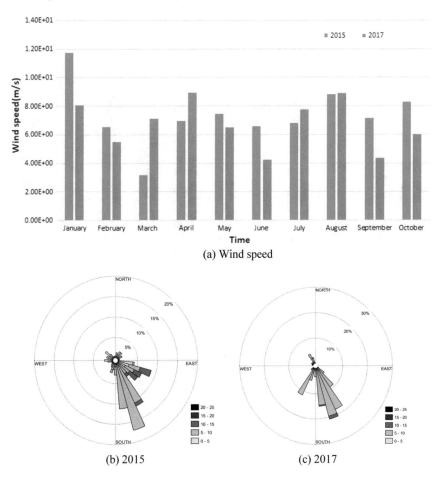

(a) Wind speed

(b) 2015   (c) 2017

Fig. 5 Wind speed and wind roses for the year 2015 and 2017 at 10 m.a.g.l at point A

2017, the average wind speed is 6.6 m/s with maximum speed in April at 8.5 m/s and minimum in September at 4.2 m/s. It is noted that in 2015, the maximum wind speed was recorded in January and minimum in March. But after the installation of the wind farm in 2017, the maximum wind speed was observed in April and minimum in September. Same observation was made for point A. This indicates that wind speed at 3 km upstream the wind farm has also changed.

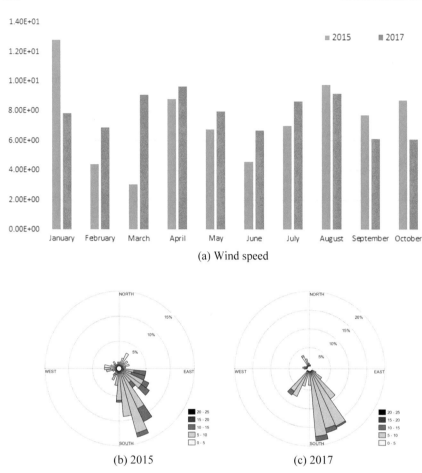

Fig. 6 **a** Wind speed and **b** wind rose for the year 2015; **c** wind rose for the year 2017 at 10 m.a.g.l at point B

4.3 Precipitation Before and After Installation of the Wind Farm

Rainfall data during the day at 2 m.a.g.l for the years 2015 and 2017 were also extracted from simulated output files of WRF at locations A and B and the extracted data are represented in bar charts as shown in Fig. 7.

It can be seen that there was less rainfall in 2017 as compared to 2015. This can possibly be attributed to the effects of the wind farm. Though a more in-depth analysis of the data for a longer period (decades) needs to be done to be able to confirm this argument.

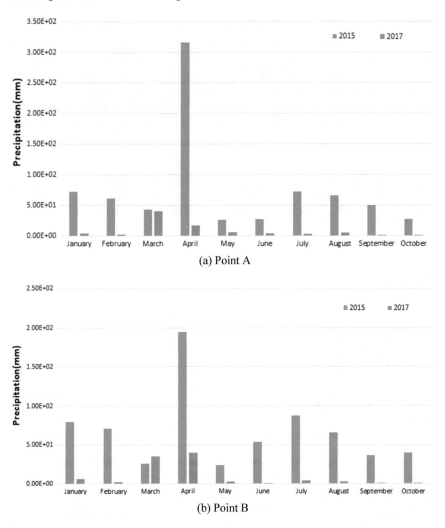

Fig. 7 Precipitation for the year 2015 and 2017 at 2 m.a.g.l at **a** Point A and **b** Point B

5 Conclusion

In this study, the WRF model has been used to study the effects of wind farming on the local weather. In the first instance, the WRF model has been validated by comparing the simulated results with observed data obtained from meteorological stations for the period February 2016–September 2017. The statistical analysis has shown that for the entire period of the simulation there was strong correlation between the simulated data and observed data. To study the effect of the Roches Noires wind farm, installed in 2016, on the local weather, simulations were run for the year

2015, i.e. one year before the installation of the wind farm and the year 2017, i.e. one year after the installation of the wind farm. Wind data were extracted from the model at two locations in the vicinity of the wind farm. One location was chosen upstream and another one was chosen downstream of the Roches Noires wind farm. The quantitative analysis of the extracted data was presented in bar charts. Wind roses were also plotted to see the trend in wind direction before and after installation of wind turbines. It was noted that the occurrence of maximum and minimum wind speed as well as precipitation was different before and after the installation of the wind farm.

Acknowledgements The authors would like to thank the University of Mauritius and the University des Mascareignes for providing facilities for this research. Special thanks are extended to the Tertiary Education Commission for supporting the work of Z. Dhunny in the form of a Postdoctoral fellowship.

References

1. AWS: Wind Resource Assessment Handbook Fundamentals for Conducting a Successful Monitoring Program (1997)
2. Baidya Roy, S., Pacala, S.W., Walko, R.L.: Can large wind farms affect local meteorology? J. Geophys. Res. Atmos. **109**(D19) (2004)
3. Bilal, M., Solbakken, K., Birkelund, Y.: Wind speed and direction predictions by WRF and WindSim coupling over Nygardsfjell. J. Phys. Conf. Ser. **753**(8), 082018 (2016)
4. Byrkjedal, O., Berge, E.: The use of WRF for wind resource mapping in Norway. In: 9th WRF Users Workshop (2008)
5. Carvalho, D., Rocha, A., Gómez-Gesteira, M., Santos, C.: A sensitivity study of the WRF model in wind simulation for an area of high wind energy. Environ. Model Softw. **33**, 23–34 (2012)
6. Carvalho, D., Rocha, A., Gómez-Gesteira, M., Santos, C.S.: WRF wind simulation and wind energy production estimates forced by different reanalyses: comparison with observed data for Portugal. Appl. Energy **117**, 116–126 (2014)
7. Chen, S.: Are wind farms changing the weather in China. South China Morning Post (2010)
8. ElTahan, M., & Magooda, M. (2017). Evaluation of different WRF microphysics schemes: severe rainfall over Egypt case study. https://arxiv.org/abs/1711.04163
9. Evans, J., & McCabe, M. (2010). Regional climate simulation over Australia's Murray-Darling basin: a multi-temporal assessment. J. Geophys. Res. D Atmos. **115**(D14)
10. Fiedler, B., Bukovsky, M.: The effect of a giant wind farm on precipitation in a regional climate model. IOP (2011)
11. Guo, Z., Xiao, X.: Wind power assessment based on a WRF wind simulation with developed power curve modeling methods. Abstr. Appl. Anal. **2014** (2014)
12. GWEC: Wind in numbers | GWEC. [Online] Gwec.net. Available at: http://gwec.net/global-figures/wind-in-numbers/. Accessed 5 March 2018 (2017)
13. Habtezion, B.L.: Wind Resource Assessment in Dragash-kosovo. UNDP, New York (2013)
14. Hahmann, A.N., Lennard, C., Badger, J., Vincent, C.L., Kelly, M., Volker, P.J., Argent, B., Refslund, J.: Mesoscale modelling for the wind atlas of South Africa (WASA) project. DTU Wind Energy **0050**, 80 (2014)
15. Horvath, K., Koracin, D., Vellore, R., Jiang, J., Belu, R.: Sub-kilometer dynamical downscaling of near-surface winds in complex terrain using WRF and MM5 mesoscale models. J. Geophys. Res. Atmos. **117**(D11) (2012)

16. Larsén, X.G., Badger, J., Hahmann, A.N., Mortensen, N.G.: The selective dynamical down-scaling method for extreme-wind atlases. Wind Energy **16**(8), 1167–1182 (2013)
17. Larsén, X.G., Larsen, S., Hahmann, A.N.: Origin of the waves in a case-study of mesoscale spectra of wind and temperature, observed and simulated: Lee waves from the Norwegian mountains. Q. J. R. Meteorolog. Soc. **138**(662, Part A), 274–279 (2012). https://doi.org/10.1002/qj.916
18. Liming, Z., Yuhong, T., Somnath Baidya, R., Chris, T.: Impacts of wind farms on land surface temperature. Nat. Clim. Change **2**, 539–543 (2012)
19. Roy, S.B.: Simulating impacts of wind farms on local hydrometeorology. J. Wind Eng. Ind. Aerodyn. **99**(4), 491–498 (2011)
20. Skamarock, W., Klemp, J., Dudhia, J., Gill, D., Barker, D., Duda, M., Huang, X., Wang, W., Powers, J.: A Description of the Advanced Research WRF Version 3, NCAR Technical Note, Mesoscale and Microscale Meteorology Division. National Centre for Atmospheric Research, Boulder, Colorado, USA (2008)
21. Storm, B., Dudhia, J., Basu, S., Swift, A., Giammanco, I.: Evaluation of the weather research and forecasting model on forecasting low-level jets: implications for wind energy. Wind Energy **12**(1), 81–90 (2009)
22. Wang, C., Prinn, R.G.: Potential climatic impacts and reliability of very large-scale wind farms. Atmos. Chem. Phys. **10**(4), 2053–2061 (2010)
23. Zhou, L., Tian, Y., Roy, S.B., Thorncroft, C., Bosart, L.F., Hu, Y.: Impacts of wind farms on land surface temperature. Nat. Clim. Change **2**(7), 539 (2012)

Driving Behaviour Analysis Using IoT

Aadarsh Bussooa and Avinash Mungur

Abstract This paper addresses the significant problem of dangerous driving and road accidents in Mauritius by constantly monitoring the driver's behaviour and by detecting dangerous driving patterns which could lead to road accidents. The two dangerous driving patterns that were monitored and detected are speeding and overtaking on solid white line. When these patterns are detected, the driver, as well as authorities are alerted. A gyroscope sensor and Global Positioning System (GPS) sensor, connected to a Raspberry Pi, were used to gather data about the motion of the vehicle. An algorithm known as Dynamic Time Warping (DTW) was used to identify where overtaking occurs in real time. The vehicle's speed was obtained from the GPS sensor. These data were sent to a server for processing. The server would subsequently decide whether the detected motion was an offence or not and the client device would be informed in order to alert the driver of an offence being committed.

Keywords Driving behaviour analysis · Real time · Dynamic time warping
Internet of things · Global positioning system · Gyroscope · Raspberry Pi

1 Introduction

Almost every week, radio stations in Mauritius broadcast news of road accidents and the tragic loss of lives that follows as a result. On 28 August 2017, the 100th victim of road accidents, since the start of the year 2017, made headlines in the news, prompting people to drive safely. Speeding, drink driving, indiscipline, by not abiding by traffic regulations, and distracted driving, through the use of mobile

A. Bussooa · A. Mungur (✉)
Department of Information and Communication Technologies, University of Mauritius, Reduit, Moka, Mauritius
e-mail: a.mungur@uom.ac.mu

A. Bussooa
e-mail: aadarsh.bussooa@umail.uom.ac.mu

© Springer Nature Singapore Pte Ltd. 2019
S. C. Satapathy et al. (eds.), *Information Systems Design and Intelligent Applications*, Advances in Intelligent Systems and Computing 863,
https://doi.org/10.1007/978-981-13-3338-5_22

phones while driving, are bad driving habits that Mauritian drivers have developed. Thus, if the bad driving behaviour is detected and corrected at an early stage, then that bad driving behaviour might never become a habit. The main objective of this paper is to constantly monitor the driver's behaviour, gathering data where dangerous driving patterns have been observed and alert the driver as well as relevant authorities of those patterns. This would deter dangerous driving. This effect would be similar to the presence of a police officer on the road while a person is driving.

With the presence of a police officer, drivers are affected with fear, respect and concern [6]. The first instinct of driver is to apply the brakes and slow down. They follow good driving habits which last until the police officer is out of sight. Being under constant surveillance and threatening to alert authorities, the good habit could become long-lasting or even omnipresent. In order to provide a constant monitoring, the concept of Internet of Things is used to gather essential data on the driver's behaviour. The data is processed and analysed by the DTW algorithm to detect whether the vehicle performed an overtaking motion. The processed data is sent to the server for further processing and returns information about the offences committed, mainly speeding and overtaking on solid white line.

This paper is organised as follows: Sect. 2 provides brief literature reviews. Section 3 provides an overview of the proposed system. Section 4 presents the evaluation result of the system and Sect. 5 concludes this paper.

2   Literature Review

This section provides an overview of the related work along with a brief critical analysis of the surveyed paper.

In a survey [3], different methods for detecting driving behaviour were discussed. Researchers have made use of sensors found inside the car, built in smartphones and deployed along roads. GPS, accelerometer, gyroscope, microphone and camera are some of the sensors that were used in their implementation. Their proposed methodologies were to use threshold values, algorithms, pattern recognition techniques, statistical models or neural networks to detect driver's behaviour from the sensor data. Nowadays, smartphones possess a wide array of sensors. It enables the effortless implementation of projects based on driving behaviour analysis, as surveyed by the paper. However, the mere thought of using smartphones to detect dangerous driving seems counter-intuitive. Indeed, using a mobile phone while driving is termed as dangerous driving. Therefore, its use to deter dangerous driving may not be effective since a smartphone would receive calls and notifications that would potentially distract the driver. A possible alternative would be to create a completely separate system such that there is no reliance on a mobile phone, like the use sensors found inside vehicles and on-road sensors, as discussed in the survey. It would ensure dangerous driving prevention and compliance with traffic regulations regarding mobile phone usage.

In a published research article [5], the researchers identified three types of dangerous driving behaviour to be detected namely speeding, irregular driving direction change and abnormal speed control. Speeding detection made use of GPS and accelerometer coupled with a novel speed estimation algorithm to estimate the vehicle's speed. Irregular driving direction change detection used a gyroscope and microphone to record angular speed and audio beeps of the turn signal to detect the irregular events. Abnormal speed control used an accelerometer to detect sudden acceleration and braking. The underlying determination of erratic behaviour is based on a threshold scheme for all the three categories. One dangerous driving behaviour, whose significance is mostly undermined, is the failure to use the turn signal. An audio detection algorithm is used to detect the turn signal sound using the smartphone's microphone. The algorithm coupled with data from the gyroscope sensor determines whether the driver used the turn signal while the vehicle was turning. The paper gives an interesting insight on the use of a microphone sensor in driving behaviour analysis. The paper could be improved by using the microphone to detect engine acceleration sounds and excessive horn use which would show that the driver is dangerous. Moreover, noise level in the car could be sensed to check the driver's distractedness.

In a project thesis [7], driving behaviour analysis using an application developed on an Android smartphone was proposed. The detection of driving behaviour was based on a threshold scheme derived from sample test data. The concept of "gamification" was used to give a score to the driver. However, it concluded that the accuracy of the system using fixed thresholds was low when tested on different road conditions. It proposed the use of an algorithm known as Dynamic Time Warping (DTW) as a solution. The project thesis demonstrated that threshold values could be used to detect left turns, right turns and lane change. However, the use of threshold values is problematic since the detection of driving events might not work with different road layouts. A more thorough approach to the use of threshold values could have been devised. The system could have stored different threshold values for different locations and road conditions. This would have provided a more meaningful usage of the GPS sensor which is being used solely to detect speed.

In the paper [2], smartphone sensors were used to detect dangerous driving. The system, known as MIROAD, relied on the Dynamic Time Warping (DTW) algorithm to detect events. They argued that the algorithm was suitable since vehicles have a limited range of movement. The system was able to detect turns, U-turns, lane changes, acceleration, braking and speeding. It concludes that the algorithm is valid for the purpose of detecting dangerous driving. Experimentation was carried out with different vehicles and different drivers on various road conditions. It concluded that 97% of dangerous driving events were detected. This consolidates the use of the DTW algorithm for the purpose of driving behaviour analysis. However, the paper mainly detected aggressive driving behaviours. Non-aggressive driving behaviours like typical lane changes or overtaking was not detected by their proposed system. Overtaking a vehicle on a pedestrian crossing is a dangerous driving behaviour, but not necessarily an aggressive one. They were unable to distinguish natural lane change movements from noise, claiming that the movement did not generate enough force on the sensor. Moreover, they have put emphasis on the fact that all processing is done on a smartphone. While it preserves privacy and ensures uptime, since no external processing is done and no communication network is required, it lacks some desirable features like the possibility to gather essential data on the behaviour of drivers and to improve the overall system by updating some parts of the program logic.

3 Proposed Solution

This section provides an overview of the proposed architecture and the use of the DTW algorithm to detect driving behaviour. Figure 1 shows the architectural diagram of the proposed system.

1. Both the GPS sensor and the gyroscope sensor were connected to the Raspberry Pi. Data from the GPS sensor was passed through the location parser to extract meaningful information from the raw values of the sensor. Data from the gyro-

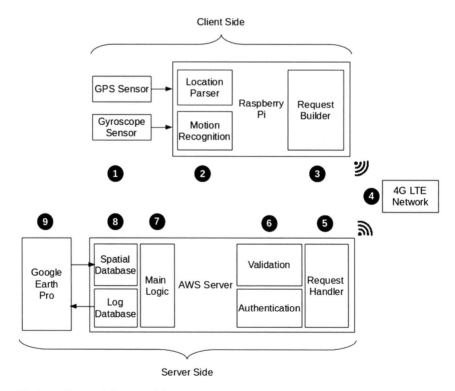

Fig. 1 Architectural diagram of the system

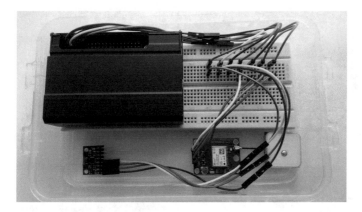

Fig. 2 Client-side hardware implementation

scope sensor was passed through the motion recognition algorithm to identify the motion of the vehicle.

2. The location parser gave out data about the location and speed of the vehicle. Location was in the form of GPS coordinates, namely longitude and latitude. The motion recognition algorithm output whether the vehicle overtook on solid white line or not.

3. The interpreted data, namely the longitude, latitude, speed and overtaking, were passed through the request builder. These information were included in a request which was then transmitted to the server via the wireless 4G modem.

4. The service provider's 4G LTE network provided fast connection speeds to the server. It routed the request via the network to the intended server.

5. The server listened on the default web traffic port 80 for requests. Once a request was received, the server started processing the request.

6. The data sent in the request was validated to prevent invalid data or harmful code injections attempts from being processed by the server. The sender of the request was authenticated with their id and password.

7. The main logic of the processing included the decision-making on whether the parameters received constituted an offence or not. Databases were queried to determine this fact.

8. The spatial database, known as Tile38, was queried to know whether the vehicle was inside a speed zone or a no-overtaking zone. The MySQL log database recorded the offences which have occurred and alerted relevant authorities.

9. Google Earth Pro is the tool which enabled authorities to view the recorded offences. Moreover, it was used to update speed zones and no-overtaking zones with the server's spatial database.

Figure 2 shows the client-side hardware implementation. The device was placed on the floor of the vehicle.

```
try {

    while(true){

        longitude = l.getLongitude();
        latitude = l.getLatitude();
        speed = l.getSpeed();
        overtaking = m.getOvertaking();

        System.out.println(longitude +" :::: "+ latitude +" :::: "+ speed +" ::

        responseArray = r.sendRequest(longitude, latitude, speed, overtaking);
        speeding = responseArray[0];
        overtakingProhibited = responseArray[1];

        System.out.println(responseArray[0] +" :::: "+ responseArray[1]);

        if((speeding == true) && (previousSpeeding == false)){
            Function.alertDriver(true, false);
        } else if((overtakingProhibited == true) && (previousOvertakingProhibit
            Function.alertDriver(false, true);
        }

        previousSpeeding = speeding;
        previousOvertakingProhibited = overtakingProhibited;

        TimeUnit.MILLISECONDS.sleep(500);

    }

} catch (InterruptedException ie){
    System.out.println("[\033[31mError\033[0m] (InterruptedException occurred d
}
```

```php
//////////////////////// Client Response ////////////////////////
$speeding = false;
$overtakingprohibited = false;

if($response1_array["count"] == 0 && $response2_array["count"] == 0){

    // Do nothing

}else if($response1_array["count"] == 1 && $response2_array["count"] == 0){

    if($overtaking === 'true'){
        $overtakingprohibited = true;
    }

} else if($response1_array["count"] == 0 && $response2_array["count"] == 1){

    $speed_limit = $response3_array["fields"]["speed"];

    if($speed > $speed_limit){
        $speeding = true;
    }

} else if($response1_array["count"] == 1 && $response2_array["count"] == 1){

    if($overtaking === 'true'){
        $overtakingprohibited = true;
    }

    $speed_limit = $response3_array["fields"]["speed"];

    if($speed > $speed_limit){
        $speeding = true;
    }

} else {

    $json_array = array("ok" => false, "error" => "Invalid count obtained");
    echo json_encode($json_array);
    exit(0);

}
```

Fig. 3 Client-side software implementation (left). Server-side software implementation (right)

Figure 3 (left) shows the main logic of the client-side software implementation. The client-side device continuously gathered data from the sensors and sent them to the server. It then waited for a response to decide on whether to alert the driver of an offence committed.

Figure 3 (right) shows the main logic of the server-side software implementation. Based on queries made to the spatial database and the responses that followed, the server determined whether an offence was committed and sent a response back to the client.

3.1 Motion Recognition Algorithm (Client-Side)

A gyroscope sensor was used to detect overtaking. However, the values of the gyroscope sensor would have to be interpreted, to extract meaning or features and attempt to identify whether overtaking occurred at a particular time. Such interpretation was achieved through time-series analysis. More specifically, the system measured the similarity between a live data stream sequence from the sensor and a pre-recorded data sequence of an overtaking motion.

The ability of the DTW algorithm, to match time-series sequences that are misaligned and stretched in the time domain [1] is especially useful in the context of detecting overtaking. Indeed, it allowed the detection of overtaking no matter how fast or slow the vehicle is moving.

The DTW algorithm can be summarised as follows:

1. Consider a time-series sequence A of x amplitude values $(x_1, x_2, x_3, \ldots)$
2. Consider another time-series sequence B of y amplitude values $(y_1, y_2, y_3, \ldots)$

3. Construct an x times y matrix M where $M_{i,j} = | x_i - y_i | + \min (M_{i-1,j}, M_{i-1,j-1}, M_{i,j-1})$

4. $M_{x,y}$ is the DTW distance or dissimilarity score, with value zero being a perfect match.

The proposed solution used an implementation of the DTW algorithm from the Gesture and Activity Recognition Toolkit [4] which contains development tools for building gesture-based applications.

3.2 Real-Time DTW Algorithm Design (Client-Side)

An extra layer over the DTW algorithm enabled the use of the algorithm in real time as new data was recorded by the system.

A queue was created to store the motion sensor values. A queue has the ability to en-queue and de-queue values from itself. This was useful in creating a First-In First-Out (FIFO) stream.

An example of how the real-time DTW algorithm works is illustrated in Fig. 4. With each iteration of the client-side main logic, old data is polled from the start of the queue, the most recent data is added to the end of the queue and the DTW algorithm is applied to the data which is now found in the queue.

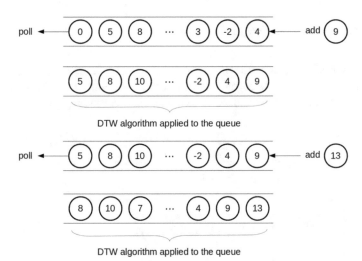

Fig. 4 The real-time DTW algorithm using a queue

3.3 Method of Operation (Client-Side)

The sensor was placed on the floor of the front passenger area of the vehicle. The sensor was placed in such a way that its axes of measurement aligned with that of the vehicle. However, only the z axis was considered in this paper. Therefore, the gyroscope sensor only measured how much the vehicle steered left or right.

The z-value of the gyroscope sensor was used to record the overtaking motion of a vehicle for a duration of 5 seconds. The recorded pattern, as shown in Fig. 5, was saved in a file that was accessible by the client-side software. The real-time DTW algorithm matched incoming data from the gyroscope with that of the recorded pattern. If a dissimilarity score, below an arbitrarily defined threshold, was obtained, it concluded that overtaking did occur. The motion recognition algorithm therefore output either true, if overtaking occurred, and false, if overtaking did not occur.

To determine the speed and location of the vehicle, a Global Positioning System (GPS) sensor was used. The sensor was interfaced with a Raspberry Pi. Raw GPS data requires processing to obtain relevant information. A location parser, in the form of a Linux software package known as gpsd, was used to parse the raw GPS data.

Furthermore, request parameters (in the form of an HTTP request), were sent to the server roughly every second containing the credentials, the Boolean showing whether overtaking occurred, the location of the vehicle and the speed at which the vehicle is moving.

3.4 Method of Operation (Server-Side)

Although the GPS coordinates of the boundaries of the no-overtaking zones and speed zones could have been recorded manually by going at the location of the zones, Google Earth Pro was used to ease that time-consuming process. Satellite

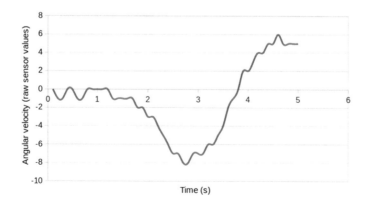

Fig. 5 Graph of recorded overtaking motion

Fig. 6 Creating a speed zone using the polygon tool in Google Earth. The speed limit of the speed zone is specified in the description

imagery provided by the software was used to identify speed zones, lane markings, speed breakers and pedestrian crossings.

The zones were created using the polygon tool in Google Earth Pro, as shown in Fig. 6. The software stores the coordinates of all the zones in a Keyhole Markup Language (KML) file which is a type of Extensible Markup Language (XML) file. The KML file was uploaded to the server where a script parsed the file and stored the GPS coordinates in the Tile38 database.

A spatial database known as Tile38 was used to store and manipulate the GPS coordinates. The database stored the current location of the vehicle together with multiple GPS coordinates, forming a polygon, describing the perimeter of the different zones. To determine whether the vehicle lay in a zone, the "INTERSECTS" command was used and the zone that the vehicle intersected with was returned.

Moreover, for each HTTP request sent from the client, a response was issued by the server. The response contained two Booleans stating whether the overtaking, on solid white line, offence occurred or the speeding offence occurred.

4   Results

The proposed architecture was evaluated in road traffic conditions. The test road was suitable for the evaluation since it did not have much traffic, was wide enough and had only two lanes, one for outgoing traffic and the other for oncoming traffic. Multiple accident-prone areas were identified and recorded in Google Earth Pro. As the vehicle drove along the road, the driver was alerted to reduce their speed

Fig. 7 Vehicle overtaking on solid white line (left). Corresponding offence, displayed as yellow dot, shown on Google Earth Pro among other offences (right)

Fig. 8 Vehicle approaching pedestrian crossing too fast (left). Corresponding offence, displayed as red dot, shown on Google Earth Pro among other offences (right)

when approaching the identified accident-prone areas. Since it would have been an offence to overtake moving vehicles on solid white line, the proposed system was instead evaluated by overtaking stationary vehicles. Figures 7 and 8 demonstrate two scenarios of where the system worked and how the offence was shown on Google Earth Pro.

When the vehicle crossed the solid while line and performed a right lane change, the DTW algorithm matched closely with the pre-defined template, as shown in Fig. 5, and a low dissimilarity score was obtained. The overtaking parameter returned true and an HTTP request was sent to the server. The script on the server updated the location of the vehicle in the spatial database and determined that the vehicle was in a no-overtaking zone. The incident was logged in the database which was subsequently displayed on Google Earth Pro, as shown in Fig. 7, while the driver was alerted by an audio signal at the same time.

When the vehicle approached the pedestrian crossing, the GPS sensor calculated the speed of the vehicle at about 35.3 km h^{-1}. An HTTP request was sent to the server and the location of the vehicle was updated. The server script determined that the vehicle was in a speed zone and retrieved the speed limit. Since the vehicle's speed was greater than the speed limit, the incident was logged in the database and displayed on Google Earth Pro, as shown in Fig. 8, while the driver was alerted by an audio signal at the same time.

5 Conclusion

This paper addresses the significant problem of dangerous driving and road accidents in Mauritius. It gives insights on trends in car safety and ongoing research in the field. This paper has investigated and evaluated the use of different methods and tools to analyse driving behaviour. An elaborate system was designed and evaluated in road traffic conditions. The positive results reinforced the validity of the DTW algorithm in driving behaviour analysis. This paper has also demonstrated how the algorithm could be used in the context of Internet of Things (IoT) where some of the program logic is stored in the cloud.

The proposed system in this paper is able to constantly monitor and alert the driver of offences they are committing. A possible application of the proposed system is in driving schools where the driving skills of learners are assessed. Moreover, if the proposed system is equipped on vehicles driving on everyday road traffic, the live feed of incidents on Google Earth Pro would enable authorities to identify regular road traffic offenders and strategic locations for the effective planning of police checkpoints.

References

1. Berndt, D., Clifford, J.: Using dynamic time warping to find patterns in time series. In: Workshop on Knowledge Knowledge Discovery in Databases, vol. 398, pp. 359–370. Available at: http://www.aaai.org/Papers/Workshops/1994/WS-94-03/WS94-03-031.pdf (1994)
2. Johnson, D.A., Trivedi, M.M.: Driving style recognition using a smartphone as a sensor platform. In: ITSC, Proceedings of the IEEE Conference on Intelligent Transportation Systems, pp. 1609–1615. https://doi.org/10.1109/itsc.2011.6083078 (2011)
3. Kalra, N., Bansal, D.: Analyzing driver behavior using smartphone sensors: a survey. Int. J. Electron. Electr. Eng. 7(7), 697–702 (2014)
4. Lyons, K., Brashear, H., Westeyn, T.: GART: the gesture and activity recognition toolkit. Electronics 4552, 718–727 (2007). https://doi.org/10.1007/978-3-540-73110-8_78
5. Ma, C., et al.: DrivingSense: dangerous driving behavior identification based on smartphone autocalibration. In: Mobile Information Systems. https://doi.org/10.1155/2017/9075653 (2017)
6. Paulas, R.: The Psychological Impact of Driving Among Police Cars, Pacific Standard. Available at: https://psmag.com/news/bad-boys-bad-boys-what-you-gonna-do. Accessed: 3 Nov 2017 (2015)
7. Stoichkov, R.: Android smartphone application for driving style recognition. In: Lehrstuhl fur Medientechnik, Technische. Available at: http://www.eislab.fim.uni-passau.de/files/publications/students/Stoichkov-Projektarbeit.pdf (2013)

Exploring the Internet of Things (IoT) in Education: A Review

Dosheela Devi Ramlowat and Binod Kumar Pattanayak

Abstract Over the years, with the evolution of the global Internet, global communication among people of the world has become a reality. Nevertheless, with the emergence of the Internet of Things (IoT), human-to-human communications have been transformed into communication between everything and everything that refers to human-to-human, human-to-machine and machine-to-machine communications altogether. As a result of it, IoT has been able to capture all spheres of human life like security and privacy provisioning, health care, recycling, environmental monitoring and so on. In this paper, we address the significance of IoT technology in the field of education in improving the efficiency of teaching and learning. Here, we discuss its implementation from various aspects of education like computer science education, medical education, distance education, consumer green education and so on.

Keywords IoT · Networks · Communication · Education

1 Introduction

The Internet's evolution and growth are said to be exponential. In the past 25 years, the Internet has expanded itself thereby linking people of the world through computers, laptops, smartphones and other devices. Today with the evolution of the global Internet, a wide range of devices such as home appliances, cars, different electrical equipment as well as varieties of smart devices could communicate using the services of the Internet as well thus creating the Internet of Things (IoT). IoT makes it possible for different real-world objects referred to as "things" to communicate among themselves on the global Internet using Internet Protocol (IP) enabled ser-

D. D. Ramlowat (✉)
Department of Software Enginering, Universite des Mascareignes, Rose-Hill, Mauritius
e-mail: dramlowat@udm.ac.mu

B. K. Pattanayak
Siksha 'O' Anusandhan Deemed to be University, Bhubaneswar, India
e-mail: binodpattanayak@soa.ac.in

© Springer Nature Singapore Pte Ltd. 2019
S. C. Satapathy et al. (eds.), *Information Systems Design and Intelligent Applications*, Advances in Intelligent Systems and Computing 863,
https://doi.org/10.1007/978-981-13-3338-5_23

vices with the help of wired or wireless communication networks. These things are capable of sensing the environment around them and acting upon it autonomously, thus transforming the physical world around them into a very large information and knowledge base. Several terms have originated from the literature on IoT technology namely the Internet of Everything (IoE), the Internet of Anything (IoA), the Web of Things, Machine-to-Machine communication or the Industrial Internet of Things (IIOT).

Moreira et al. [1] stated in their paper that IoT has no unanimously recognised specific definition and provided some interesting definitions as defined by various authors available in the literature. Ray et al. [1] defined IoT "as a set of connected physical objects—or things" consisting of electronics, software, sensors that permit the "things" to collect and disseminate data. The authors go further and present a more systematic definition of IoT and refer to it as "an ecosystem" that is able to scale and exploit the existing infrastructure of "embedded and connected devices". Bota et al. [1] further extend the above definition by stating that the idea behind this is the "ubiquitous presence of objects/things" in the user's environment and which can act autonomously in the environment in which they are utilised. Perera et al. [1] present IoT as an omnipresent "technology that allows individuals and things to be connected" via any available "network or service". O'Brien [1] refer to IoT as "a technology that allows, through sensors, to connect objects with the Internet". This further triggers data collection from different sources, which can be processed to provide feedback and better monitoring of the environment in which these "objects" operate. Atzori et al. [1] considering the interdisciplinary nature of the technology, present a definition based on the cross-referencing of three paradigms namely an "Internet"-oriented paradigm (middleware), a "Things"-oriented paradigm (sensors), and a "Semantics"-oriented paradigm. Sundmaeker and Saint-Exupéry [1] defined IoT as "a dynamic global network infrastructure with self-configuring capabilities based on standard and interoperable communication protocols". Objects or devices operate seamlessly within this "dynamic global network" and possess "identities, physical attributes" and use "intelligent interfaces". According to Xia et al. [1], IoT refers to the "networked interconnection of everyday objects, which are often equipped with ubiquitous intelligence".

IoT covers most of the spheres of human life. In this paper, the focus is on reviewing the implementation of IoT in the field of education and the challenges associated with it. Section 2 includes a description of IoT protocol architecture. In Sect. 3, some examples of IoT applications are detailed. Implementation of IoT in the field of education in different paradigms are covered in Sect. 4. Section 5 concludes the paper with possible future work.

2   IoT Architecture

The IoT architecture in Fig. 1 consists of a set of layers which determine the specific technologies used from the lower layers responsible for managing the sen-

Fig. 1 IoT architecture

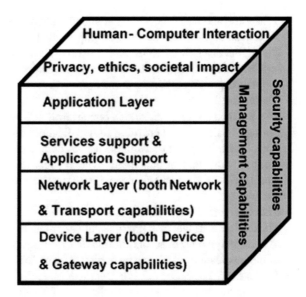

sors/actuators and network operations to the upper layers responsible for application and presentation of data. The architecture also consists of some cross-cutting vertical layers and horizontal layers such as security management and human–computer interaction, respectively. The application layer represents an interface to the IoT applications. Service support and application support layer assist some generic as well as some specific applications. Network layer performs the tasks associated with network and transport layers in the context of the Open Systems Interconnection (OSI) model of communication. Device layer deals with the functionality of the physical layer.

3 IoT Application Domains

As the technology of IoT matures, processing as well as communication capabilities tend to be much more accessible and at the same time, versatile. This gives rise to increasing interconnectivity of devices and multidisciplinary applications of IoT. Furthermore, the decreasing size and cost of sensor technology coupled with improved Internet connectivity and other networking technologies will fuel the development and implementation of IoT applications in the future.

- *Smart Home*
  Energy consumption can be measured by IoT systems and transmitted to the energy provider electronically. A smart home can be set up by using a combination of sensors and actuators to measure temperature, humidity, light and other parameters. One example is the optimisation of energy usage for the purpose of house cooling

or heating, and which hence automatically adjusts the comfort level of the house. Moreover, the smart home will be equipped with smart appliances such as a fridge having the capabilities to reorder groceries automatically or a person entering a building with a smartphone that can subsequently activate his/her preference for setting the profile pertaining to the home [2].

- *Healthcare*
  The healthcare sector can largely benefit from IoT technology in the remote monitoring of patients or in enabling ad hoc diagnosis in the case of emergency situations. There is a range of available technologies and wearables that can be used to measure vital signs of patients like blood sugar level, body temperature, cholesterol level and so on and hence facilitate storage and access of health records of a patient [2].
- *Pharmaceutical*
  The use of smart labels in pharmaceutical products would not only allow patients to know about dosage and expiration date of drugs but also verify the authenticity of the medication which can be achieved using IoT services [2].
- *Retail, Logistics, Supply Chain Management (SCM)*
  The benefits that IoT can bring in the field of Retail and SCM are enormous. With the implementation of RFID-equipped items and smart shelves, it would be easier for retailers to track the stock availability and replenishment of goods in real time or even detection of shoplifting. Other advantages that IoT would bring to the retail sector would be effective check-outs in retail shops, customised shopping experience for customers and personalised marketing strategies [2].
- *Safety, Security and Privacy*
  Applications of IoT in the field of security are numerous. Some examples are environment surveillance with regard to natural disasters, monitoring of water leaks, gases, vibrations, fire, unauthorised entry, vandalism in buildings and in secure equipment surveillance and payment systems [2].
- *Environment Monitoring*
  IoT can contribute in developing environmentally friendly programs, for example by the integration of RFID and Wireless Sensor Networks (WSNs) in applications such as remote data sensing in forest fire, earthquake detection, or air pollution [2].

In addition to the above, IoT can also be usefully implemented in the field of education which is detailed in the next section.

4   Related Work on IoT in Education

The field of education also is expected to undergo massive transformation due to IoT. The education sector is considered as a strong candidate for the application of IoT systems. In recent years, IoT technology has led to the consolidation of educational resources in the form of scalable, rich-media content repositories. Furthermore, a full spectrum of research work has been conducted in the field of IoT in education.

In this section, some research applications on the impact of IoT in education are presented. The domain of the various research work conducted, and method-ologies/tools used, have been analysed. Implementation results of the various IoT educational projects as well as future scope have been discussed.

4.1 IoT Courses and Tools in Computer Science Education

Burd et al. [3] highlighted the main motivations and challenges that post-secondary educators in the field of computer science and information technology might face in teaching IoT. IoT technology is described as being pervasive in nature as it touches various contemporary computing topics such as embedded systems, communication networks, client–server architecture, cloud computing, service-oriented architecture-based systems and human–computer interaction. IoT is predicted to have a significant impact on today's major industrial applications. Hence, it is imperative that students learn to analyse, develop and maintain such systems in ethical and secure ways by getting exposure to new knowledge areas.

The authors listed a set of challenges that need to be considered with regard to teaching IoT. In their paper, the authors focused on the following issues:

- How can the concepts of IoT and the related technologies be incorporated into existing curricula?
- How can the software and hardware concepts related to IoT technology be integrated together?

Various methods and tools were used for data collection. The scope of the search was restricted to formal education in the development of IoT systems at post-secondary level education, especially at the undergraduate level. Online courses and aggregators (e.g. Coursera, Class-Central) as well as online articles posted on Google Scholar, the ACM Digital Library and the IEEE Xplore Digital Library were searched for the identification and reviewing of several courses from U.S., Swedish and Norwegian universities. In an attempt to obtain qualitative data that would give a deeper insight into the subject, six semi-structured interviews were conducted. In addition to this, analysis of course documents and other related academic document were carried out. After analysis of data collected, the authors came to the conclusion that, first, a broad introduction to IoT concepts in a single course and secondly, a focused course in the capacity of an IoT specialisation, would be the natural way of integrating a completely new topic with the existing curriculum. Some other issues that must be addressed are students' existing knowledge in IoT, topic and tools selection for teaching, lab space and technology challenges including network access. Future challenges that need attention in the field of IoT education are namely needed for training and teamwork and model curricular guidelines, democratisation of access to IoT equipment and a rapidly evolving technology.

4.2   Current Status and Overview of IoT in Education

Moreira et al. [1] also identified in their work the potential challenges of IoT in education. They discussed the integration of IoT and open data in school books and which was referred to as "smart books" in the paper. Smartbooks would be the driver behind personalisation of education by transforming the conventional school textbook from a mere resource to a service capable of facilitating other methods of interaction like "augmented reality or mobile computing" thereby enriching the educational contents belonging to the book. The authors further presented a set of open data sources along with possible scenarios in order to integrate these data into the manuals of institutions. As claimed by the authors, Johnson et al. refer to the NMC Horizon report 2012 in which it was stated that IoT would take the evolution of "intelligent objects" to the next level as the boundary between physical objects and digital information would be blurred by their interconnectivity. This ease of interconnectivity in IoT can be harnessed in the field of education by allowing smaller electronic equipment to be connected in a very subtle way and then use them to track, monitor, maintain and record data on the objects. Schools can use the data captured to monitor, track and perform inventory related activities of the facilities and objects and at the same time, grant access to specific locations for students, teachers and other staff members. The NMC Horizon Report 2015 coined the term "hyper situation", which is described as the "ability to amplify knowledge based on the user's location, contextualising it from its geo-location". The authors lay emphasis on the importance of creating a social and technical ecosystem which would comprise of hardware, data and associated contents along with relevant services. Such an ecosystem would facilitate access to and interpretation of data, thus empowering students to act on these findings and solve real-world challenges. Most researchers agree that the main challenges arising from the application of IoT in education would be security and privacy especially with regard to the increased accessibility to credentials, location and learning history of students as well as data storage and costs associated to management of resources such as data, hardware, and software as well as scholar equipment. Training of teachers cannot be neglected as they would be the key instruments of change and they would be called on to put in place new educational practices and guiding principles for effective use of IoT technology. For an effective exploitation of the advantages that IoT present and also, better prepare all stakeholders to face the challenges related to IoT, it is desired that an awareness must be created at different levels, namely government, educators and society in general so as to increase access to connectivity, networks and technology.

4.3   IoTFLiP Platform for Medical Education

Maqbool et al. [4] proposed in their paper an "IoT-based Flipped Learning Platform (IoTFLiP)" that is principally intended for medical learning which could also be extended to other domains as well. IoTFLiP uses IoT resources and infrastructure for supporting Case-Based Learning (CBL) in a flipped learning environment. IoT-FLiP is compared to Interactive "Case Based Flip Learning Tool (ICBFLT)", a flipped case-based learning platform running in a cloud environment. Whilst ICBFLT was designed with security, privacy and delivery of personalised data in mind, it lacked support for common and conventional cases of patients. Therefore, the main objective of IoTFLiP was to overcome this main weakness of ICBFLT whilst extending the services that are mostly on-demand, sharing of knowledge and preserving the privacy of data. The IoTFLiP platform consists of eight layers that are conceptually split into two separate blocks with respect to the mode of communication and relevant resources. The blocks are called the local and cloud processing blocks. The local block consists of four layers namely Data Perception, Data Aggregation, Local Security, and Access Technologies which is capable of handling the communication as well as the respective resources locally whilst the cloud block includes Cloud Security, Presentation, Application and Service, and Business layer. IoTFLiP has three access levels namely Administrator, Expert and Student. The medical expert records patient history. The patients' vital signs such as blood pressure, blood glucose are recorded using IoT wearable devices and transmitted via a virtual network. The medical expert is then able to visualise and analyse the patients' vital signs after which a CBL case is generated and fed into the ICBFLT tool. On the other hand, medical students have access to ICBFLT to interpret and solve CBL case. The medical expert evaluates students' CBL solutions and provides feedback accordingly. The medical knowledge is saved in the CBL Case Base and finally, feedback is sent to medical students so that they are able to improve their skills and knowledge.

4.4   Consumer Green Education and Green Marketing via IoT

Tu et al. [5] used IoT as a supporting technology to determine, whether green marketing has an impact on people's thoughts and behaviour and consequently, on green education. Green education targets at teaching people about environment protection and sustainable development. Green marketing allows an enterprise to target its business marketing strategies whilst at the same time promoting sustainable environmental protection. Interactive advertising includes an element of feedback to its target audience thus allowing products or services to be more effectively customised. Qualitative analysis was conducted in the first instance using case study and focus group. The statistical results of questionnaires were then analysed using factor based analysis along with a regression in order for determining the effectiveness of an IoT-based green marketing on consumers' green education. As the results of regression analysis depict, there is the high influence of advertising attitude upon the purchasing intentions of a consumer. Green attitude has the same impact to purchasing intention as advertising on purchasing attitude. The following hypotheses held using single factor variance analysis were valid:

1. Education is the most important factor that significantly affects the consumers' green attitude.
2. Green attitude is supposed to leave a high impact on consumers' purchasing intentions.
3. Advertising attitude of a consumer can significantly affect the purchasing intentions with respect to green products.
4. Green marketing using the services of IoT would be the new approach towards green education.

4.5 Interrelations of Cloud Computing, Connectivism and IoT

Sarıtaş [6] investigated the relationships between three emerging technological innovations namely, Cloud Computing (CC), Internet of Things (IoT) and the learning theory "Connectivism". Connectivism, refers to "a new learning theory" proposed by Siemens (2005) that is presented as an "alternative and effective theory" using cloud computing technology to support learning and teaching. The potential impact of the above three emerging technologies in education is also discussed here. Cloud computing combined with IoT has given rise to new learning paradigms and teaching experiences. It shifts from the traditional practice of teaching to a peer-reviewed transparent process, which is more appropriate to the new generation of students who are extremely conversant with the use of digital media and technologies. However, there are few challenges which need to be addressed. Cloud Computing combined with IoT would demand new teaching methodologies which need to be effective and at the same time, compatible with newly explored networked learning activities. The authors recommended that educational institutions prepare a long-term strategy capable of focusing on curricula development, professional development of teachers, educational philosophy, data security, legal and political issues, and transformation of resources and infrastructure, in order to address the various challenges that would come with new technological revolution.

4.6 IoT Manpower Training Using HOPPING and ESIC

Cha et al. [7] argued that the potential of IoT technology can only be harnessed if professional manpower is available and which would only be possible through education and developing the appropriate skills of future generations. The aim of this study was to design a systematic educational program called "HOPPING", which is expected to provide effective and specialised training courses in IoT. The HOP-

PING curriculum was structured into firstly, a *Hierarchical course* which aimed at maximising the efficiency of education as well as the convergence of it by virtue of the adoption of systematic curricula, step by step learning process and horizontal education operating system. The second component was an *Ordered course* which determined the course design and project topics based on the industry demand. This approach enabled students to develop both their theoretical and practical skills by working on real-world projects. Thirdly, a *Practical course* aimed at filling in the mismatch between educational programs and real needs of the industry making use of field training along with start-up operations. "ING" provided a common platform among prospective employees, potential start-up companies and other stakeholders for innovation and idea generation. "ESIC" program was split into four separate types namely, Education (E), Start-ups (S), Incubator (I), fieldwork and Career (C) and provided support from experts in each field. Trainees joined the ESIC program and went through the different stages which prepared them to adapt to the workplace effectively. After selection of an enterprise, the job support management team and education program evaluation team used the PTS (Project Training System) and the ECMS (Employment Curriculum Management System) to target job-seekers and business start-ups. Evaluations were carried out in various forms such as quiz, interview, examination, practical test and results were fed into a feedback system. The Feed-Back System (EFBS) aimed at improving the learning ability of the trainees, reducing the drop-out rate and number of inefficient learners. The evaluation system broadly rated the academic level of students, the ability for problem-solving, collaboration and teamwork skills.

4.7 IoT in Distance Education

The authors in [8] address the issue of construction of distance education classroom based on IoT. The basic components involved in it are ZigBee/GPRS wireless technology, sensor networks, embedded systems along with web distributed software as well as database systems. A scheme of distance teaching experiment system platform based on IoT proposed by the authors is oriented around an experiment terminal group (ETG) that comprises of various terminal systems in the capacities of nodes in a ZigBee network. Here a dedicated teaching server is made to communicate with ETG via the GPRS network. A student client and teacher client systems can communicate with the teaching server either by a wired or a GPRS based wireless network. As claimed by the authors, this IoT-based system is associated with the following advantages:

1. Scores of students can be significantly enhanced for the reason that students can achieve vivid as well as visual learning that makes them acquire the contents with a high level of understanding.
2. This system can convincingly enhance the effect of teaching from teachers' point of view for the reason that it incorporates multiple methods of information trans-

mission, remote interaction with students and convenient data as well as resource sharing facilities.

4.8   IoT in Education Model

Bagheri et al. [9] address an education business model wherein they focus on four important aspects such as energy management along with ecosystem monitoring, students' healthcare monitoring, access control of classroom and improvement of teaching–learning processes. In their work, the authors carry out an extensive comparison of methods and processes involved in traditional and modern education business models and conclude that in modern scenario with the introduction of IoT technology into this business model, a set of advantages like reduction in cost, enhancement of comfort, saving time, enhancement of safety, exploring personalised learning and increasing student collaboration can be achieved.

4.9   Green IoT in Engineering Education

The aspects relating to the implementation of green IoT methodology in engineering education in order for deploying smart classrooms are addressed by authors in [10]. A focus is made on the sustainability of resources of IoT in engineering education thereby encouraging green behaviour to be exhibited by all the stakeholders of the institution. In order to achieve this, a set of tasks need to be taken up as claimed by the author. These tasks are: (i) optimal utilisation of resources; (ii) disposal, reuse and recycling of IoT resources and, (iii) creation of awareness of green IoT among the stakeholders of educational institutions. Performing these tasks would significantly enhance the sustainability of future IoT resources.

5   Conclusion and Future Work

In this paper, the variations in the implementation of IoT technology in the field of education have been addressed. Emphasis was set on the innovations proposed by various authors relating to the implementation of IoT resources in the fields of education such as computer science education, distance education teaching and learning, medical science education as well as consumer green education. The interrelations between IoT and cloud computing paradigm in the education sector has also been discussed. It can be concluded that using the resources as well as facilities of IoT technology, the efficiency of teaching and learning process can be significantly enhanced. Furthermore, investigations can also be carried out in the context of some innovative tools specific to IoT that can be useful in education as a whole.

References

1. Moreira, F.T, Magalhães, A., Ramos, F., Vairinhos, M.: The power of the internet of things in education: an overview of current status and potential. In: Proceedings of Conference on Smart Learning Ecosystems and Regional Development, pp. 51–63 (2017)
2. European Research Projects on the Internet of Things (CERP-IoT): Internet of Things Strategic Research Roadmap (2009)
3. Burd, B., Barker, L., Divitini, M., Perez, F.A.F, Russell, I., Bill, S., Tudor, L.: Courses, content, and tools for internet of things in computer science education. In: Proceedings of the ITiCSE Conference on Working Group Reports, pp. 125–139 (2017)
4. Maqbool, A., Hafiz, S.M.B., Muhammad, A.R., Jawad, K., Sungyoung, L., Muhammad, I., Mohammad, A.T.C., Soyeon, C.H., Byeong, H.K.: IoTFLiP: IoT-baseFlip learning platform for medical education. In: Proceedings of Digital Communications and Networks, pp. 188–194 (2017)
5. Tu, J., Chen, Y., Chen, S.: The study of consumer green education via the internet of things with green marketing. In: Proceedings of International Conference on Applied System Innovation (ICASI), pp. 6133–6145 (2017)
6. Sarıtaş, M.T.: The emergent technological and theoretical paradigms in education: the interrelations of cloud computing (CC). Connectivism Internet Things (IoT) **12**(6), 161–179 (2015)
7. Cha, J.S., Kang, S.K.: The study of a course design of IoT manpower training based on the HOPPING education system and the ESIC program. Int. J. Softw. Eng. Appl. **9**(6), 71–82 (2015)
8. Yang, Y., Kanhua, Y.: Construction of distance education classroom in architecture specialty based on internet of things technology. Int. J. Emerg. Technol. Learn. (IJET) **11**(5), 56–61 (2016)
9. Bagheri, M., Movahed, S.H.: The effect of internet of things (IoT) on education business model. In: Proceedings of 12th IEEE Conference on Signal-Image Technology and Internet-Based Systems, pp. 435–441 (2016)
10. Maksimovic, M.: Green internet of things (IoT) at engineering education institution: the classroom of tomorrow. Infotech-Jahorina **16**, 270–273 (2017)

Evaluation of Cloud Computing Adoption Using a Hybrid TAM/TOE Model

Aatish Chiniah, Avinash E. U. Mungur and Krishnen Naidoo Permal

Abstract With the expansion of businesses and the advent of the Internet, the moderately simple client–server architecture evolved and became more complex. The evolution saw the birth of tier levels; namely tier-two, tier-three and tier-four architectures. As a result, it is sound to assume that the cost of implementation and maintenance of such system would definitely grow exponentially. Many companies and businesses thought about having the IT part relegated to backend so that they may concentrate fully on developing their core-business. But unfortunately IT usually consumes a major part of business activity and capital since most business nowadays require the IT medium to compete on the market. Although cloud computing adoption is still very slow and restricted to certain businesses who are willing to make the shift, many companies believe that cloud computing may offer a new competitive model that may reduce costs and complexity while increasing operational efficiency. In this work, we aim at evaluating the already known factors for cloud adoption/non-adoption by the ICT sector of Mauritius. To this end, we opt to use a Hybrid Technology Acceptance Model (TAM) and Technology-Organization-Environment Model (TOE) as they complement each other. We surveyed 93 ICT related companies/organizations. We also developed a Cloud Computing Adoption Tool that will help any organization identified that will help them determine the type of cloud service most suitable for them, identify local or international cloud service providers and finally the ROI for the Cloud Adoption.

Keywords Cloud computing · Cloud adoption · TAM · TOE

A. Chiniah (✉)
Faculty of Information Communication and Digital Technologies, University of Mauritius, Reduit, Moka, Mauritius
e-mail: a.chiniah@uom.ac.mu

A. E. U. Mungur · K. Naidoo Permal
University of Mauritius, Reduit, Moka, Mauritius
e-mail: a.mungur@uom.ac.mu

K. Naidoo Permal
e-mail: krishnen8@gmail.com

© Springer Nature Singapore Pte Ltd. 2019
S. C. Satapathy et al. (eds.), *Information Systems Design and Intelligent Applications*, Advances in Intelligent Systems and Computing 863,
https://doi.org/10.1007/978-981-13-3338-5_24

1 Introduction

In Mauritius, there are several cloud service providers proposing a variety of services to the local industry, goCloud [1] operational since January 2014 has been among the pioneer and among the first local company to supply public cloud. Data Communication Limited, Harrell Mallac Technologies, Emtel, FRCI and more recently Orange (Mauritius Telecom) have all invested into the cloud business. On the international level, we find huge institutions like Amazon with its AWS (Amazon Web Service) being a leader in cloud services on every charts across the globe. Not far behind Rackspace and VMWare are good examples for private cloud technology providers. Though private cloud is quite expensive compared to public cloud, it requires huge initial investment from the company willing to acquire such technology, many businesses are still more interested in signing for a private cloud due to the fact that they have direct control over the infrastructure and have the ability to ensure the security of their data.

As a matter of fact, there is one alternative to Public and Private model that is triggering a sudden interest from enterprises. This model is genuinely interesting and effective in the sense that it combines the best parts of Public and Private cloud into a single architecture that satisfies two important criteria; cost-effectiveness and data security. Hybrid Cloud [2] is getting thumbs up by more and more CEOs. As stated in the 2016 RightScale prediction survey [3], Hybrid cloud's adoption has increased by a net 71% within a year. Some of the main actors in the Hybrid Cloud business are Microsoft with Azure, VMW are vCloud Air, EMC, Cisco Fabric, HP Helion, IBM Bluemix and Rackspace RackConnect.

2 Cloud Computing Adoption

There are three major cloud computing service models available; Software as a Service, Platform as a Service and finally Infrastructure as a Service. There is a fourth model, which is less common but slowly being adopted by the IT landscape; Recovery as a Service (RaaS). In January 2016, RightScale conducted a survey on the latest cloud computing trends [RightScale 2016]. The survey asked 1060 IT professionals about their views on the adoption of cloud technologies.

Cloud computing as mentioned earlier has the ability to reduce cost a company needs to spend on IT infrastructure. Additionally, it also reduces the complexity linked to running traditional systems within Data Centre walls. Table 1 refers to the 2016 RightScale survey, the results are self-explanatory; cloud computing adoption is on the rise. Out of the 1060 companies surveyed, 42% have a working personnel of over 1000 employees, showing that big companies are also getting more and more involved in the cloud business as a mean to make more profit and reduce associated cost with traditional systems.

Table 1 Cloud computing adoption rate [3]

Cloud computing adoption survey 2016: conducted by RightScale with approx. 1060 respondents						
2015	2016	Remarks				
Public	Private	Hybrid	Public	Private	Hybrid	With the adoption rate expected to continue to be rising
65%	63%	58%	70%	77%	71%	

2.1 Challenges for Cloud Adoption

Cloud computing has been a game changer for IT companies. The way businesses manage and build their infrastructure has drastically evolved. Previously, it could be that a company is investing massively in data centre and having to incur recurrent maintenance and upgrade expenses, now it is all about hosting the same data on a cloud platform. The real challenge is to actually transfer the operations from one's data centre to a cloud environment, the task requires high technical and organizational skills due to the fact that IT is very complex in nature. Since Cloud is a relatively new concept whereby it is difficult to define its essence and singularity; it is made up of several components each one more complex than the other, as a result moving towards the cloud-based environment is full of complexity and risks. These altogether creates adoption barriers, because the enterprise is unable to obtain tangible and viable data about the possible outcome of investing in cloud computing. According to a recent survey conducted by Forbes (see Fig. 1), cloud computing implementation is subject to a number of critical variables. Security, cost uncertainty and loss of control are among the main reasons behind the reluctance to adopt cloud.

2.1.1 Security

Security is one of the topmost concerns for companies in today's technological landscape. Being able to protect data within an internal network from tampering and unauthorized access has become a colossal challenge for companies. Attackers threaten company integrity from different sources, insider threats, cyber-espionage, disgruntled employees; outsider threats social engineering, hacking, etc.

2.1.2 Cost of Platform

There are different cloud providers with considerably different service portfolio. Each one of these, locally or on the international level, have their own pricing strategy based

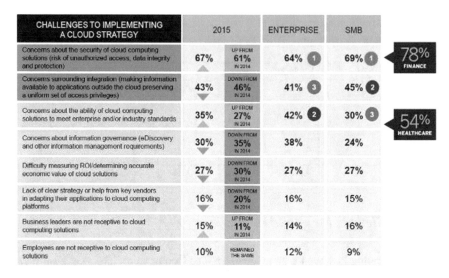

Fig. 1 Survey on challenges for cloud adoption [4]

Table 2 DCL SME boost package offering [6]

Package type basic	Details	Comment
Internet	512 kbps	
Email account	1 user	
Website	3 pages	
IP phone	1	
ERP account	1 user	
Storage capacity	5 GB	
International calls	Rs 300	Free
Installation and setup	Rs 3000	Offered
Training and support		

on their market share or target group. For instance, Amazon's cheapest solution, the t2.nano, is available for $3.29/month [5]. Locally price range varies depending on the service purchased; the lowest price observed so far is around Rs. 2175.00 which is being provided by DCL in the form of a boost package, as shown in Table 2.

2.1.3 Loss of Control

Moving data or even business to cloud usually implies losing a particular subset of the company. This could be tangible or intangible in nature but still is one major factor that prevents companies from moving to a cloud platform. So far, through this paper, we have been able to ascertain one point; companies like and want to remain in control, especially when it comes to their assets. Therefore, losing control is a de-motivation factor, which can be regrouped into two main categories; technical and organizational control loss.

2.1.4 Service-Level Agreement (SLA)

A proper SLA, as mentioned above, should at least include the following; details about the service, details about how will the service be delivered to the customer, a description of the nature of the service to be provided to the customer, the location where shall the service be delivered, customer premise or any secondary location, to whom shall the service be provided and when will the service be provided to the customer.

2.1.5 Performance

Most cloud providers or service provider, in general, will tend to guarantee infrastructure uptime and availability and provide customers with performance in best effort. Let us say one particular company is hosting its application server and middleware on the cloud platform. Application services usually require high performance and throughput because they are directly related to the user experiences. While setting up the SLA one needs to ensure that the supplier, cloud provider, does guarantee a certain level of service.

2.1.6 Compliance

Security for enterprises is very important. As such, companies handling customer data find it even more important to ensure that they comply with international standards such as ISO 27k, PCI-DSS, SOX, HIPAA and for European Union Countries, the European DPA. For instance, in Mauritius, we have the Data Protection Act 2004, ICT Act 2000, Electronic Transaction Act 2000 and Computer Misuse and Cybercrime Act 2003 that are employed by the legal framework to ensure compliance. Additionally, companies locally, usually make it a must to adhere to ISO 27k and PCI-DSS when dealing with customer data and monetary transaction using credit cards.

2.1.7 Lock-in Problem

Different service providers on the market will provide to customers a unique dashboard to control and operate their respective cloud platform. For instance, customers subscribing to Platform as a Service and Software as a Service are faced with one

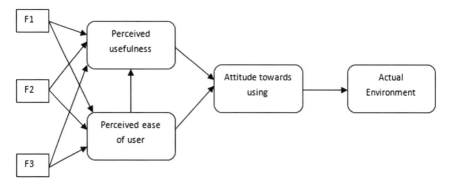

Fig. 2 Technology acceptance model; Davis 1989 [8]

major challenge; mastering and learning how to operate their environment via a predefined dashboard given to them by the supplier.

3   Research Model

3.1   TAM (Technology Acceptance Model)

First published in 1986 by F. Davis et al., the Technology Acceptance Model, or TAM, mission was to answer one very important question; will technology be accepted or rejected by users. TAM can be considered a derivation of Fishbein and Ajzen's 1975 Theory of Reasoned Action (TRA). TRA was devised in order to study and analyse human behaviour towards a wide array of domains. Though TRA could be applied to studying user interaction and acceptance of technology, it was very broad and general in nature. To be able to provide better insight and more accurate results when studying technology acceptance, Davis 1986 refined the model and came up with TAM (Fig. 2), which was more centred towards computers and technology. The first variant of TAM consisted of defining the actual architecture with regards to the available options and features and the user motivation to use the architecture. Then with further research Davis was able to propose an improved TAM model, which addressed three critical aspects of user acceptance towards technology; Perceived Ease of Use, Perceived Usefulness and Attitude towards using the system [7]. These factors would then be subject to the influence of the dynamics of the external factors F^*, which represents the architecture design and features.

Further research saw the birth of new variants and improvement of TAM; Davis came up with the first version as shown in Fig. 2.

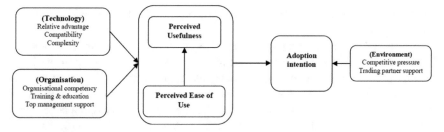

Fig. 3 Hybrid TAM/TOE model [13]

3.2 TOE (Technology-Organization Environment)

The Technology-Organization-Environment framework as described by Tornatzky and Fleisher [9], involves a process-based approach that companies use in order to implement a particular technology. Though TOE can be considered similar to TAM, it addresses different technology acceptances from a different angle. Technology adoption and acceptance, under TOE, will be assessed using three different criteria; firstly, from a technological perspective, secondly from an organizational perspective and finally from an environmental perspective [10].

Another aspect that is also very important is the organization itself. Certain key variables should be accounted for before deciding to invest in new technologies, is the company size and income big enough in order to sustain investment and adoption of new technologies. While adopting new technologies the manpower available needs to be assessed; do the people have the necessary expertise and are they dynamic enough to adapt to changes [11] and resulted in a model as presented in [12].

3.3 Hybrid TAM/TOE

A conceptual framework (Fig. 3) was developed [13] using technological and organizational variables of TOE framework as external variables of TAM model while environmental variables were proposed to have a direct impact on cloud computing adoption.

3.4 Survey

Data collection for research can be conducted in various ways, in this particular two methods will be necessary, a quantitative approach will be used to collect primary data from a survey sent to companies in the IT and non-IT business, while a qualitative approach will be used to interpret the data from past researches and existing surveys on the Internet.

Ethics are norms of behaviour that leads to moral choices about a person's behaviour. Respondents were informed of the purpose of the survey in the introductory part of the questionnaire and complete confidentiality of respondents was maintained throughout the study. In order to be able to verify the soundness of the hypothesis in Mauritius and possibly other technologically developing countries, a web-based survey consisting of 43 questions was designed. The survey has been divided into three sections:

- **Section 1**: Background Study and Demographic about the type of company we are assessing. It should be noted that this survey will be extended throughout the IT and Non-IT landscape as well as SMEs. Section 1 consists of eight questions.
- **Section 2**: Designed with the five-point Likert scale system, this section will tackle the framework proposed, TAM and TOE. Specific questions have been prepared in order to obtain critical data about the factors affecting the adoption of cloud in the country. Section 2 is made up of 32 questions.
- **Section 3**: Open-ended section containing only three parts that will allow the respondent to express suggestions.

In order to distribute the survey questionnaire to potential respondents, two methods were deployed. First, since the survey is mainly web based; the link to the survey was sent by email to groups of people with proper explanation with regards to the work. Second, Social media like Facebook and LinkedIn were used as a vector in order to reach a larger number of people.

4 Data Analysis

4.1 Demographics

Organization Size—Out of the 93 respondents, 43% represents organization between the range 51 and 250, while organization between that are more than 250 in size amount to 41.9%. From this table, it is clear that the majority of enterprises that responded to this survey can be categorized as mid-size and large companies. Additionally, this is indicative of the use of IT facilities within an organization compared to smaller companies with a response rate of less than ten percent (10%).

Organization Sector—One very important observation during this research concerns the companies that responded to the survey. Sixty-five per cent are directly related to the IT sector, which shows that IT is very much present in the local business landscape. Another very interesting observation that will definitely be beneficial to the cloud computing adoption cause in Mauritius, is the percentage of respondents who know what cloud computing means. 96.8% of respondents to the survey, are

Table 3 Cronbach's alpha for perceived benefits

Reliability statistics		
Cronbach's alpha	Cronbach's alpha based on standardized items	No of items
0.850	0.872	12

aware of the cloud computing concept, which proves very significant for this research that aims at improving the adoption of cloud computing in Mauritius.

Cloud Computing Knowledge—a combined 75% of the respondents claim they have "Good" and "Very Good" knowledge of Cloud Computing Technology.

4.2   Perceived Benefits of Cloud Computing

Table 3 shows the result of reliability test, Cronbach's alpha [14], for the perceived benefits of cloud computing. Based on previous research by Nunnally [15], a result of 0.7 is indicative of adequate internal consistency. The result being 0.850 shows that the data is 85% reliable with a 5% acceptable error margin.

There were 12 questions related to perceived benefits of Cloud Computing and we note that the different companies that participated in the survey demonstrated a positive attitude towards the usage of cloud computing. One point can be argued; some respondents did not agree to the fact that cloud computing facilitated access to the latest technology. Cloud customers are limited to the platform that service providers have, eucalyptus works with Red Hat Linux and CentOS, Windows Azure is Windows proprietary. While Amazon supports nearly everything [5].

4.3   Perceived Ease of Use

Similar to the previous section, reliability statistic, Cronbach's alpha, for the perceived ease of use of cloud computing returns a positive result. Though there is one exclusion in the result set, the obtained 85.6% alphas with a 5% margin of error provide ample reliability to the test results.

In Mauritius, it is quite easy to find local cloud providers as well as international cloud service providers; on the local market we have companies like DCL, Emtel, HM-Technologies that are among the market leaders while on an international level, Amazon, Rackspace, Windows Azure, SalesForce are example of providers. As a result, the response shows that 50.54% strongly agreed and 31.18% agreed that finding a cloud service provider is relatively easy on the local market. Similarly, obtaining proper support for acquired services is very important, 40.86% strongly agree and 36.56% agreed that finding appropriate support for cloud service is easy in Mauritius.

One concern that was observed during this research is that respondents were predominantly against the idea that it was easy to import data to cloud platform. 51.09% disagreed with the statement that important apps and data to cloud was without difficulty. A sound reasoning could be that importing data from a native platform to cloud is dependent on various components; licenses, distribution and version of software, platform and base OS among others.

As a whole, from the response of the 10 questions related to perceived ease of use, we can conclude that the majority of respondents do find cloud computing easy to use.

4.4 Attitude Towards Cloud Computing

Cloud computing involves the storage and processing of company data at a remote location, typically where the cloud provider data centre is located. As a result, before moving to cloud platform companies will thoroughly assess and consider options.

So far, we have noticed that the surveyed companies in Mauritius are very positive with the use of cloud computing. But the attitude towards technology is also very important. Privacy of data is very important, with 91.40% stating it is very important to consider privacy before moving. 8.602% still considers privacy as important.

Respondents to this survey have highlighted the importance of non-repudiation, it is easy to deny being the author of an electronic transaction, especially when data is not under direct control of the owner. Tsai 2002 [16], relates to the need and importance of non-repudiation in practice and the mechanism needed by organizations in order to prevent fraud. 69.89% defined repudiation as very important, 21.51% considered the variable as important. While it is interesting to see that 1.075% saw repudiation as not important and 7.527% were undecided.

From the 10 questions related to Attitude towards Cloud Computing, the majority of the respondents are very concerned with the security aspects and the availability of their data.

4.5 Hypothesis Testing

Several hypotheses were formulated and tested using chi-square Test, and outcome of Hypothesis accepted are as follows:

(1) H1: There is no significant relationship between technology readiness and organization sector.
(2) H2: There is no significant relationship between Cloud computing adoption and Information Security.

Table 4 Model summary regression analysis of Hypothesis 1

Model summary[b]

Model	R	R square	Adjusted R square	Std. error of the estimate
1	0.58[a]	0.003	−0.008	0.283

[a]*Predictors* (Constant), organization sector
[b]*Dependent Variable* Has your company implemented—or is considering or planning on implementing—cloud-based services?

(3) H3: There significant relationship between technology barriers and migration towards cloud computing.
(4) H4: There is no significant relationship between Firm size and adoption of cloud computing.
(5) H5: There is no significant relationship between org. sector and Cloud computing adoption.
(6) H6: There is no significant relationship between cloud computing and competitiveness of the industry.

4.6 Regression Analysis

We also tested Hypothesis 1 (Table 4) with Regression Analysis.

5 Cloud Adoption Tool

We developed a Cloud Adoption website with an integrated Cloud Service Provider recommendation feature. The website provides details about the Cloud, Cloud Services, and Cloud Services Providers locally and internationally. For the recommendation tool, users have to enter some information as laid out in Fig. 4. Then the tool will match the requirements of the user with the data crawled from the internet (cloud service providers and their respective prices.) and the recommendations will be shown as in Fig. 5.

6 Conclusion

Cross-tabulations and Pearson chi-square tests have been conducted to verify the relationship between variables. So far, only 1 hypothesis out of the 6 possible has

Fig. 4 Cloud adoption tool

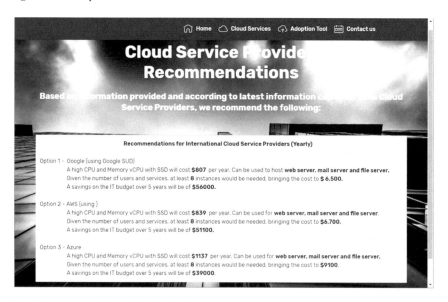

Fig. 5 Recommendation of cloud adoption tool

been accepted. The reason being that cloud computing has had several leaps forward in a short lapse of time.

Security is no more the major concern for cloud adoption; rather companies are more interested in the benefits cloud can provide to their business. Increased advantage, access to new technology, competitiveness are the variables with the highest scores.

References

1. goCloud Mauritius.: Providing Cloud Services to SME. Retrieved 12 February 2018, from http://www.gocloud.mu/ (2018)
2. Yousif, M.: Hybrid clouds. IEEE Cloud Comput. **3**(1), 6–7 (2016). https://doi.org/10.1109/mcc.2016.9
3. Weins, K.: Cloud Computing Trends 2016. RightScale. Retrieved 12 January 2018, from http://www.rightscale.com/blog/cloud-industryinsights/cloud-computing-trends-2016-state-cloud-survey (2016)
4. Forbes Welcome: Forbes.com. Retrieved 12 January 2018, from http://www.forbes.com/sites/louiscolumbus/2015/01/24/roundup-of-cloud-computing-forecastsand-market-estimates-2015/#460eafbc740c (2018)
5. Amazon EC2 Pricing—AWS. Amazon Web Services, Inc. Retrieved 6 January 2018, from https://aws.amazon.com/ec2/pricing/ (2018)
6. Boost SME—Small Medium Enterprise Mauritius: Boost.mu. Retrieved 16 November 2017, from http://www.boost.mu/ (2017)
7. Davis, F.D.: Perceived usefulness, perceived ease of use, and user acceptance of information technology. MIS Q., 319–340 (1989)
8. Davis, F.D., Bagozzi, R.P., Warshaw, P.R.: User acceptance of computer technology: a comparison of two theoretical models. Manage. Sci. **35**(8), 982–1003 (1989)
9. Tornatzky, L.G., Fleischer, M., Chakrabarti, A.K.: The Processes of Technological Innovation. Issues in organization and management series. Lexington Books. Available at http://www.amazon.com/Processes-Technological-Innovation-Organization/Management/dp/0669203483 (1990). Accessed 10 June 2013
10. Tan, M., Lin, T.T.: Exploring Organizational Adoption of Cloud Computing in Singapore (2012)
11. Misra, S.C., Mondal, A.: Identification of a company's suitability for the adoption of cloud computing and modelling its corresponding return on investment. Math. Comput. Model. **53**(3–4), 504–521 (2011)
12. Baker, J.: The technology–organization–environment framework. In: Information Systems Theory, pp. 231–245. Springer, New York, NY (2012)
13. Gangwar, H., Date, H., Ramaswamy, R.: Understanding determinants of cloud computing adoption using an integrated TAM-TOE model. J. Enterp. Inf. Manage. **28**(1), 107–130 (2015)
14. Bland, J.M., Altman, D.G.: Stat. notes: Cronbach's alpha. Bmj **314**(7080), 572 (1997)
15. Nunnally, J.: Psychometric Methods (1978)
16. Tsai, W.: Social structure of "coopetition" within a multiunit organization: Coordination, competition, and intraorganizational knowledge sharing. Organ. sci. **13**(2), 179–190 (2002)

Digital Transformation from Leveraging Blockchain Technology, Artificial Intelligence, Machine Learning and Deep Learning

N. Chandrasekaran, Radhakhrishna Somanah, Dhirajsing Rughoo,
Raj Kumar Dreepaul, Tyagaraja S. Modelly Cunden
and Mangeshkumar Demkah

Abstract These are exciting times as new software development paradigms are fast emerging to cope up with the shift in focus from "mobile first" to "AI first" approach being adapted by Google, Facebook, Amazon and others. This can mainly be attributed to the stability of the cloud computing platform and the developments in search capabilities which have extended from traditional text and web pages to achieving voice and vision recognitions relating to images and videos. Continued research focus has brought the error rate in image recognition by machine to converge sharply with that of the human. Apart from developments in big data analytics, artificial intelligence, machine learning and deep learning, break throughs in peer to peer distributed ledgers with a blockchain technology platform, which incorporates multiple levels of strong encryptions, have created massive developmental interests. Most of the "popular apps" that we use today, are being built using AI algorithms. To achieve this, changes are being incorporated to computational architecture to make them compatible with "AI first" data centers equipped with AI driven features. Tensor Processing Unit (TPU), which powered Google's developments in ML and AI, has now become part of cloud computing service. Anticipating cost related issues, new hardware developments are focusing on moving from the cloud to the edge with the new "Edge TPU". Digital transformation is further augmented by the fact that block chain platforms, which are built on de-centralized tools and technology, are exhibiting greater maturity by the day. The paper highlights several blockchain applications to deliver on several of the promises. The paper also discusses the fundamentals of Neural network to demonstrate how well these concepts that are incorporated in deep learning have decreased error rates by tenfold compared to previous technologies.

N. Chandrasekaran · R. Somanah (✉) · D. Rughoo · R. K. Dreepaul · T. S. M. Cunden
M. Demkah
Université des Mascareignes, Rose Hill, Mauritius
e-mail: dsomanah@udm.ac.mu

R. Somanah
University of Mauritius, Reduit, Mauritius

© Springer Nature Singapore Pte Ltd. 2019
S. C. Satapathy et al. (eds.), *Information Systems Design and Intelligent Applications*, Advances in Intelligent Systems and Computing 863,
https://doi.org/10.1007/978-981-13-3338-5_25

1   The Uniqueness of Blockchain Technology

It is inconceivable that at a time when tremendous breakthroughs are occurring in technologies relating to cloud computing, big data, artificial intelligence, etc. attention is being drawn by technology disruptors to the new technology platform that is fast emerging, by highlighting the importance of Blockchain Technology, vis-à-vis, the distributed nature of the platform and the high degree of encryptions incorporated at multiple levels. Since it operates peer to peer, the centralized nature of the existing banking system among others are being challenged. The digital economy has till today relied heavily on controls and operations by banks and numerous other centralized agencies to handle the finances of people and institutions. This paradigm extends to all aspects that touch human life. Examples include dealing with property documents, personal records, etc. The over-reliance of the asset owners on the banks, governments, etc. as custodians have made them extremely powerful today. Blockchain technology makes it possible to dispense with these third parties by making available an alternative technology platform that operates peer to peer with a potential to remove all shackles.

Due to its powerful peer-to-peer architecture, this technology, which emerged initially to deal with digital currency transactions, is being exploited by the developers to extend their capability to several other wide-ranging new areas involving ledgers. Designed with the original objective to trace the history of payments of every digital currency in the system, this particular feature came handy for the developers to devise numerous other wide-ranging applications. Another attractive feature that allowed it to operate under different circumstances, is the ability to provide proof of a transaction with a history of ownership and time stamps. What is important is that this technology has perfected the concept of distributed ledgers. These ledgers are publicly accessible as they reside on networks of numerous clients across the world. The architecture ensures openness and provides for secure access. The success of the operation of a distributed ledger depends on ensuring that all the nodes in the network collectively agree with the contents of the ledger. This is guaranteed by a combination of complex mathematics and a built-in "consensus mechanism". This mechanism defines the process by which the nodes agree on how to update the blockchain whenever assets transfer takes place between individuals [1].

2   Mathematical Concepts

The blockchain technology applications are numerous and encompass ownership rights pertaining to intellectual property, digital wallets, neighbourhood microgrids, etc. In a nutshell, each block represents one particular transaction and a chain of blocks are created by connecting a series of such transactions/blocks. The concept of encrypted "distributed ledgers" ensures that the chain of blocks is always secure. As can be seen from Fig. 1, an algorithm is used to generate "hash value" for each

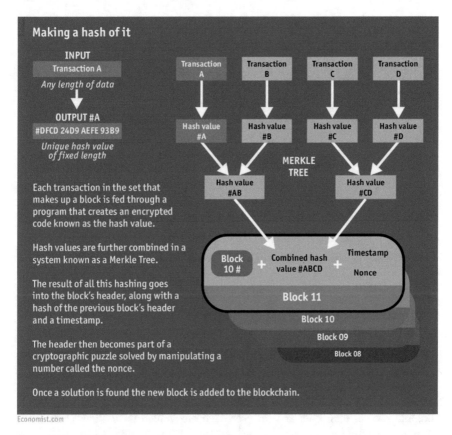

Fig. 1 The use of hashing algorithms to create blocks [1]

transaction. A combination of these hash values leads to the creation of "Merkel Tree". The sequence of mathematical operations that joins the encrypted blocks together is seen in Fig. 1.

As can be seen from Fig. 1, "Hash functions" computes hash values taking as inputs differing in lengths of data representing varying transactions. In other words, each transaction can be defined with varying string lengths. Hash value outputs are always unique and of fixed length irrespective of the input string lengths. This process of conversion can only be performed one way, implying that while it is possible to get the hash value from the data, it will not be possible to generate the data backward given the hash value. An analogy is the juice making process, which is irreversible, as whole juice can be made from the fruit one can never get the fruit from the juice. Even though the hash does not contain the data, it is still unique to them as is juice to the fruit. Even if a small variation in a transaction is introduced, due to alteration, the hash would automatically become different, refer to Carnegie Mellon University lecture notes on "Introduction to Hashing" [2] and hence can be detected at once.

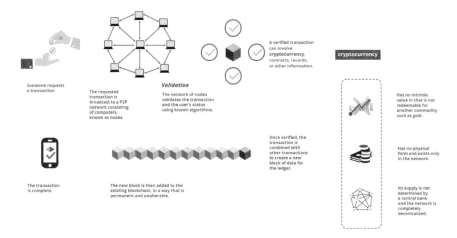

Fig. 2 Schematic describing the formation of "Blockchain" resulting from each peer-to-peer transaction [3]

3 How Does This Technology Work?

In the previous section, we have described in detail the formation of chain of blocks by linking a number of individual blocks. It has been stated before that the financial service providers like big banks or the public service providers like the Government, currently employ centrally controlled servers or cloud computing environments to store and protect their data and services. The blockchain technology architecture conceptually offers an alternative mechanism as it stores info across millions of computers across the world. The encryption process is quite rigorous in the sense that each block of data is encrypted with a very high level of sophistication. That is one of the major reasons that the time taken for each transaction is currently disturbingly high.

Figure 2 demonstrates the peer-to-peer nature of the transactions as well as the blockchain that is being created with every transaction.

In a recent announcement, Bank of England has announced a revamped payments system based on the blockchain technology [4]. Earlier this year, the bank working in association with a number of financial technology companies, proposed a revamp of RTGS in a clear boost to the distributed ledger technology.

The potential for this technology is so huge that Stanford computer scientists have recently announced the creation of a Center for Blockchain Research [5]. According to this bulletin, "since blockchain allows traceability, security and transparency, many companies are exploring ways it might be used to improve supply chain management, expedite real-estate transactions and the transfer of deeds, or modernize voting technology". A partial financial support has been provided by Vitalik Buterin, who has developed the "Ethereum Blockchain", which is a virtual machine and cloud platform housing user-crafted legal contracts. Ethereum allows new projects to be developed that include creating an alternative to Wall Street. It can also help to create a new environment for controlling the functioning of a virtual model of democratic governance. The transparent and open nature of the platform allows citizens to take direct control of the entire process of government functions by ensuring that the politicians are fully accountable to citizens.

The network always remains in a consensus state. For instance, we have talked about the hashing of all transactions, which always results in a unique output. These results are also verified by other nodes on the network, which re-run the same hashing algorithms (mathematical equations). Obviously, the output hash values have to match exactly all the time. If there is a mismatch, this is a clear indication that the inputs (the transactions) don't match, either due to an error or fraud. The transaction then has to be ignored. The process works exactly the same way for validating single transactions or for validating new blocks. The data cannot be corrupted as altering any unit of information on the blockchain would mean using a huge amount of computing power to override the entire network.

4 Some Typical Applications of Blockchain Technology

Steps to create a digital wallet account are described in [6]. As its name implies, "Wallet app" functions like a virtual wallet to store virtual currencies as well as for using these currencies for commerce. The strong feature of this app is the process of verification of identity and the security features that have been built into link transactions with personal identity. Identity management becomes part of this powerful technology platform and hence can become a powerful tool in this field in the future.

Another interesting app is "Smart Contracts" that can be used to execute business contracts. The app facilitates the drawing up of a contract after it ensures that certain conditions are adequately met.

Another important app is the "Sharing Economy", which is a virtual equivalent of Uber and Airbnb, the services that are flourishing in the real-world sharing economy. In the current environment, Uber, like a centralized bank in the case of financial transactions, must be relied upon to provide the taxi sharing service. As an alternative, an app should be able to establish direct contact between the user and driver, thus enabling personal interaction. This can then permit decentralized direct payment to be made between concerned parties, see Fig. 3.

A whole range of other applications have been built using this technology including "OpenBazaar", which allows the user to transact with vendors without paying any transaction fees, "Crowd Funding" and "Governance" apps, "Board Room" app, which, as its name implies, is intended to bring Board room capabilities to this technology platform.

"Supply Chain Auditing" is another important application, which is exhibiting enormous maturity. This particular app serves the important purpose of providing certification as to how original and genuine the seller's products are. It is heartwarming to see a similar app being used in Indian Education sector to identify false

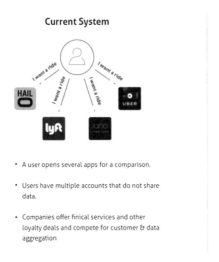

Current System

- A user opens several apps for a comparison.
- Users have multiple accounts that do not share data.
- Companies offer finical services and other loyalty deals and compete for customer & data aggregation

Future System

- Rider has one interface that aggregates all offers available at that moment.
- Riders, drivers, and service providers have one account that aggregates data from all transactions.
- Companies optimize to connect peers in the best way using reputation calculators and

Fig. 3 Futuristic peer to peer taxi-sharing schema

academic qualifications. Needless to point out that this feature allows the buyer to make better-informed decisions. The solar energy produced by rooftop solar panels, may at times be in excess or in shortfall being not able to meet the demands of the household. Blockchain apps can create a community based microgrids' system to judiciously buy or sell the renewable energy generated. Contracts can be executed using Ethereum-based "smart contracts".

Most of the applications are still emerging as few mishaps have exposed some of the vulnerabilities in this technology. Adria conferences [7] provide a clear perspective on the developments by clubbing the new applications under four areas. For instance, developments relating to healthcare, title records, ownership, voting and IP rights are covered under "record keeping", see Fig. 4. Several other fields are grouped under three other categories.

5 Basic Characteristics of Artificial Intelligence

On the other end of the spectrum resides the massive developments that are occurring in the field of Artificial Intelligence (AI). As the name suggests, AI algorithms are designed to artificially mimic the intelligence of humans, vis-à-vis how we learn and understand to perform several behavioural tasks. In the past few decades, AI has evolved considerably from just being used to play games and in the development of expert systems to teach computers to interpret natural language and in neural networks.

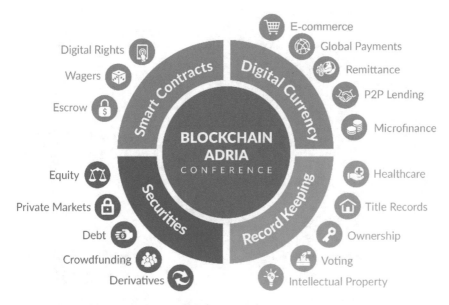

Fig. 4 Broad areas of applications of blockchain technology [7]

Once the human inquiry process has been utilized to set up the system, AI can reliably perform most tasks without fatigue normally associated with humans. Most of the products that we use on a day-to-day basis, tend to perform better when AI adds intelligence to these products. For instance, Alexa becomes a far superior product capable of performing a wider range of tasks, when AI tools are integrated. AI uses powerful datasets to facilitate progressive learning as it lets the data to execute the programming. To enable the algorithm to acquire a new set of skills, AI searches to locate structure and regularities in data. The algorithm then performs the classification or prediction function. For instance, Amazon may use this feature to recommend the type of product that you are likely to buy next while online. Similar strategy is adopted to predict (say) the next move that can be made in Gaming.

As can be seen in Fig. 5, deep learning is a subset of machine learning, while the latter is a subset of artificial intelligence. All the definitions are also provided in the figure.

6 Historical Perspective of Evolution Leading to AI and Their Components

The first generational changes began with developments in "data analytics", which mainly involved simple analysis, followed by a depiction of the analytical results in graph form, etc. The second generation saw the emergence of the field of "data

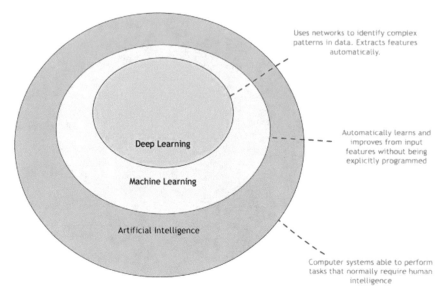

Fig. 5 Deep learning (DL) and machine learning (AI) in the world of AI

science" (or "clever analytics"), when scientific principles and methodologies were applied. Data science ensured that complex patterns in data can clearly be identified and the features are automatically extracted. As in Fig. 5, this represents Deep Learning (DL). The third generational changes brought in a new field called "machine learning". Basically, ML is just data science with a feedback loop. Robotics is also a major field related to AI. Robots require intelligence similar to the humans to perform tasks relating to navigational capabilities and manipulation of objects.

7 Breast Tomosynthesis Facilitated by AI

As computer algorithms get better at reading and detecting cancerous cell growth, physicians need not necessarily have to spend an enormous amount of time interpreting mammograms. Detection of lesions can be automated and reliably identified using Computer-assisted detection (CAD) AI tools. Public digital mammography datasets (INbreast) helps the AI model to detect a large percentage of cancers and that too with precise locations. It should be mentioned that with repeated training (and with a wider availability of larger datasets), false positive detection of an image has been considerably brought down to negligible proportions. Janet Burns [8] in a Forbes article has emphasized the fact that when the mammogram indicates that there is likely to be a potential problem, women frequently undergo breast biopsies

Fig. 6 Neural network

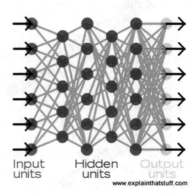

Input units   Hidden units   Output units

www.explainthatstuff.com

to take a closer look. With the use of AI tools, these steps can be completely avoided as the algorithms will be able to provide more in-depth insight.

8 Fundamentals of Neural Networks

The importance of incorporating vision capabilities to AI cannot be overstressed. The concept of neural networks is based on enabling a computer to learn from observational data. Deep learning employs this concept to learn to generate a powerful set of techniques. These techniques are mainly responsible for many of the advancements that are currently occurring in the fields of natural language processing, image and voice recognitions.

Till recently, algorithms were taught to recognize patterns using human supervised process employing labelled data sets. Scaling up becomes an issue, as can be seen from the fact that Facebook is currently using 3.5 billion users shared Instagram images. The number of hashtags used is in tens of thousands. Since attributes can now be accurately defined, say using PhotoDNA, Facebook is now, for instance, able to detect and monitor pornography posts. Google Photos use DL techniques to offer suggestions to fix or edit your photos.

The major objective of a neural network is to get the computer to learn things, recognize patterns, and make decisions in a humanlike way [9]. Several iterations are required to be performed before the computer can mimic the brain function of learning. Human brain has a far denser structure compared to neural networks. The computer simulation techniques have various other huge limitations.

Let us illustrate the process by an example, see Fig. 6, which depicts a fully connected neural network. The main units of the network consist of input and output as also the intermediary hidden units. The whole process is dynamic and constantly evolves as the network continually learns from the iterations.

9 Feedback Process for Learning by Neural Network

It is important to point out that, deep learning models can recognize objects on images better when they are shown more examples. They have the ability to "learn", by refining the parameters of subsequent filtering steps. This is what we meant by the incorporation of the feedback loop into data science. In certain visual tasks, deep learning has decreased error rates by tenfold compared to previous technologies.

Akin to the functioning of the closed-loop feedback control systems in electronics, the output is constantly compared with the desired output to generate the error data. The weights of the connections between the units in the network are modified in the depending upon this error. The technique is aptly named as "backpropagation" as the algorithm works backward in the reverse direction from the output units to the input units and through the hidden units. During the learning process, the calculations are repeated until complete convergence takes place (error becomes close to zero).

10 Machine Learning Example

The MNIST database (Modified National Institute of Standards and Technology database) is the most extensively employed to train and test ML algorithms. MNIST contains a huge number of handwritten digits, which can be helpful for training purposes. Figure 7 shows one of the handwritten codes, whose pattern is being analysed.

A partial screen capture of a Python code, which contains several mathematical definitions for acquiring machine learning capability is shown in Fig. 8.

11 Understanding Machine Learning Algorithms

There are numerous hands-on procedures that have been outlined by various authors, see [10] for one such listing. The algorithms can be broadly classified into three types.

1. Supervised Learning: This is the approach that has been employed in a large number of algorithms. The main objective is to learn the mapping function, which expresses the output variable as a function of the input variable. The aim is to define the mapping function in such a fashion that the function can compute an accurate output value for any new input data too. Training datasets are used for the supervised learning process. The training process continues until the model achieves the desired level of accuracy on the training data.

2. Unsupervised Learning: For a given input data, there is no definite corresponding output variable exists. There are no correct answers and hence there is no supervision. The primary objective is to model the underlying structure or distribution in the data to learn more about the data. By themselves, algorithms discover the

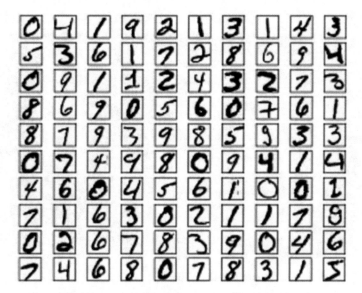

Fig. 7 Handwritten code for the computer to read

```
import numpy as np
class Network(object):
    def __init__(self, sizes):
        self.num_layers = len(sizes)
        self.sizes = sizes
        self.biases = [np.random.randn(y, 1) for y in sizes[1:]]
        self.weights = [np.random.randn(y, x)
                        for x, y in zip(sizes[:-1], sizes[1:])]
    def feedforward(self, a):
        for b, w in zip(self.biases, self.weights):
            a = sigmoid(np.dot(w, a)+b)
        return a
    def SGD(self, training_data, epochs, mini_batch_size, eta,
            test_data=None):
        if test_data: n_test = len(test_data)
        n = len(training_data)
        for j in xrange(epochs):
            random.shuffle(training_data)
            mini_batches = [
                training_data[k:k+mini_batch_size]
                for k in xrange(0, n, mini_batch_size)]
            for mini_batch in mini_batches:
```

Fig. 8 Part of the python code to impart machine Learning capability

structure in the data. Unsupervised learning problems can be further grouped into clustering and association problems. In the absence of any outcome variable to predict or to estimate, the output is clustered into different groups.

3. Reinforcement Learning: These are goal-oriented algorithms and they learn how to meet certain complex objectives or goals. They always start with a blank slate and with learning can achieve superb performance. Reinforcement is achieved

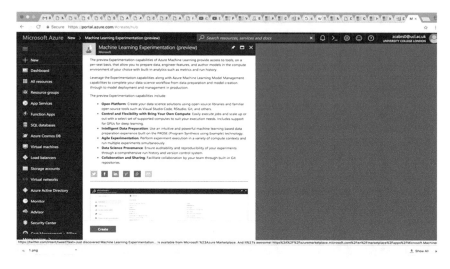

Fig. 9 Microsoft machine learning experimentation

by providing incentives when right decisions are made and punished for wrong decisions.

A wealth of information and tools are available at Microsoft Azure portal, to get trained in using Visual Studio Code and the AI Extension with Azure Machine Learning Work Bench, see Fig. 9.

12 Digital Transformation of a Nation

There exists a strong correlation between the strength of the economy of a nation and the nature of the adoption of sophisticated technologies like the ones outlined in this paper. According to some estimates, even the most advanced countries have been able to deploy only 6% of the solutions that are currently being investigated.

The following provides a list of FANG businesses, whose efforts relating to AI work are bringing in top dollars:

Facebook (F) has access to trillions of data points, has an understanding of the users to increase the user activity and advertisement revenue.

Amazon (A) is in possession of enormous data relating an extensive range of products. It also has data on purchase history and hence is engaged in applying AI for business development.

Netflix (N) has a tremendous amount of user data pertaining to movies.

Google (G) had been dominant in the text- and web-based search, is increasingly becoming the best in the world as it introduces AI into a number of its products. It also makes available many of the tools to the public.

These technology giants are growing at a faster rate because they are able to exploit the advancements in ML, AI, etc. much better than even many of the nations. There is no reason why any nation or business should relegate the technology impacts just to these tech giants [3].

13 Concluding Remarks

By some estimates, in about 4 years' time, the job scenario will drastically change to create about 9% new jobs that do not exist today. Both the Artificial Intelligence (including the Machine Learning and Deep Learning subsets) will act as a catalyst during the transformation phase as also the peer to peer technologies like blockchain. While AI techniques have broader connotations, Machine Learning (also known as Applied AI) involves specific scientific methods currently in vogue for building AI. Apart from data analytics capabilities, ML employs complex models and algorithms that lend themselves to the prediction to enable the emergence of the field of predictive analytics.

References

1. https://www.economist.com/briefing/2015/10/31/the-great-chain-of-being-sure-about-things
2. https://www.cs.cmu.edu/~guna/15-123S11/Lectures/Lecture17.pdf
3. https://blockgeeks.com/guides/what-is-blockchain-technology/
4. https://www.reuters.com/article/us-boe-blockchain-payments/bank-of-england-says-payments-system-can-serve-blockchain-users-idUSKBN1KD2AM
5. https://engineering.stanford.edu/news/stanford-computer-scientists-launch-center-blockchain-research
6. https://www.onlinehomeincome.in/bitcoins-how-to-create-a-blockchain-wallet-account.php
7. https://blockchainadria.com/
8. https://www.forbes.com/sites/janetwburns/2016/08/29/artificial-intelligence-can-help-doctors-assess-breast-cancer-risk-thirty-times-faster/#c6964495908e
9. https://www.explainthatstuff.com/introduction-to-neural-networks.html
10. https://www.analyticsvidhya.com/blog/2017/09/common-machine-learning-algorithms/

Mobile-Based Attendance Monitoring Using GPS and Network Provider Location

Aatish Chiniah, Sahil Raghoober and Krishtee Devi Issur

Abstract Research has shown that attendance of students for lectures, tutorials, and practicals has a direct impact on their performance. As such proper monitoring of attendance provides an incentive for students to be present for their classes. In this work, we aim at designing and implementing a mobile-based attendance monitoring system and will make this process transparent and ubiquitous. The system makes use of both GPS and network provider localization information so as to balance between accuracy and power usage. The Mobile App is authenticated using fingerprint and allows automatic attendance monitoring if the app is open. All the information is stored in a NO-SQL database Firebase, on the cloud, and retrieved for statistical analysis. The results of this work show the enhanced features in terms of localization and authentication compared to existing works and also in terms of the execution time of the different features to provide a seamless user experience while using the system.

Keywords Attendance monitoring · Mobile app · GPS · Network provider
Geofence · Fingerprint

1 Introduction

Two studies [1, 2] conducted at the University of Mauritius have shown the direct impact of attendance on the results of students. Maintaining an attendance log is vital and as far as we remember universities have been taking attendance of students

A. Chiniah (✉) · S. Raghoober · K. D. Issur
Department of Digital Technologies, Faculty of Information Communication and Digital
Technologies, University of Mauritius, Reduit, Moka, Mauritius
e-mail: a.chiniah@uom.ac.mu

S. Raghoober
e-mail: sahil.raghoober@umail.uom.ac.mu

K. D. Issur
e-mail: krishtee.issur@umail.uom.ac.mu

© Springer Nature Singapore Pte Ltd. 2019
S. C. Satapathy et al. (eds.), *Information Systems Design and Intelligent
Applications*, Advances in Intelligent Systems and Computing 863,
https://doi.org/10.1007/978-981-13-3338-5_26

285

by either passing the register to the students for them to enter their signatures or lecturers call out the names of the students one by one to mark their presence. Since everything has to be done manually, this process is perceived as time-consuming and error prone, that is, when lecturers forget to bring their attendance log book, they pass a paper around for students to input their names, ID and signatures, but very often the lecturers forget to update their registers and thus will not be aware of who attended the class. Some lecturers can also lose that piece of paper which will result in a similar problem as mentioned above. Another factor when lecturers pass over the attendance book to the class is that some students sign for their friends, who are not in the class or they can make other amendments like putting someone absent who was present.

It can be seen that taking attendance manually wastes a lot of time and energy. Therefore, to overcome these issues, we suggest altering the existing system by an android mobile app which will be a real-time tracking system. The Android mobile app will use the Internet (Wi-Fi) in order to work with GPS which will be used to track students and will also use fingerprints to stiffen security. It will be a user-friendly app which will be useful and straightforward for lecturers, and can thus check the attendance accordingly. As for the students, they will input their fingerprints on their phone's sensor and if their fingerprints match with the one stored in the database then they will be marked as present. They will also use this app to view timetables, view facilities within the university (mostly for Year 1 students) and obtain their overall rate of presence and absenteeism. The main purpose of the "Mobile-based Attendance Monitoring System" is to discourage the use of paper as we are all moving towards an eco-friendly world. The rate at which everything is becoming computerized nowadays there is always a technological push, this is why we feel that it is the right moment to develop a "Mobile-based Attendance Monitoring System", which will facilitate both students and lecturers. It will also give the University of Mauritius a chance to alter their current attendance methods and become an example to others.

2   Related Works

Attendance monitoring plays a vital role in schools, universities and workplaces. Numerous research has been carried out to demonstrate how attendance is conducted at the above-cited places. Let us see why attendance system is important and how it is carried out according to the following researchers.

The owner of Great House Screen Printing [3] in San Diego implemented the biometric facial recognition system in his company to capture attendance of his employees. Every morning employees must smile at a biometric device in order to mark their presence or time of arrival. Once the device has detected the person, it will forward the data to a cloud-based time and attendance software program. Shawn Greathouse then stated that he was paying his staff only for the number of hours that they have worked as the biometric facial recognition enlightened him of which employees attended work and their arrival and departure times. Journal TICOM Vol. 1 No. 3 [4] of the University of Selangor from Malaysia conducted a research on the use of a barcode scanner to track the attendance of students in the institution in Malaysia. Students make avail of their student card to deem themselves present. So, when a student will present his card to the barcode scanner, the reader will scan the barcode number of the card and it will transmit the data to a student attendance system. The data will then be saved on a database where the lecturers can access the system at any time to print the student attendance report. Bringing technology into schools: NFC-enabled school attendance supervision, another research made by Ervasti et al. [5] of the University of Oulu from Finland asserted that the students of a primary school in Finland make use of a smart card called "Robo". All they had to do is pass the "Robo" card over a smart card reader or use the card to touch their teacher's NFC-enabled mobile phone. The students also had to do this process before leaving school. The NFC card was also being used for early departure; if ever a student was unwell or had other commitments. Teachers accessed the system in order to view or modify the records. Moreover, if ever students forgot to pass their cards on the reader, they will automatically be deemed as absent. Furthermore, parents were also included to the system and were receiving phone notifications (SMS) or were given access to the "citizen's portal", which is an online system to obtain information about their children's presence, lateness or absence. Radical Global [6] introduces the use of RFID for attendance monitoring system in schools. The RFID card reader operates by reading the card that the student presents when he enters and leaves the school. The RFID reads the code of the card and sends the data to the cloud server. The server saves the information to a database and then communicates to the SMS gateway in order to deliver attendance messages to their parents. Lecturers or teachers will have a website from where they will be notified about the attendance recordings of their students and can thus make necessary amendments where required. QR-code-based systems is a study from the Development of the Online Student Attendance Monitoring System [7] from Malaysia which demonstrates how attendance is performed in some of the institutions in Malaysia. Students are given a QR-code which they have to access through their phones. Those students who do not have a mobile phone will get their QR-code as a hard-copy format. When students will go for their lectures, the teachers will make use of their mobile phones or laptops or tablets to scan the QR-codes. After that, the QR-codes that have been scanned will be transmitted to the student attendance monitoring system where the lectures will get the results of the students who are present in the class.

3 Design and Algorithm Proposed for Mobile-Based Attendance Monitoring

When the user uses their mobile device to access the attendance app, any activity such as login, register, etc. must be passed through Wi-Fi and the Internet, in order to interact and communicate with the SQL Server database and also if the mobile phone is used to access the website, it will again have to use Wi-Fi and Internet so that it can interact with the IIS web server and the SQL Server database.

The following algorithm will automatically take the attendance for each student. This algorithm will be able to distinguish each student based on their StudentID. If the student is outside the boundaries of the Geofence when attendance is being checked, then the system will set them as 'Absent'.

```
Start
```

1. Student login (get StudentID and Programme of studies)
2. Attendance Monitoring

    2.1 Check if GPS is turned on
      - If GPS is on
        - Load map
      - Else
        - Student is navigated to the location settings page
        - Go to 2.1

3. Search for student's GPS location
4. Lecture starts, student receives a push notification as a reminder that their class has started
5. Boundary of classroom is created
6. Check if student is within the Geofence (boundary)

    6.1 Check student's location 3 times

7. Halfway of a lecture, student must input their fingerprint

    7.1 Check in database if fingerprint matches the student that logged in
      - If fingerprint matches and student is within the Geofence 2 out of 3 times
        - Set the student as 'Present' in the database
        - Notify the user about his state
        - Go to 3
      - Else
        - Set the student as 'Absent' in the database
        - Notify the user about his state
        - Go to 3

```
End
```

4 Mobile App and Backend

The implementation of the mobile and backend system is important for the proper functioning of the app. Let us have a look at the tools and environment we have used to build the system.

Implementing the application (Mobile system) refers to coding in Java. So, to build the system we will opt for Android IDE because it already consists of predefined plugins which will be useful for us since we will have the necessary tools in that platform itself. Both the Mobile and the Backend System comprises of different components and in order to create a relationship and link those tables, we will use SQL Server database. For our system, the Admin (Backend system) needs to obtain information about students, timetables, or lecturers and for this to be possible, we have chosen to use IIS web server which will fetch data from SQL Server database and will then display those data to the admin.

The Mobile App is made up of different modules. These are as follows:

- **View Statistics**: A student has to select a module from the drop-down box and click on a button to obtain his rate of presence for that module.
- **View Timetable**
- **Attendance monitoring**: When a student clicks on the "ATTENDANCE MON-ITORING" button, then the app will display a map with the student's current location and will check if the student is within the given Geofence. If the app finds that the student is within the right Geofence at the right time then halfway a lecture they will be asked to insert their fingerprint. If the fingerprint matches the one in the database, then the app will insert the student as present. The app will check for the student's location several times during that lecture and will subsequently mark the student as present or absent.
- **Facilities within University of Mauritius**
- **View Students' Attendance**: Lecturers can either select a program to check the attendance of a class or they can input the ID of a student to verify the attendance of that student.

5 Results

The table below compares the different technologies used in different implementations that were mentioned in Sect. 2 (Related works) and the ones used for this project (Mobile-Based Attendance Monitoring) as well as the differences between them (Table 1).

Table 2 gives an overview of the features that the app consists of.

In order for the system to operate effectively, it is important for all the modules to function in a suitable way so that they can deliver the correct output at the right time. The performance of some components of the app is provided in Table 3.

Table 1 Comparison between the different technologies

Project title	Employees turn to biometric technology to track attendance	Journal TICOM Vol. 1 No. 3 bar code scanner based student attendance system	Bringing technology into school: NFC-enabled school attendance supervision	Mobile-based attendance monitoring
Technology	Biometric facial recognition	Barcode scanner	Smart card reader. NFC-enabled mobile phone	GPS. Network Provider. Fingerprints
Security	Secure because of the sensors ability to detect individual facial traits and patterns and based on this, a student will be marked as present	Less secure as someone else can take the card and scan the barcode of that card and this will mark the student as present even if he did not attend the class	Less secure as another person can take the smart card and pass it to a card reader or NFC-enabled mobile phone to mark that student as present	More secure as the sensor will detect only the fingerprint stored in the database and it will also depend on GPS, Geofence and network provider to mark a student as present
Cost	Expensive as the biometric device needs to be purchased and implemented	The barcode scanner has to be bought, therefore, it is expensive to implement	Expensive as a smart card reader has to be purchased and should also provide the teachers with an NFC-enabled mobile phone	No additional equipment has to be purchased, hence, it is cheaper

Table 2 Features with different technologies that make up the app

Technologies	GPS	Network provider	Fingerprints
Proximity	Approximately 6–10 m	Approximately 25 m depending on the cell size	Requires physical touch of the finger
Accuracy	High accuracy	The smaller the cell, the higher the accuracy	High accuracy
Performance	20 s (timer checks location every 20 s)	20 s (timer checks location every 20 s)	368 ms

Table 3 Performance of the different components in the system	Components	Performance (ms)
	Fingerprint	368
	Registration	304
	Login	201
	Push notification	161
	Take attendance	272
	Statistics (percentage)	306
	Statistics (table)	412
	Bar graph	332

5.1 Screenshots of Interfaces

Student's must place their fingerprint on their phone's respective scanner during lectures (see Fig. 1a). When Attendance Monitoring is clicked and GPS is on, the user will be taken to the map's page with the mobile device geolocation (see Fig. 1b, c).

When "Statistics" is clicked (see Fig. 1b), then the user is taken to the statistics page (Fig. 2a) where they can select a module and obtain their rate of presence. If a student clicks on the "TABLE" button (see Fig. 2b) then they can view their attendance in tabular form (Fig. 3).

Lecturers can view the percentage of attendance per week based on their modules.

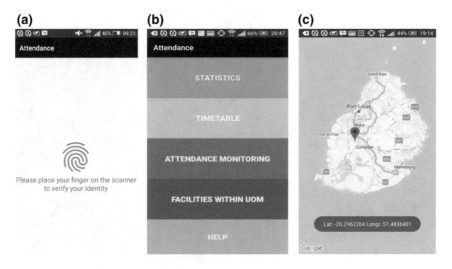

Fig. 1 a, b, c Screenshots of the mobile app

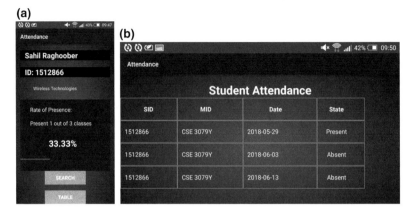

Fig. 2 a, **b** Statistics page

Fig. 3 Bar graph

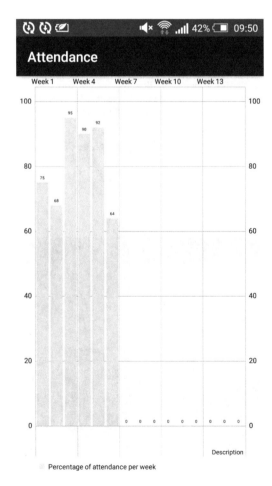

6 Conclusion

It can be asserted that the Android mobile app will be more sophisticated than the traditional system and it will result in less time for taking attendance as well as more reliable outputs will be obtained. Moreover, the mobile app will respond quickly to users' requirements such that if a lecturer wants to view the attendance of a student, they will get the result in just a click of a button. Security will also be maintained and the students will not be able to make corrections to cheat the attendance for the simple reason that the system uses fingerprint recognition and only when the inserted fingerprint will be identified as a valid one then the system will detect whether someone was present or absent for a lecture.

References

1. Pudaruth, S., Nagowah, L., Sungkur, R., Moloo, R., Chiniah, A.: The effect of class attendance on the performance of computer science students. In: 2nd International Conference on Machine Learning and Computer Science (IMLCS'2013), pp. 9–15 (2013)
2. Pudaruth, S., Moloo, R., Chiniah, A., Sungkur, R., Nagowah, L., Kishnah, S.: The Impact of Continuous Assessments on the Final Marks of Computer Science Modules at the University of Mauritius. Available at: https://www.researchgate.net/publication/305032754_The_Effect_of_Class_Attendance_on_the_Performance_of_Computer_Science_Students (2010)
3. Sarah Fister Gale Employees Turn To Biometric Technology To Track Attendance. Available from: http://www.workforce.com/2013/03/05/employers-turn-to-biometric-technology-to-track-attendance (2013)
4. Subramaniam, H., Hassan, M., Widyarto, S.: J. TICOM 1(3) (Mei-Bar Code Scanner Based Student Attendance System). Available from: https://www.researchgate.net/publication/245025631_Bar_Code_Scanner_Based_Student_Attendance_System_SAS (2013)
5. Ervasti, M., Isomursu, M., Kinnula, M.: Bringing Technology into School: NFC-Enabled School Attendance Supervision. Available from: https://www.researchgate.net/publication/38289342_Bringing_technology_into_school_NFC-enabled_school_attendance_supervision (2009)
6. Radical-global.com. RFID Automatic School Attendance System|Radical Global. [online] Available at: http://www.radical-global.com/rfid-student-attendance/ (2017)
7. Rahni, A., Zainal, N., Adna, M., Othman, N., Bukhori, M.: Development of the Online Student Attendance Monitoring System Based On QR-Codes And Mobile Devices. [ebook] Available at: http://jestec.taylors.edu.my/Special%20Issue%20UKM%20TLC%202013_2/UKMTLC%202013_6_2015_2_028_040.pdf (2015)

Hand Gestures Categorisation and Recognition

Maleika Heenaye-Mamode Khan, Nishtabye Ittoo
and Bonie Kathiana Coder

Abstract In this digital era, the focus is now on the development of applications that allow human beings and machines to interact directly. Up to now, many hand gesture recognition systems have been developed for different applications such as sign language recognition and smart surveillance. In recent years, researchers have shown interest in the development of hand gesture recognition applications for dancing movements, which involve dynamic hand gestures. However, there are still various challenges such as extraction of invariant factors, automatic segmentation, the transition between gestures, mixed gestures issues, nature of dynamic hand gestures and occlusions that need to be addressed. This research work aims at developing an application to categorise and recognise the classical "Bharatanatyam" dance hand gestures. Since no online database of "Bharatanatyam" gestures is available to the public for research purposes, a customised database has been built with 900 images, consisting of 15 instances for each hand gesture. In this work, Chain Codes and Histogram of Oriented Gradients (HOG) are proposed for the feature representation of the hand gestures. For the classification of the images, Support Vector Machine (SVM) and K-Nearest Neighbor (KNN) are explored. From the experiments conducted, Chain codes with SVM provide a recognition rate of 99.9% and a false rejection rate of only 0.1%, which is a promising technique for the deployment of movement recognition applications.

Keywords Hand gestures · Chain codes · HOG · SVM · KNN

M. Heenaye-Mamode Khan (✉) · N. Ittoo · B. K. Coder
University of Mauritius, Moka, Mauritius
e-mail: m.mamodekhan@uom.ac.mu

N. Ittoo
e-mail: manishaittoo@outlook.com

B. K. Coder
e-mail: bonie.coder@gmail.com

© Springer Nature Singapore Pte Ltd. 2019
S. C. Satapathy et al. (eds.), *Information Systems Design and Intelligent Applications*, Advances in Intelligent Systems and Computing 863,
https://doi.org/10.1007/978-981-13-3338-5_27

1 Introduction

Gestures are bodily movements which help a person to express himself easily without words. Hand gesture is defined as gestures or movements produced by hands or arms combined to be able to express a person's intention. These movements can be a natural communication between a human being and a machine [1]. Hand gestures can be categorised as static or dynamic gestures. In static gesture, the movement of the hand is not taken into consideration. As for dynamic movement, the sequence of hand gestures is taken over a period of time. Gestures have improved the lives of people, for instance applications such as sign language helps visually impaired people to interact with computers through hand movements. Dance hand gesture recognition is another area that is gaining popularity. This type of application aims at recognising the hand movement and tries to determine the meaning of the gestures. Bharatanatyam is a popular dance that is originated from India, performed by dancers all over the world. It consists of elegant hand movements known as mudras. Many people want to learn classical dance without a dance teacher and thus many choose to learn it by watching some videos and try to mimic the hand movements. However, they may be doing the movement incorrectly as there is no current gesture recognition system to recognise the gestures. Hence to address this obstacle, it is important to deploy automated dance movement verification systems, which can be possible with the growing advancement in the field of computer vision. This application will automatically determine whether the user is doing the correct hand gesture and will help in the self-assessment. In this work, only the hand movements are being considered as the hand is the part of the body that conveys more meaning in the Bharatanatyam dance style. To develop such applications, the most relevant frames have to be extracted from the videos and the appropriate techniques should be adopted at all the development phases.

1.1 Research Gap and Motivation

So far, there are only a few dance hand gestures applications that have been deployed. The hand is one of the smaller objects that form part of the entire human body and has many complex articulations that is subject to segmentation errors [2]. Dynamic image recognition system is facing many challenges like the extraction of invariant features, transition between gestures, automatic filtering and segmentation of features, matching techniques, mixed gestures recognition and complex backgrounds amongst others [3]. Up to now, applications that have been implemented consist of images that are mostly captured from the central viewpoint only. Consequently, the correct gestures performed from different orientations like side views and skewed frontal views are not identified correctly. This implies that even if a user is doing a correct gesture but in a different orientation, the system will fail to recognise the hand gestures. In addition, it is noticed that hand gestures application recognise static images and not sequences of images captured from videos. The sequence of movement is very important in

Bharatanatyam dance and thus have to be taken into consideration. Image classification is yet another important aspect that has to be explored when developing a hand gesture recognition system. Since there is no Bharatanatyam database available online for research purpose, a customised Bharatanatyam database was built from various viewpoints. In this research, a dynamic approach is used to take the sequence of movement into consideration. The paper is organised as follows: Sect. 2 provides a brief literature review, the design of the proposed application and the development of the techniques are discussed in Sect. 3, followed by the experimental results in Sect. 4.

2 Literature Review

Development of hand gesture recognition systems consists of four main parts namely: image capture, preprocessing of hand gestures, representation of features, image matching and image classification. Different researchers have deployed different stages by applying existing image processing techniques or by devising new suitable algorithms. In [3], the authors proposed a static hand gesture recognition system using an Intel Pentium processor and MATLAB R011b. Their dataset comprises 140 images and for 28 mudras, 5 instances were kept for each mudra. In [4], the authors have used clustering technique to classify the images for the development of a dynamic hand gesture recognition application. The input to the system was a video file which considered of 25 or 29 frames per second (fps) depending on video type which was separated into frames by a frame grabber. In the preprocessing phase in [3], the hand image is first converted from RGB to grayscale. Texture-based segmentation was applied to the images to segment the background and foreground using a threshold of 0.7. Sobel edge technique was then applied to obtain the hand boundary. To smooth the uneven thickness of the boundary, morphological shrink operation was adopted. The slopes of the boundary were calculated and approximated based on this straight line using the concept of the decagon. In [4], YCbCr skin colour was used to detect skin regions in each frame. The two morphological operations namely erosion and dilation were applied to remove holes from the thresholded image. The authors in [5] have segmented the image using a threshold of 80. After image segmentation, contours were detected and polygon approximation was adopted to approximate the curves of the images.

In the feature extraction phase, [3] have applied the chain code algorithm. To obtain equal length of chain code, zero padding of the chain codes were carried out. The authors mentioned that standard length of chain code makes the technique simpler and helps in the matching phase. Fourier descriptor , which is invariant to rotation, scaling, translation and starting point of the hand, was adopted to extract features from the hand region in [4]. On the other hand, the authors in [5] have applied the three techniques of chain code descriptor: Freeman Chain Code, Vertex Chain Code and Chain Code Histogram. PCA, which is a dimen-

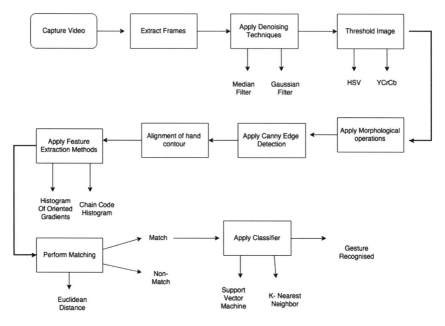

Fig. 1 Design of the proposed hand gesture application

sion reduction technique, was then applied to project the feature vectors of the hand gesture.

To match hand gestures with the gestures found in the database, similarity function was preferred in [3]. The proposed solution resulted in obtaining an accuracy rate of 89.3% and an average time of 3.312 s for the recognition time. In [4], the k-means clustering algorithm was applied on the feature vector to create a distinct class on each gesture and to label them. The system obtained an accuracy rate of 97.03% and a time complexity of 53 ms. In [5], KNN, SVM and Naive Bayes Classifiers for classification. The literature shows that several techniques have been applied to preprocess and classify hand gestures. However, there is a need to explore feature representation techniques to represent the movement of hand gestures and improve the recognition rate of such gestures recognition systems.

3 Deployment of Hand Gesture Application

Figure 1 shows the architectural design of the hand gesture recognition application.

3.1   Creation of Hand Database

An experimental setup was devised at the University of Mauritius to capture hand movements of dancers. Three video cameras were placed at three different positions: front, left and right. It is to be noted that variables such as illumination, the distance between subjects and camera and the position of the cameras have been taken into considerations. Many experiments were conducted to obtain optimal videos. Professional "Bharatnatyam" dancers volunteered to pose as subjects. The subjects were requested to perform the twelve hand movement in the image capture setup. Frames were extracted from the recorded videos to obtain each gesture. Fifteen instances for each gesture were kept in the database as the same gesture may vary slightly as hands have a degree of freedom. The 15 samples of each gesture consisted of 5 for the centre view, 5 for the right view and 5 for the left view. The database was built with 900 images; 450 for the training set and 450 for testing set.

3.2   Preprocessing of Hand Images

First of all, frames are extracted from the videos at a rate of 8 fps. After obtaining the hand gesture, the captured image needs to be preprocessed to obtain the features that will be used for the identification of the hand gestures. Several noise filtering namely Salt and pepper, Median filter, Gaussian filter and Wiener filter were applied on the hand images. From the experiments conducted, the Median filter is able to clean the hand images. This filter is a non-linear smoothing technique which reduces blurring of edges and removes impulse noise [6]. During the filtering phase, every single pixel has been taken into consideration. The pixel value is then replaced with the median of its neighbouring pixel values. The median is computed by first sorting all the neighbouring pixel values in numerical order. The current pixel is then replaced with the computed median. If there are two similar values, the average of these two values is used as median. The resulting hand image is then thresholded. There are two thresholding techniques that can be used namely *HSV* and YCbCr. *HSV* uses hue, saturation and value to define a colour where, hue is the wavelength, saturation is the domination of hue and value refers to the lightness of the colour. Inspired by the work conducted in [7] the skin region is being detected using *HSV* colour space with the following threshold values: $0.05 < H < 0.17$, $0.1 < S < 0.3$, $0.09 < V < 0.15$. As for YCbCr, this colour space is a variation of YIQ colour model, which is a popular colour model normally used for the video signal. YCbCr colour model is used for digital video transformations where Y refers to the luminosity and CbCr refers to the difference in the red and blue chrominance of the picture [8]. The skin region can be detected by using this colour space since an RGB image can easily be converted into YCbCr using the following equation:

$$Y = 0.299R + 0.57G + 0.114B \qquad (1)$$

Note: $Cr = R - Y$ and $Cb = B - Y$

To enhance the output of the thresholding process, morphological operations were used to fill in remaining holes found in the object to obtain blobs. The two basic operators are erosion and dilation [9]. The aim of erosion is to remove the white pixels that usually appears on the foreground of the image. The aim of dilation is to enlarge the white pixels at the borders areas that are normally the foreground pixels. In this way, these operations make the area of foreground pixels become larger in size and makes the holes within those areas smaller [10]. It is noticed that hand recognition system encounters problems with regards to the alignment of the image, which consequently has an impact on the performance. To avoid this issue, the centre of gravity of the hand image has been aligned to the centre of the frame using translation. Canny edge detection is then applied on the image to get the outline of the hand gesture.

3.3 Hand Feature Representation Using Chain Code Histogram (CCH)

Chain code is a technique that is commonly used for image encoding and feature extraction. It extracts feature by representing an object boundary by a connected sequence of integers of specified length and direction. The direction code is either a 4 direction chain code or an 8 direction chain code. To obtain the chain code, the object is being scanned from left to right. The object is being traversed from the starting point till the end pixel [5]. CCH counts the frequency occurrence of each code within the contour and normalises it by dividing it with the total number of codes. Chain codes are a method that provides lossless compression and preserves all topological and morphological information which will be useful in the analysis of line patterns in terms of speed and effectiveness [11]. Inspired by the fact that CCH is translation and scale invariant and starting point invariant, this method is adopted to represent the hand gestures in this research work.

The outline of the hand gesture is used to compute the chain code. The code is obtained by keeping track of the pixel as it moves from one contour to the other in a counterclockwise direction. The direction of each segment is then coded by using a numbering scheme. This means that there is a difference of number in the code whenever the number of direction changes between two adjacent elements of the code. The number of changes in the direction that occur between two pixels is thus recorded. These direction changes are then used to create the feature vector. Thus, the chain code of the boundary is stored in a vector $C1$. The frequency of each code is counted and is stored in a vector $C2$. The chain code histogram $N1$ is then normalised. The chain code histogram is then stored.

3.4 Hand Feature Representation Using Histogram of Oriented Gradients (HOG)

HOG is a technique used for the detection of object [12]. HOG is a local descriptor which can be used to represent hand gestures. This technique is capable of producing the same feature vector for the same hand at any location in the image and it is invariant to lighting. The images are divided into cells, where the histogram of oriented gradients of each cell is computed. A feature descriptor is obtained for each cell. The principle of a feature descriptor is to generalise the object in such a way that the same object produces as close as possible to the same feature descriptor when viewed under different orientations or conditions. This technique is very robust and preferred for classification of images. This method uses the sliding detection window technique, where it is moved around the image. At each position of the detector window, a HOG descriptor is computed for the detection window [12].

In this research work, the hand image is cropped to a 64×128 window. To calculate the HOG descriptor, an 8×8 pixel cells is used within the detection window. The gradient vector at each pixel within the cell is then computed. Thus, for each cell 64 gradient vectors are obtained, which are then placed into a 9-bin histogram. The purpose of the histogram is to quantize the vectors. The mean and the standard deviation of the feature vector is then computed. The normalised HOG is then stored in a matrix.

3.5 Hand Feature Classification Using SVM and KNN

Classification is the phase where each gesture is categorised with respect to its features. Each gesture is assigned to a unique label. In this work, two classification methods namely, support vector machine (SVM) and K-Nearest Neighbor (KNN) are proposed. SVM is a supervised learning algorithm which is normally adopted for classification and regression analysis problems. SVM is usually applied for two-class problems; the first one in which the training data is linearly separable in the input space and the second one in which the training data is not linearly separable to the input. SVM separates the training set into two classes with a hyperplane that maximises the margin between them. A good separation is achieved by a hyperplane when it has the largest distance to the nearest training data points of any class [13]. In this work, many experiments have been conducted to obtain an optimal margin. Note that in the application, different views of the image have been considered and thus making the system more complex. However, the error rates of the applications were carefully computed so that an appropriate hyperplane is obtained to maximise the chance of recognition. KNN is a non-parametric classifier which classifies a test data by identifying its k-nearest neighbor. This technique computes the distance between the objects.

Table 1 Performance of the hand gesture application

Techniques	Recognition rate (RR), %	False acceptance rate (FAR), %	False rejection rate (FRR), %
Chain codes with SVM	99.9	0	0.1
Chain codes with KNN	98.8	0	1.2
HOG with SVM	95.6	2	2.4
HOG with KNN	92.2	1.5	6.3

Table 2 Comparison of similar applications

Papers	Techniques	Performance
[3]	Chain codes	RR: 89.3%
[5]	Chain codes for hand shape	RR: 80–93%
[14]	Hidden Markov model	RR: 90%

4   Experimental Results and Evaluations

The captured video is broken into frames and stored. After the preprocessing phase, the features are extracted. After the application of the CCH, a histogram was generated for each frame. These histograms are then compared with the gestures templates to find matches. It is to be noted that the dance consisted of 12 movements. Different levels of achievement, that is, level 1 for a beginner, level 2, for intermediate and level 3 for professional, were set depending on the number of correct dancing movements. Several sets of experiments were conducted and the average was taken. The three metrics; recognition rate (RR), false acceptance rate (FAR) and false rejection rate (FRR) were used to determine the performance of the gesture recognition system. Several test sets were used to avoid biased results. The results are summarised in Table 1.

From the experiments, chain codes with SVM outperform the other techniques by achieving a recognition rate of 99.9%. The system was also tested with various views of the image, that is, frontal view and side view. It was able to correctly recognise the gestures in spite of the views, making the application a robust one. This is an indication that chain codes with SVM are a promising technique that can be used for motion detection. Though HOG with SVM has a recognition rate of only 95.6, it can still be adopted in various motion detection application. The proposed techniques have been compared with other existing hand gesture applications as shown in Table 2:

Other similar applications have produced a recognition rate between 80 and 93%. Our proposed work has brought some improvement in the recognition rate where applications are able to correctly identify gestures without human intervention.

5 Conclusion

In this work, a hand gesture application, using the Bharatanatyam hand gestures, has been developed. Up to now, many hand gesture applications exist. Nevertheless, the recognition rate was still not satisfactory and at some point in time, human intervention was required to justify the performance of the system. Thus, more effort is required to develop techniques that could enhance the performance of the gesture recognition system. The main challenge in this type of application is to determine the change in movement and capture the appropriate features. In addition, the required frame representing the image has to be extracted correctly. In this research work, various techniques have been studied and analysed. We have created a hand gesture database by setting up an optimal environment taking into consideration the lighting and environment interferences. After feature extraction, we have applied Chain codes and HOG to represent the features. SVM and KNN were the two classifications techniques that have been adopted in the development of the application. A recognition rate of 99.9% was obtained for the chain codes with SVM.

Declaration We have taken the required permission of dataset, images used in this work from the respective authorities. We are solely responsible if any problem that arises in the future.

References

1. Xu, Y., Dai, Y.: Review of hand gesture recognition study and application. Contemp. Eng. Sci. **10**(8), 375–384 (2017)
2. Malima, A.K., Özgür, E., Çetin, M.: A fast algorithm for vision-based hand gesture recognition for robot control. In: IEEE 14th Signal Processing and Communications Applications, Antalya (2006)
3. Saha, S., Konar, A., Gupta, D., Ray, A., Sarkar, A., Chatterjee, P., Janarthanan, R.: Bharatanatyam hand gesture recognition using polygon representation. In: Control, Instrumentation, Energy and Communication (CIEC), 2014 International Conference on, IEEE, pp. 563–567 (2014)
4. Panchal, J.B., Kandoriya, K.P.: Hand gesture recognition using clustering based technique. Int. J. Sci. Res. (Ijsr). **4**(6), 1427–1430 (2015)
5. Fating, K., Ghotkar, A.: Performance analysis of chain code descriptor for hand shape classification. Int. J. Comput. Graph. Anim. **4**(2), 9 (2014)
6. Sonka, M., Hlavac, V., Boyle, R.: Image Processing, Analysis, and Machine Vision. Cengage Learning, 4th edn (2014)
7. Dongare, Y.B., Patole, R.: Skin color detection and background subtraction fusion for hand gesture segmentation. Int. J. Eng. Res. Gen. Sci. **3**(4) (2015)
8. Hearn, D., Baker, M.P., Baker, M.P.: Computer Graphics with OpenGL, vol. 3. Pearson Prentice Hall, Upper Saddle River, NJ (2004)
9. Sarkar, A.R., Sanyal, G., Majumder, S.: Hand gesture recognition systems: a survey. Int. J. Comput. Appl. **71**(15) (2013)
10. Soille, P.: Erosion and dilation. In Morphological Image Analysis (pp. 63–103). Springer, Berlin, Heidelberg (2004)
11. Andrew, M.: Introduction to Digital Image Processing with Matlab, USA: Thomson Course Technology, pp. 353 (2004)

12. Dalal, N., Triggs, B.: Histograms of oriented gradients for human detection. In: IEEE Computer Society Conference on Computer Vision and Pattern Recognition, CVPR 2005, vol. 1, pp. 886–893 (2005)
13. Zou, Z., Premaratne, P., Monaragala, R., Bandara, N., Premaratne, M.: Dynamic hand gesture recognition system using moment invariants. In: IEEE 5th International Conference on Information and Automation for Sustainability (ICIAFs), pp. 108–113 (2010)
14. Chen, F.S., Fu, C.M., Huang, C.L.: Hand gesture recognition using a real-time tracking method and hidden Markov models. Image Vis. Comput. **21**(8), 745–758 (2003)

Infect-DB—A Data Warehouse Approach for Integrating Genomic Data of Infectious Diseases

Shakuntala Baichoo, Zahra Mungloo-Dilmohamud, Parinita Ujoodha, Veeresh Ramphull and Yasmina Jaufeerally-Fakim

Abstract With the expansion of biological data sources available online, integration is a major challenge facing researchers wishing to explore this information. Users often need to integrate data derived from multiple, diverse and heterogeneous sources for investigation. This paper presents the features of Infect-DB, a data warehouse that can localize and integrate genomes of pathogenic species, retrieved from NCBI, based on information from the American Biological Safety Association (ABSA). The list of bacteria and their corresponding host specificity were programmatically accessed from ABSA and integrated into Infect-DB. The list of organisms obtained from ABSA was used to target the automated download of corresponding genomes from the NCBI FTP site. Infect-DB provides a set of analysis tools, including a comparison of genomes using local-BLAST, dN/dS analysis, multiple sequence alignment, phylogenetic analysis and visualization tools. To date, Infect-DB has integrated 854 bacterial genomes from 207 genera considered as important pathogens causing infectious diseases.

Keywords Data warehouse · Genome comparison · Infectious diseases
Analysis tools · Bacterial genomes

S. Baichoo · Z. Mungloo-Dilmohamud (✉) · P. Ujoodha · V. Ramphull · Y. Jaufeerally-Fakim
University of Mauritius, Reduit, Moka, Mauritius
e-mail: z.mungloo@uom.ac.mu

S. Baichoo
e-mail: shakunb@uom.ac.mu

P. Ujoodha
e-mail: pari_ujoodha@yahoo.com

V. Ramphull
e-mail: veereshramphull@gmail.com

Y. Jaufeerally-Fakim
e-mail: yasmina@uom.ac.mu

© Springer Nature Singapore Pte Ltd. 2019
S. C. Satapathy et al. (eds.), *Information Systems Design and Intelligent Applications*, Advances in Intelligent Systems and Computing 863,
https://doi.org/10.1007/978-981-13-3338-5_28

1   Introduction

The ultimate goal of bioinformatics is to enable the discovery of new biological insights and to create a global perspective from which unifying principles in biology can be discerned using computing technologies. DNA sequencing technologies have created astronomical amounts of data. To date, many species have been sequenced including human, rat, chimpanzee, chicken and many pathogens. As the information becomes larger and more complex, there is a need for properly organizing the data and also have relevant computational tools to sort through the massive amount of data.

This paper describes Infect-DB, a data warehouse for infection-causing bacterial species. It integrates data from the NCBI, based on the American Biological Safety Association (ABSA).

1.1   Data Warehouses in Bioinformatics

With the expansion of the biological data sources available online, integration is a new, major challenge facing researchers who wish to explore this information. For a typical research project, a user must be able to merge data derived from multiple, diverse and heterogeneous sources freely and readily.

The data warehouse approach localizes/assembles data sources into a centralized system with a global schema and an indexing system for integration and navigation. A data warehouse depends largely on relational database management systems (RDBMS) and offers the advantage of high-level standard query language (SQL). The biological data sources are quite dynamic and unpredictable, and few of the public biological data sources use structured database management systems. Developing a single data warehouse to integrate all the biological data sources would require considerable effort and a lot of storage space.

1.2   Methodology

The incremental software development model [1] was used to implement Infect-DB. Requirements of the system were obtained through interview of biologists and prospective users. Once the software requirements were obtained, the individual modules were developed and the functionalities were incrementally developed and were validated by end-users. Depending on end-users' feedback, functionalities were improved as necessary. Additional functionalities were then added in the next increments.

2 Related Works

A lot of research has been carried out in the area of biological data integration and several software tools have been implemented for performing analyses and inferring biological conclusions from the integrated data [2, 3]. The following subsections describe some state-of-the-art tools in integrative bioinformatics.

Atlas [4] is a data warehouse, integrating biological sequences, homology information, biological ontologies, molecular interactions and functional annotation of genes on a single platform. Atlas aims at providing a Bioinformatics workbench [4] for inferring biological relationships between the stored entities.

BioDWH [5] is a data warehouse toolbox which retrieves life science information from different public repositories and integrates them into a local database management system. It is mainly used for research and development.

Prokaryotic Genome Analysis Tool (PGAT) is a web application mainly for multistrain analysis of genomes [6]. The main features of PGAT include gene comparison at the sequence level, the computation of pangenome of user-selected strains, the identification of Single-Nucleotide Polymorphism (SNPs) within a set of Orthologs, the determination of the absence or presence of a set of user-defined genes in some metabolic pathways as well as their comparison [6].

IMG [7, 8] is a data warehouse with data from the three domains of life. It is mainly used to compare publicly available genomes, genomes submitted on IMG by sequencing centres, draft genomes and genomes of viruses and plasmids to answer biological queries. It provides a wide variety of analysis tools for genomes, genes and gene functions [8] and integrates third-party platforms such as VISTA [9], Dot Plot and Artemis ACT [10] to enhance the viewing of the genomes in a comparative context.

MicrobesOnline [11] is an integrative web portal integrating genomes from the three domains of life. The key features are an interactive comparison browser for comparative genomics, functional genomics, gene carts and microarrays categories. It integrates genomes available from the NCBI RefSeq database [12] and all sequences are mapped to their taxonomic lineage by making use of the NCBI taxonomy database [13].

MicroScope is a web-based integrative platform for the comparative analysis and annotation of microbial genomes [14–16]. It has been developed with the aim of integrating complete genomes and on-going genomic projects in order to compare their genomic sequences and infer biological conclusions about evolution and natural selection. Microscope offers interactive genome browsers and caters for the manual and expert annotation of genomic sequences during sequencing projects.

All the main tools that have been discussed above are general and, therefore, contain a lot of data. The aim here is to provide a similar tool but targeted only to genomes causing infectious diseases. The tool should be a stand-alone tool that can work even without an internet connection since internet connectivity can be a challenge in some parts of the world.

3   Design of Infect-DB

Infect-DB follows the general schema of a data warehouse whereby biological data
has been taken from disparate sources and integrated into one single repository for
ease of analysis and retrieval. The system architecture of Infect-DB is similar to a
typical data warehouse, as shown in Fig. 1.

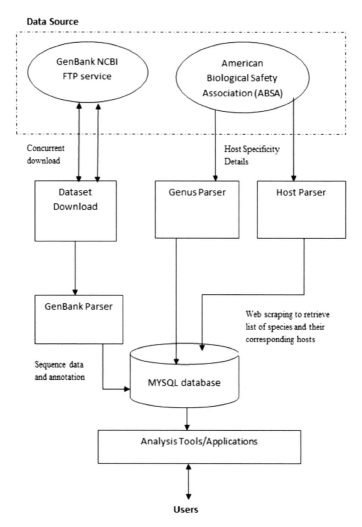

Fig. 1   Data integration model

3.1 Data Sources

The data sources used in building the data warehouse and the justifications for using these specific data sources are detailed in the following subsections.

GenBank NCBI FTP service. GenBank stores bacterial genomes from different sequencing centres and makes them publicly available via FTP service [17]. It is also a central repository which integrates genomes from the EMBL data library [18] and DNA databank of Japan (DDBJ) [19], thus ensuring a uniform distribution of genomes across all the databases.

GenBank was chosen because it is easy to download specific bacterial genomes programmatically using its FTP service. Moreover, GenBank integrates data from the main biological databases like EMBL and DDBJ thus ensuring that its data is reliable and a wide variety of bacterial genomes are available.

American Biological Safety Association (ABSA). The American Biological Safety Association (https://absa.org/) is mainly responsible for addressing the needs of biosafety professionals. With respect to the following needs, ABSA provides a Risk Group Database which contains the list of all infectious bacteria classified by their genus. The database also provides the pathogenicity and host range for all the infectious bacteria.

3.2 Modules

A number of modules have been designed to implement the different functionalities of Infect-DB. They are elaborated below.

Genus Parser. The Genus Parser module handles the extraction of the list of genera from ABSA, filters any duplicate or incorrect information and finally inserts the appropriate list of genera in the database. The list of genera for infectious diseases is already available on the HTML page of ABSA and the data was pulled and filtered directly into the MySQL database. This module ensures a dynamic population of the list of genera for infectious diseases in the data warehouse.

Host Parser. The Host Parser module is mainly responsible to identify the complete set of infectious bacteria and their respective hosts. It is dependent on the Genus Parser module as it uses the list of genera extracted from Genus Parser to automatically mine the list of species and the corresponding hosts (Human, Animal, Plant) they infect. The filtered information is thereby populated in the respective table.

Dataset Download. The Dataset Download module handles the automated downloading of GenBank files for infectious bacteria from the NCBI FTP site. This module caters for the tedious task of connecting, searching and downloading of GenBank files for each genus and species that fall into the category of infectious diseases. This module uses the list of genera and species retrieved from ABSA to search for the corresponding GenBank files and download them concurrently into a specified directory. The concurrent download is achieved by the use of Java multi-threading.

GenBank Parser. The GenBank Parser module parses GenBank files and extracts relevant information such as organism name, source information, sequence length; CDS feature tags including the gene names, locus tags, corresponding protein sequences and gene references amongst others. After the extraction of the specific data, the module performs a sequential insertion of the features of each infectious bacterium in the respective tables of the database. The insertion is performed sequentially to avoid any type of foreign key constraints conflict and to ensure that the complete set of features for a species has been inserted successfully in the database.

Feature Comparison. The Feature Comparison module compares features from a chosen set of genomes using either (1) Gene annotation, (2) Product annotation or (3) alignment of protein sequences using local-BLAST application. The gene annotation and product annotation information are obtained from the GenBank files while they are parsed.

3.3 Database Design

The database schema of Infect-DB consists of six main tables to facilitate the searching and analysis of the biological data stored. The main table is the organism table which stores the primary information such as organism name, GenBank ID, sequence length amongst others, from the GenBank files of specific strains downloaded from NCBI. The organism table is linked to the reference table such that one organism can have different publications or references stored in one GenBank file. The same principle applies to the CDS table which stores all the coding sequences and their respective features for each strain found in the organism table. The genus table and species_host table store information that has been parsed from ABSA and are linked to the organism table. The information stored in these tables are processed programmatically to perform the biological analysis.

3.4 Analysis Tools and Applications

Genomes in the data warehouse can be chosen for analysis either by selecting a list of genomes found in the data warehouse or by starting from genomes that infect a specific host namely human, animal or plant, or a combination of these.

A set of analysis tools have also been implemented and they are: comparison of genomes based on gene annotation, based on product annotation or based on the alignment of protein sequences, a genome browser and a search for genes/products in the entire data warehouse. For genes/product common among different genomes/strains, a user has the facility to: extract the protein/DNA sequences in FASTA format to be used for further analysis, align the protein/DNA sequences using ClustalW, perform a phylogenetic analysis, perform a dN/dS analysis and perform a comparative GC content analysis.

4 Implementation of Infect-DB

4.1 Tools and Libraries Used

A number of open-source tools and libraries have been used to support the development of Infect-DB. The Java programming language was used along with BioJava libraries which provides support for biological problems. The MySQL/MariaDB database was used for data integration and querying. GView, Rs2xml, Guava and Apache Commons Net libraries were used. The GView [20] Java package was used to render and display genomes in either circular or linear format. Jsoup (https://github.com/jhy/jsoup/) was used to parse the HTML pages from ABSA and retrieve the necessary host specificity information in an appropriate format to insert in the database. Rs2xml was used for mapping database queries to JTable models in accordance with the database queries. Guava (https://github.com/google/guava) was used for handling I/O, collections, string processing amongst others, hence ease the comparison of bacterial genomes at the annotation level. The Apache Commons Net library was used for handling the FTP connection and GenBank file transfer from the NCBI FTP site to the application software in the Infect-DB data warehouse. Additionally the BLASTp functionality of the local-BLAST (Basic Local Alignment Search Tool) [21] application was used for comparing protein sequences, TreeView [22] was used to read the Newick format and display the phylogenetic trees, JCoDA [23] was used to rapidly screen for genes and regions of genes under selection using PAML and to perform dN/dS analysis.

4.2 The Infect-DB Application Evaluation

Infect-DB was evaluated by comparing the genomes/strains of the Mycoplasma species which infects both humans and animals. Infect-DB provides the option of performing the comparison of genomes using either annotations or results of BLASTp. In cases where the annotations are consistent, the results of the comparison are meaningful. However, annotations are not consistent in a lot of cases, hence the comparison based on the alignment of sequences gives a more accurate result. For the same four (4) Mycoplasma genomes selected, a comparison based on product annotations gave only 5 common features while comparison based on alignment gave 576 common features.

5 Conclusion

Although a number of tools exist in the field of integrative bioinformatics, the main achievement of Infect-DB is that it is specific to infectious diseases. A novel mechanism to fetch the whole list of infectious bacteria from ABSA and automatically fetch and store their complete genomic data from NCBI has been implemented. The database is regularly updated with newly sequenced genomes.

Moreover, the host specificity information which was absent from all the existing comparative databases was complemented in Infect-DB. Web scraping was used to retrieve and store the information in a standardized format to facilitate querying and searching of host specificity data about a particular infectious organism. Using local-BLAST application, it was possible to find common features (genes/products) in genomes, more accurately than using gene/product annotations. Multi-threading has also been used to cope with the time-consuming BLASTp process. If a powerful server with a number of cores is available, the processing time will be further shortened.

A new algorithm was developed to enhance the visualization of the phylogenetic tree for the comparison of a single gene from multiple species of the same genus, by displaying the labels associated with each organism correctly in TreeView.

The tool aims at providing easy access to whole genomes for researchers in microbiology: food microbiology, plant pathology and medical microbiologists.

6 Future Work

Some improvements can be brought to the implemented system. The existing algorithm for finding common features can be enhanced by comparing protein domains and gene enrichment using the gene ontology. The DataSet download module can be improved to fetch updated information from NCBI upon addition of new or update of existing genome files are.

Additional data sources like KEGG can be used to extract further information about biological pathways or gene functions so as to categorize genes into functional units and perform more efficient analysis of the bacterial genomes. A tool to generate the pangenome of bacterial species using the available strains can be added to the existing analysis tools of Infect-DB, similar to that of PGAT. Lastly, WebAct genome comparison tool [24] can be integrated to conduct whole genome comparisons of existing bacterial genomes display the results graphically.

References

1. Benediktsson, O., Dalcher, D., Reed, K., Woodman, M.: COCOMO-based effort estimation for iterative and incremental software development. Softw. Qual. J. **11**, 265–281 (2003)
2. Triplet, T., Butler, G.: A review of genomic data warehousing systems. Brief. Bioinform. **15**, 471–483 (2014)
3. Ramharack, P., Soliman, M.E.S.: Bioinformatics-based tools in drug discovery: the cartography from single gene to integrative biological networks. Drug Discov. Today (2018)

4. Shah, S.P., Huang, Y., Xu, T., Yuen, M.M.S., Ling, J., Ouellette, B.F.F.: Atlas—a data warehouse for integrative bioinformatics. BMC Bioinf. **6**, 1–16 (2005)
5. Topel, T., Kormeier, B., Klassen, A., Hofestadt, R.: BioDWH: a data warehouse kit for life science data integration. J. Integr. Bioinform. **5** (2008)
6. Brittnacher, M.J., Fong, C., Hayden, H.S., Jacobs, M.A., Radey, M., Rohmer, L.: PGAT : A Multistrain Analysis Resource for Microbial Genomes, vol. 27, pp. 2429–2430 (2011)
7. Markowitz, V.M.: The integrated microbial genomes (IMG) system. Nucleic Acids Res. **34**, D344–D348 (2006)
8. Markowitz, V.M., Chen, I.M.A., Palaniappan, K., Chu, K., Szeto, E., Pillay, M., Ratner, A., Huang, J., Woyke, T., Huntemann, M., Anderson, I., Billis, K., Varghese, N., Mavromatis, K., Pati, A., Ivanova, N.N., Kyrpides, N.C.: IMG 4 version of the integrated microbial genomes comparative analysis system. Nucleic Acids Res. **42**, 560–567 (2014)
9. Mayor, C., Brudno, M., Schwartz, J.R., Poliakov, A., Rubin, E.M., Frazer, K.A., Pachter, L.S., Dubchak, I.: Vista: visualizing global DNA sequence alignments of arbitrary length. Bioinformatics **16**, 1046–1047 (2000)
10. Carver, T., Berriman, M., Tivey, A., Patel, C., Böhme, U., Barrell, B.G., Parkhill, J., Rajandream, M.A.: Artemis and ACT: viewing, annotating and comparing sequences stored in a relational database. Bioinformatics **24**, 2672–2676 (2008)
11. Dehal, P.S., Joachimiak, M.P., Price, M.N., Bates, J.T., Baumohl, J.K., Chivian, D., Friedland, G.D., Huang, K.H., Keller, K., Novichkov, P.S., Dubchak, I.L., Alm, E.J., Arkin, A.P.: Microbesonline: an integrated portal for comparative and functional genomics. Nucleic Acids Res. **38**, 396–400 (2009)
12. Pruitt, K.D., Tatusova, T., Maglott, D.R.: NCBI reference sequences (RefSeq): a curated non-redundant sequence database of genomes, transcripts and proteins. Nucleic Acids Res. **35**, 501–504 (2007)
13. Sayers, E.W., Barrett, T., Benson, D.A., Bolton, E., Bryant, S.H., Canese, K., Chetvernin, V., Church, D.M., Dicuccio, M., Federhen, S., Feolo, M., Geer, L.Y., Helmberg, W., Kapustin, Y., Landsman, D., Lipman, D.J., Lu, Z., Madden, T.L., Madej, T., Maglott, D.R., Marchler-Bauer, A., Miller, V., Mizrachi, I., Ostell, J., Panchenko, A., Pruitt, K.D., Schuler, G.D., Sequeira, E., Sherry, S.T., Shumway, M., Sirotkin, K., Slotta, D., Souvorov, A., Starchenko, G., Tatusova, T.A., Wagner, L., Wang, Y., John Wilbur, W., Yaschenko, E., Ye, J.: Database resources of the national center for biotechnology information. Nucleic Acids Res. **38**, 5–16 (2009)
14. Vallenet, D., Engelen, S., Mornico, D., Cruveiller, S., Fleury, L., Lajus, A., Rouy, Z., Roche, D., Salvignol, G., Scarpelli, C., MeDigue, C.: Microscope: a platform for microbial genome annotation and comparative genomics. Database **2009**, 1–12 (2009)
15. Vallenet, D., Belda, E., Alexandra, C., Cruveiller, S., Engelen, S., Lajus, A., Le Fevre, F., Longin, C., Mornico, D., Roche, D., Rouy, Z., Salvignol, G., Scarpelli, C., Smith, A.A.T., Weiman, M., Medigue, C.: MicroScope—an Integrated Microbial Resource for the Curation and Comparative Analysis of Genomic and Metabolic Data, vol. 41, pp. 636–647 (2013)
16. Vallenet, D., Calteau, A., Cruveiller, S., Gachet, M., Lajus, A., Josso, A., Mercier, J., Renaux, A., Rollin, J., Rouy, Z., Roche, D., Scarpelli, C., Medigue, C.: Microscope in 2017: an expanding and evolving integrated resource for community expertise of microbial genomes. Nucleic Acids Res. **45**, D517–D528 (2017)
17. Wheeler, D.L., Barrett, T., Benson, D.A., Bryant, S.H., Canese, K., Chetvernin, V., Church, D.M., Dicuccio, M., Edgar, R., Federhen, S., Feolo, M., Geer, L.Y., Helmberg, W., Kapustin, Y., Khovayko, O., Landsman, D., Lipman, D.J., Madden, T.L., Maglott, D.R., Miller, V., Ostell, J., Pruitt, K.D., Schuler, G.D., Shumway, M., Sequeira, E., Sherry, S.T., Sirotkin, K., Souvorov, A., Starchenko, G., Tatusov, R.L., Tatusova, T.A., Wagner, L., Yaschenko, E.: Database resources of the national center for biotechnology information. Nucleic Acids Res. **36**, 13–21 (2008)
18. Kanz, C., Aldebert, P., Althorpe, N., Baker, W., Baldwin, A., Bates, K., Browne, P., van den Broek, A., Castro, M., Cochrane, G., Duggan, K., Eberhardt, R., Faruque, N., Gamble, J., Garcia Diez, F., Harte, N., Kulikova, T., Lin, Q., Lombard, V., Lopez, R., Mancuso, R., McHale, M., Nardone, F., Silventoinen, V., Sobhany, S., Stoehr, P., Tuli, M.A., Tzouvara, K., Vaughan, R., Wu, D., Zhu, W., Apweiler, R.: The EMBL nucleotide sequence database. Nucleic Acids Res. **33**, 29–33 (2005)

19. Miyazaki, S., Sugawara, H., Ikeo, K., Gojobori, T., Tateno, Y.: DDBJ in the stream of various biological data. Nucleic Acids Res. **32**, D31–D34 (2004)
20. Petkau, A., Stuart-Edwards, M., Stothard, P., van Domselaar, G.: Interactive microbial genome visualization with GView. Bioinformatics **26**, 3125–3126 (2010)
21. Altschul, S.F., Gish, W., Miller, W., Myers, E.W., Lipman, D.J.: Basic Local Alignment Search Tool. http://www.ncbi.nlm.nih.gov/pubmed/2231712%5Cn, http://www.cmu.edu/bio/education/courses/03510/LectureNotes/Altschul1990.pdf (1990)
22. Page, R.D.M.: Visualizing phylogenetic trees using Treeview. Curr. Protoc. Bioinformatics. Chapter 6, Unit 6.2 (2002)
23. Steinway, S.N., Dannenfelser, R., Laucius, C.D., Hayes, J.E., Nayak, S.: JCoDA: a tool for detecting evolutionary selection. BMC Bioinformatics **11**, 1–9 (2010)
24. Abbott, J.C., Aanensen, D.M., Rutherford, K., Butcher, S., Spratt, B.G.: WebACT–an online companion for the artemis comparison tool. Bioinformatics **21**, 3665–3666 (2005)

Measuring the Influence of Moods on Stock Market Using Twitter Analysis

Sanjeev K. Cowlessur, B. Annappa, B. Kavya Sree, Shivani Gupta
and Chandana Velaga

Abstract It is a well-known fact that sentiments play a vital role and is an incredibly influential tool in several aspects of human life. Sentiments also drive proactive business solutions. Studies have shown that the more appropriate data is gathered and analyzed at the right time, the higher the success of sentiment analysis. This paper analyses the correlation between the public mood and the variation in stock prices towards companies in different domains. For each tweet, scores are assigned to eight predefined moods namely "Joy", "Sadness", "Fear", "Anger", "Trust", "Disgust", "Surprise" and "Anticipation". A regression model is applied to the mood scores and the stock prices dataset to obtain the R-squared score, which is a metric used to evaluate the model. The paper aims to find the moods that best reflect the stock values of the respective companies. From the results, it is observed that there is a definite correlation between public mood and stock market.

Keywords Mood · Lexicon · Regression · Twitter analysis · Stock market
Sentiment analysis

S. K. Cowlessur (✉)
Department of Software Engineering, Université des Mascareignes, Beau Plan, Pamplemousses, Mauritius
e-mail: scowlessur@udm.ac.mu

B. Annappa · B. K. Sree · S. Gupta · C. Velaga
Department of Computer Science and Engineering, National Institute of Technology
Karnataka, Surathkal, Karnataka, India
e-mail: annappa@ieee.org

B. K. Sree
e-mail: kavyasbhagavatula@gmail.com

S. Gupta
e-mail: shivaniarchana@gmail.com

C. Velaga
e-mail: chandanavelaga1@gmail.com

© Springer Nature Singapore Pte Ltd. 2019
S. C. Satapathy et al. (eds.), *Information Systems Design and Intelligent Applications*, Advances in Intelligent Systems and Computing 863,
https://doi.org/10.1007/978-981-13-3338-5_29

1   Introduction

Analysis of the stock market has always posed to be a challenging problem for market researchers. Researchers have come up with innumerable techniques and models [1] to predict the future prices of stocks which helps investors maintain a better portfolio. Social network analysis, which primarily focuses on mining data and analyzing sentiments from Twitter, is now gaining a lot of popularity. Twitter is a micro-blogging site and an individual user post is referred to as a tweet. Research has proved that the aggregate of millions of tweets submitted to Twitter at any given time or for a particular period of time provides an accurate representation of public mood and sentiment. Twitter represents the opinions of the general public which gives a broader perspective to the problem, whereas news articles are limited to expert opinions. With the rising popularity of Twitter analysis and increasing complexity of stock market dynamics, we aim to measure the influence of public moods on stock market prediction.

2   Related Work

Mostafa et al. [2], involves obtaining the dataset by sampling Twitter data and categorizing the words for analysis. The data sampling was done based on five categories: strong buy, buy, for hold, sell, strong sell and weights were assigned to these sentiments. Visualizations and overall sentiment scores were generated which provide an unbiased representation of customers' sentiments towards specific brands and services. A four percent increase was observed in the returns obtained, based on sentiment analysis.

Bollen et al. [3] focus on establishing a correlation between collective mood states derived from large-scale Twitter feeds and the Dow Jones Industrial Average (DJIA) indicator over time. Two mood tracking tools, OpinionFinder and Google-Profile of Mood States (GPOMS) are used to analyze the text content of daily Twitter feeds. OpinionFinder measures positive against negative mood whereas GPOMS measures mood in terms of the six dimensions: "Alert", "Calm", "Happy", "Kind", "Sure", and "Vital". It was found that the inclusion of specific public mood dimensions considerably improves the accuracy of DJIA predictions. An 87.6% accuracy was obtained in predicting the daily up and down changes of the DJIA closing values.

Gilbert and Karahaliois [4] estimate anxiety, worry, and fear from a dataset of over 20 million posts made on the site LiveJournal followed by finding how an increase in expressions of anxiety can predict downward pressure on the SP 500 index, using a Granger-causal framework [5]. The Anxiety index to identify posts as anxious ("anxious", "worried", "nervous", "fearful") or non-anxious ("happy", "angry", "confused", "relaxed", etc.) is defined. The paper concludes that anxiety slows a market climb and accelerates a drop.

Choudhury et al. [6] developed a model to analyze communication dynamics captured by blog posts and comments in the online tech community called Engadget and use this to determine correlations with stock market movement. The model defined consists of contextual features such as number of posts, number of comments, etc. The stock movement of a company and the contextual feature vectors are used in support vector regression framework. The paper concludes with the observation that the support vector regression framework superseded the two baseline methods for predicting the direction of movement.

Zhang et al. [7] predict stock market indicators such as Dow Jones, NASDAQ and SP 500 by analyzing the general mood of the masses on Twitter. The authors concluded that when people express a lot of hope, fear and worry the Dow goes down the next day and when people have less hope, fear and worry, the Dow goes up.

Gloor et al. [8] try to identify trends in three information spheres: web, blogs, and online forums and find the people launching these new trends. First, the authors calculate the betweenness centrality of a concept within the chosen information sphere. Depending on the information sphere, an actor is either a website, a blog or the poster in the online forum. The authors have also given weight to actors based on their influence. Lastly, sentiment analysis is used to classify an opinion as positive or negative using a basic bag of words approach. A combination of these three subproblems gives the web buzz index [9] where moving averages are calculated against eight indices. These indices helped to discover trends in the real world scenario.

3 Methodology

3.1 Data Collection and Preprocessing

In this paper, Twitter data as well as stock data is used. Up to 160,000 tweets on JetBlue and General Electric have been collected using Twitter search API and web crawling. Web scraping is implemented using Beautiful Soup, a Python library for parsing HTML and XML documents. The historical stock prices of the companies for the past one year are collected in CSV format from the Yahoo Finance Board. TwitterTokenizer in Natural Language Toolkit (NLTK) is used to tokenize the tweets to remove Twitter handles and stop words. We always get special events that have occurred during the time period for each of the 2 companies from Yahoo finance. We extract the Twitter and stock data pertaining to 2 weeks after the event has occurred from the original dataset in order to analyze the effects different moods have on the company stock values during such time periods.

3.2 Mood Detection

We define eight moods namely, "Joy", "Sadness", "Fear", "Anger", "Trust", "Disgust", "Surprise" and "Anticipation". Using NRC emotion Lexicon [10, 11], we create a Python dictionary with words in the lexicon and their parts-of-speech (POS) tag as the keys, and we set their values as a list of their association scores [11] on a scale of 0–3, for each of the eight moods. Every tweet is tokenized using TwitterTokenizer in NLTK. Along with tokenizing, we also obtain the POS tags of each token using POS tagger in NLTK. For each token, we obtain the association scores for the eight moods by using the token and the POS tag as keys in the aforementioned dictionary. For each tweet, the eight moods are assigned scores (0–3) taking the average of the association scores for each token in the tweet.

Negation removal, which involves handling negative words (such as "not", "never"), is done by checking each token in a tweet against the negating terms list obtained from Linguistic Inquiry and Word Count (LIWC) [12]. If a token belongs to the negative terms list, the immediate next 2 tokens' mood scores are adjusted to reflect the negation of their meaning. This adjustment is done as follows:

- "Joy" and "Trust" are considered as negative moods while "Sadness", "Fear", "Anger", and "Disgust" are considered as positive moods.
- We evaluate whether the overall emotion of the token is positive or negative by comparing the average scores of the negative and positive moods.
- If the overall emotion of the token is positive, we assign a score of 0 to all the negative moods for the token and vice versa.

Sarcasm detection is done with the help of emoticons [13–15]. The sentiment scores [16] of the emoticon and the tweet [17] are compared to detect the sarcasm. If the overall sentiment of the emoticon is opposite to that of the tweet, we negate the mood scores of the tweet using the same method mentioned above to negate the mood scores of a token.

A matrix with nine columns (eight mood scores, one date of tweet) is generated with each row pertaining to a tweet. We compress this matrix to get mood scores of tweets grouped by date and aggregated by the mean score. A natural join is performed on the "date" attribute between this matrix and the stock closing values. With 365 rows and 8 columns in the data, sufficient data for regression is available without having to under-fit. Figure 1 presents the entire workflow of the paper.

3.3 Regression Model

Using the GridSearchCV module in ScikitLearn [8], an exhaustive search is carried out over specified parameter values for an estimator. This has been used in the linear regression model applied and the value of k in k-fold cross validation used in GridSearchCV is 3. Equation 1 gives the formula used for the multiple regression of the eight moods:

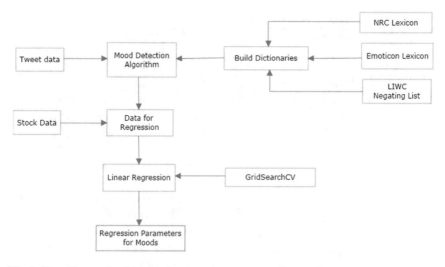

Fig. 1 Flow diagram for obtaining the regression parameters for moods

$$y = ax1 + bx2 + cx3 + dx4 + ex5 + fx6 + gx7 + hx8 + i, \qquad (1)$$

where $x1$, $x2$, $x3$, $x4$, $x5$, $x6$, $x7$ and $x8$ are the moods "joy", "sadness", "fear", "anger", "trust", "disgust", "surprise", and "anticipation" respectively with a, b, c, d, e, f, g, h and i being regression constants obtained after fitting the model on our data. The LinearRegression module in ScikitLearn is used on the regression data. The parameters for linear regression are fine-tuned using GridsearchCV, i.e., the best estimator for the data is obtained. Also, feature selection is applied to the data to eliminate the low variance features in the dataset. SelectKBest functionality in ScikitLearn is used for this purpose.

4 Experiments and Results

In this paper, we use R-squared which is a statistical measure of how close the data is to the fitted regression line.

Experiments are conducted on two chosen companies: JetBlue and General Electric. These experiments are based on three variations:

(1) Changing the lag values from 1 to 10 days.
(2) Considering special events in that time period.
(3) Changing the number of features.

Here are a few sample tweets which are used in testing the mood detection code:

(1) "Not a good flight journey @jetblue!"

Table 1 Mood scores assigned to the sample tweets

Tweet no.	Joy	Sadness	Fear	Anger	Trust	Disgust	Surprise	Anticipation
1	0.00	0.00	0.00	0.00	0.00	0.00	0.50	0.50
2	0.00	0.00	0.00	0.00	0.00	0.00	1.50	0.00
3	1.50	0.00	0.00	0.00	0.00	0.00	0.50	0.00
4	0.33	0.66	0.33	1.00	0.66	0.66	0.00	0.00
5	0.59	0.00	0.00	0.00	0.59	0.00	0.20	0.20

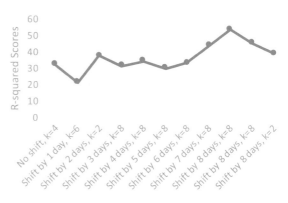

Fig. 2 Effect of lag in stock prices for JetBlue

(2) "what a wonderful flight journey @jetblue :("
(3) "so happy with @jetblue free wifi service"
(4) "i hate the new seats in the @jetblue flight!! :-("
(5) "@jetblue appreciate the effort put in providing a good service".

Table 1 shows the assigned mood scores assigned to the five sample tweets above:

It is observed that "sadness" and "fear" are the most influential features for JetBlue when the complete dataset is used, and "anticipation" and "disgust" are the least influential features. A lag of 8 days is observed to be more accurate, which is shown in Fig. 2. For the subset of data pertaining to special events, "anger" and "disgust" are the most influential features for JetBlue and "anticipation" and "surprise" are the least influential features.

It is also observed that "joy" and "sadness" are the most influential features for General Electric when the complete dataset is used, and "anger" and "disgust" are the least influential features. A lag of 5 days is observed to be more accurate, which is shown in Fig. 3.

For the subset of data pertaining to special events, "anger" and "joy" are the most influential features for General Electric and "anticipation" and "disgust" are the least influential features. In general during special events, Anger is the most significant mood while Anticipation is one of the least significant moods in both the companies.

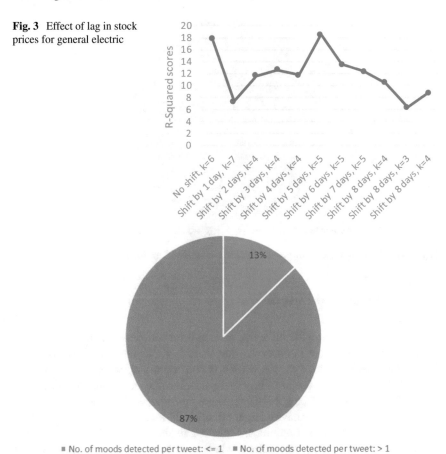

Fig. 3 Effect of lag in stock prices for general electric

Fig. 4 Mood detection per tweet JetBlue twitter data (Jan 2016–Jan 2017)

It is observed that the number of tweets with less than two moods detected, are only 13% for JetBlue whereas the value of the same metric is 32% for General Electric. This distribution of data is depicted in Fig. 4 for JetBlue and Fig. 5 for General Electric. Along with this and the comparatively less R-squared values for General Electric, it is inferred that the public is more expressive in terms of emotions when it comes to JetBlue and this is the reason for a better effect of moods on the stock market prediction.

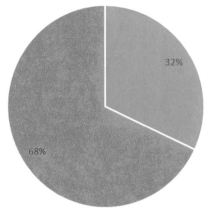

■ No. of moods detected per tweet: <= 1   ■ No. of moods detected per tweet: > 1

Fig. 5 Mood detection per tweet general electric twitter data (Jan 2016–Jan 2017)

5   Conclusion and Future Work

In this paper, we measure the influence of moods on company-specific stock prices using Twitter analysis. It is observed that there is a definite correlation between public mood and the stock market, which is consistent with the existing research. In general, "sadness" is found to be the most influential mood and "anger" is the most influential mood when special events are considered. Observations have also revealed that public moods have a larger effect on the stock market when people are more expressive about their opinions. More companies will be taken into research in the future, to generalize the results at a larger scale. Also, further research can be done in improvising the mood scoring and assigning weights to special events.

References

1. Siew, H.L., Nordin, M.J.: Regression techniques for the prediction of stock price trend. In: International Conference on Statistics in Science, Business and Engineering, Langkawi, pp. 1–5 (2012)
2. Mostafa, M.M.: More than words: social networks text mining for consumer brand sentiments. Expert Syst. Appl. **40**(10), 4241–4251 (2013)
3. Bollen, J., Mao, H., Zeng, X.: Twitter mood predicts the stock market. J. Comput. Sci. **2**, 1–8 (2011)
4. Gilbert, E., Karahaalios, K.: Widespread worry and the stock market. In: Proceedings of the Fourth International Conference on Weblogs and Social Media, pp. 58–65 (2010)
5. Granger, C.W.J.: Some recent development in a concept of causality. J. Econometrics **39**(12), 199–211 (1988)
6. De Choudhury, M., Sundaram, H., John, A., Seligmann, D.D.: Can blog communication dynamics be correlated with stock market activity? In: Proceedings of the Nineteenth ACM Conference on Hypertext and hypermedia, pp. 55–60 (2008)

7. Zhang, X., Fuehres, H., Gloor, P.A.: Predicting stock market indicators through twitter "I Hope It Is Not as Bad as I Fear". In: Proceedings of Collaborative Innovations Networks Conference, pp. 1–8 (2010)
8. Gloor, P., Krauss, J., Nann, S., Fischbach, K., Schoder, D.: Web science 2.0: identifying trends through semantic social network analysis. In: IEEE Conference on Social Computing, Vancouver (2009)
9. Gloor, P.A.: Forecasting social movements by web buzz analysis. In: Proceedings of the First ACM International Workshop on Hot Topics on Interdisciplinary Social Networks Research, ACM, New York, pp. 32–32 (2012)
10. Mohammad, S.M., Turney, P.D.: Emotions evoked by common words and phrases: using mechanical turk to create an emotion lexicon. In: Proceedings of the Workshop on Computational Approaches to Analysis and Generation of Emotion in Text, pp. 26–34. ACL (2010)
11. Mohammad, S.M., Turney, P.D.: Crowdsourcing a word-emotion association lexicon. Comput. Intell. **29**(3), 436–465 (2013)
12. Pennebaker, J.W., Booth, R.J., Francis, M.E.: LIWC2007: Linguistic Inquiry and Word Count. Austin, Texas (2007)
13. Hogenboom, A., Bal, D., Frasincar, F., Bal, M., de Jong, F.M.G., Kaymak, U.: Exploiting emoticons in sentiment analysis. In: Proceedings of the 28th Annual ACM Symposium on Applied Computing, pp. 703–710 (2013)
14. Hasan, M., Rundensteiner, E., Agu, E.: EMOTEX: detecting emotions in twitter messages. In: Proceedings of the 6th ASE International Conference on Social Computing, Academy of Science and Engineering (2014)
15. Kralj, N.P., Smailovi, J., Sluban, B., Mozeti, I.: Sentiment of emojis. PLoS One **10**(12), 1–22 (2015)
16. Gupta, N., Abhinav, K.R., Annappa, B.: Fuzzy sentiment analysis on microblogs for movie revenue prediction. In: International Conference on Emerging Trends in Communication, Control, Signal Processing and Computing Applications, The Oxford College of Engineering, Bangalore, IEEEXplore, pp. 1–4. https://doi.org/10.1109/c2spca.2013.6749350 (2013)
17. Nidhi, R.H., Annappa, B.: Twitter-user recommender system using tweets: a content-based approach. In: International Conference on Computational Intelligence in Data Science(ICCIDS), Chennai, pp. 1–6. https://doi.org/10.1109/iccids.2017.8272631 (2017)

ERP and BI Integration for Municipalities

Priyanka Ramessur, Yajma Devi Kaidoo and Bibi Zarine Cadersaib

Abstract Enterprise Resource Planning (ERP) systems and Business Intelligence (BI) are currently being used in municipalities in both developed and developing countries. Based on the literature review, it was found that ERP and BI bring along opportunities and challenges, which may depend on the type of city or town. From a fact finding process, it was found that the municipalities in Mauritius do not make efficient use of existing resources and through a citizen survey, citizens have put forward for many improvements in the complaint service. Therefore, the aim of this paper is to present an ERP framework, complemented with BI and Twitter Sentiment Analysis to address complaints. A brief evaluation was carried out which showed that the framework could be a possible solution for enhancing the efficiency of municipalities.

Keywords ERP · BI · Municipality · Twitter · Stanford core NLP

1 Introduction

Over the years, ERP systems and BI technologies have become strategic tools for organizations to improve business functions. Although these technologies are more prevalent in the private sectors, the value they can bring to local governments must not be overlooked. There are five municipalities in Mauritius [1] which have citizen-centered objectives [2]. They are concerned with delivering high-quality services to citizens and ultimately rely on effective decision making processes. All the municipalities have the same departments which include administration, pub-

P. Ramessur · Y. D. Kaidoo · B. Z. Cadersaib (✉)
University of Mauritius, Reduit, Moka, Mauritius
e-mail: z.cadersaib@uom.ac.mu

P. Ramessur
e-mail: priyanka.ramessur@umail.uom.ac.mu

Y. D. Kaidoo
e-mail: yajma.kaidoo@umail.uom.ac.mu

© Springer Nature Singapore Pte Ltd. 2019
S. C. Satapathy et al. (eds.), *Information Systems Design and Intelligent Applications*, Advances in Intelligent Systems and Computing 863,
https://doi.org/10.1007/978-981-13-3338-5_30

lic infrastructure, finance, land use and planning, welfare and public health. Their functions include administrative and financial control, managing taxes, constructions and maintenance of road infrastructure and public institutions, environment and public safety. The complaint service spreads over all the departmental activities and provides a means for municipalities to work towards citizen needs. Therefore, extracting knowledge from complaints and taking decisions backed by these facts, can help municipalities in providing better services to citizens. Despite being already equipped with an ERP system, most municipal services in Mauritius run on disparate systems thereby providing minimal synchronization between systems, data retrieval issues, time wastage and loss or duplication of data. For the complaint service, the various sources of complaints include the local government portal, the municipality websites, phone calls, letters and from the information desk center. However, basic tools and human expertise are used to address the complaints and only complaints from the website is easily tracked. BI can be integrated with the ERP to enhance decision making for complaint service and provide better services to citizens. The research questions are therefore set as follows in an attempt to evaluate the feasibility of integrating ERP and BI in view of improving service delivery.

RQ1. Can ERP and BI be used to improve the operations of municipalities?
RQ2. How does the municipal complaint service in Mauritius operate?
RQ3. Can an ERP integrated with BI be used to improve quality of service delivery by analyzing complaints data?

The following sections address each research question. Section 2 addresses RQ1, Sect. 3 addresses RQ2 and finally RQ3 is addressed in Sects. 4 and 5.

2   ERP and BI in the Context of Municipalities

A systematic literature review (SLR) [3] has been carried out to identify the current status of research being done relating to ERP and BI in the context of municipalities. However, due to space constraints, this section has been limited to only the main results of the SLR. The analysis was based on a final selection of 17 papers [4–20] out of an original pool of 836 papers from digital libraries including Google Scholar, IEEE Xplore, Research Gate and Science Direct. The major findings are provided in Table 1. ERP and BI are present in municipalities and local governments in both developed and developing countries and provide a myriad of opportunities, despite being associated with challenges. Therefore, it can be concluded that ERP and BI do have their place in the context of municipalities provided that the public authorities have the resources to handle the challenges associated.

Table 1 Main findings of the use of ERP and BI in the context of municipalities

• ERP and BI/analytics adoption are more dominant in private sectors
• Limited presence is noted in local governments of both developed and developing countries
• Implementation challenges include organizational culture, technical complexity and cost (especially for developing countries)
• Beneficial include improved operational performance, transparency and decision making process
• Complaint resolution is faster with integration of services

3   Context Analysis

Municipal interviews and a citizen survey have been conducted to understand the current complaint service provided by the municipalities. Citizens of different gender, age ranges, literacy levels and work background were considered to avoid bias. A total of 79 citizen responses were collected after scraping the answer of those who did not live in the city or town. This still represented an error margin of 11.02% by considering a total of 62,523 adults [21], a confidence level of 95% and a margin of error of 5%. The results are summarized in Table 2.

4   Proposal

Section 2 showed that ERP integrated with BI solutions are successfully being used in both developed and developing countries and from Sect. 3, main issues that are expected to be addressed by the proposed framework have been identified. These are summarized in Table 3. The case provides a unique ID for each identified issue such that Mu represents an issue identified by the municipal perspective, Ci represents an issue identified from the citizen perspective and Mu_Ci represents an issue identified by both citizens and municipal representatives.

Based on the outcomes of Sects. 2 and 3, a proposed framework is designed which consists of three main components namely; the complaint dashboard, BI reporting and sentiment analysis. The technologies used include *Odoo ERP, Microsoft Power BI, Apache Spark Streaming and Stanford Core NLP*. *Odoo erp* is an open-source ERP system, whereby the source codes are easily available and customized. To have a more customized view for the specific municipality, a new custom module (complaint dashboard module) is built in *Odoo* which provides a view of all information related to the complaints captured from the website. This module consists of the BI reporting component where the visual representations of the complaints data are published. The representations are created using *Microsoft Power BI*, whereby the data extracted from the *PostgreSQL* database are easily represented visually into bar charts, line charts and other forms of representations. To maintain accuracy and relevancy of

Table 2 Findings of the current status of the municipal complaint service

General findings	Positive aspect	Negative aspect
Perspective of complaint officer on complaint service		
• Manual validation of complaints • Letters are used to communicate with complainants • Complaints from the citizen support unit (CSU) portal are sent to municipalities	• Citizens are always notified • Complaints are directly traced on CSU website by citizens • Statistics are received through CSU	• Delayed process due to red tape • Difficulty in tracing back complaints • Difficulty in using complaints for analysis
Perspective of citizens on complaint service		
• Not aware of online complaint submission facility (59.7%) • Preferred making complaints in person (56.9%) or through calls (39.7%)	• Inform complainants when complaints are addressed (56.7%) • Willing to answer queries (71.8%) • Can submit complaints, requests and queries on CSU portal (48.6%)	• Latency in addressing complaints (65.8%) • Lack of transparency on how the complaints are addressed (44.3%) • Unaddressed complaints (19%) • Make other public authorities responsible (15.2%)
Perspective of citizens on municipal social media interaction		
• Are on social media platforms (93.7%) • Facebook (94.6%), YouTube (85.1%) • Use social media over more than 2 h daily (48.6%)	• Agreeable to use (58.2%), somewhat agreeable (29.1%), not agreeable (12.7%) • Easy interaction (72.2%) • Quicker (55.6%) • Flexible (72.2%) • Innovative idea (43.1%)	• Unsure that the authorities would actually make use of it (50.6%) • Privacy (35.6%) and public posts (26.6%) • Making complaints in person is better (12.7%)

the citizen data and to improve the quality of the proposed framework, *Twitter* has been used as a second source of data. Although the citizen survey revealed that 94.6% preferred *Facebook*, *Twitter* has been used due to the privacy restriction on Facebook data. *Apache Spark Streaming*, which allows for processing of streaming data, has been used to streamline the tweets in near-real time. Keywords such as the name of the city and service have been used to filter through the tweets to get the most relevant tweets. The tweets are then analyzed using the *Stanford Core NLP* libraries, to get the sentiments, which are stored in the database. This information can then be viewed in the complaint dashboard module. The latter is essentially the primary view of the system, whereby complaint details, user details, BI visualizations and Twitter sentiments are displayed for the officers to analyze and use for decision making purposes. The proposed framework can be seen in Fig. 1.

Table 3 Addressing issues identifies by municipal representative and citizens

Case	Identified issue	Proposed solution	Expected outcome
Mu_Ci_1	Citizens are updated only if the complaint is addressed or closed; latency in addressing complaints	Use of automatic emails notification for complaints (using ERP)	Complainant is informed of any update made by the receiving authority in real time
Mu_1	Statistics' reports from CSU are in the form of images and printed copies	Use dynamic reports which are updated in real-time (using BI)	Dynamic visualizations
Mu_2	Delayed process due to red tape	Department wise access to system to manage complaints and view reports (using ERP)	Complaints are received, stored and distributed automatically across departments
Mu_3	Difficulty in tracing back complaints	Use of email for complainants and filtering for complaint officers (using ERP)	Complainants have a copy of the complaint details via email; officers can search for complaints based on specific needs
Mu_Ci_2	Complaint officers do not close or provide feedback on addressed complaints	As complaints are addressed by the officers, all updates including the officer's identity are tracked (using ERP auditing)	Based on the changes made, norms can be introduced to ensure everybody abides to it
Mu_4	Following rigid procedures to address each complaint	Introduce criticality of complaint (using filters in ERP)	Officers can address the complaints based on the severity
Ci_1	Expected social media presence for municipalities	Twitter sentiment analysis (using spark streaming and stanford core NLP)	Officers can view sentiments of citizens for each service

5 Evaluation

The proposed system was evaluated according to its identified key components and the expected outcomes as well as by a municipal representative. The results are provided in Table 4. The following score were used: 0 if the actual outcome did not coincide with the expected outcome, 0.5 if the actual outcome partly coincided with the expected outcome, and 1 if the actual outcome completely coincided with the expected outcome.

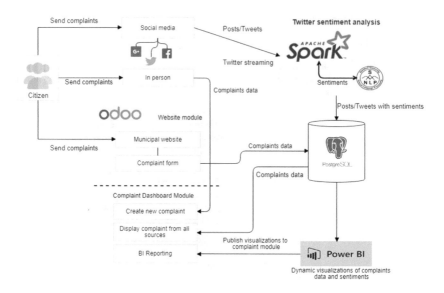

Fig. 1 Proposed framework for ERP and BI integration solution

From Table 4, it is observed that the total score per case is greater than 1 except for Ci_1 (0.5). However, the perspective of only one user and one municipal staff is not enough to be conclusive. Hence, the framework should be evaluated on a larger scale to provide a more accurate representation of its actual value in handling municipal complaints.

6 Limitations and Future Work

The framework should be assessed by several municipalities and different staff to ensure views from different angles are considered to provide a more holistic evaluation. Local data could not be gathered for the twitter sentiment analysis and thus an open dataset was used to test the framework. As part of the future works, the proposed framework shall be tested by using Business Analytics to give predictive reports to help the municipality plan for future happenings and using Facebook instead of Twitter. Other areas include updating the framework based on the evaluation results as well as exploring cloud based options.

Table 4 Results of the evaluation of the proposed framework

Case	Authors' score	Municipal rep score	Authors' comment	Municipal rep comment
Mu_Ci_1	1	0.5	Automated email improve the response time of the authorities	Allow user to keep track of complaint on website
Mu_1	0.5	1	Other tools can be used to do advanced analytics	Good to identify certain issues with a few clicks
Mu_2	1	0.5	Each department has direct access to complaints, thereby preventing red tape	Consider hierarchy of a municipality in terms of task assignment
Mu_3	0.5	1	Identify more possibilities of tracking complaints other than emails for citizens	Provides a good way to filter complaints easily, with custom filters and grouping
Mu_Ci_2	1	1	Can help identify anomalies in handling complaints	Can prove to be an effective solution
Mu_4	1	1	Provides a flexible way of addressing complaints effectively	Can address critical complaints first
Ci_1	0.5	0	Test framework using Facebook as mauritians are more present on it	Sentiments of citizens are likely to be negative because of perception

7 Conclusion

The aim of this paper was to evaluate the implementation of ERP and BI in munici-palities. As such, interviews and surveys were conducted, while focusing mainly on the complaint service. Based on the output of the fact finding exercise, a framework was proposed in an attempt to solve the issues identified in the context analysis. An evaluation was carried out, whereby relevant feedback and suggestions were noted. Future works entail in a thorough evaluation of the framework together with consid-eration of other sources of social media such as Facebook, which is more popular in Mauritius.

Acknowledgements We would like to thank the municipal authority and all the people who devoted their time to participate in this work in one way or another.

References

1. Pay Research Bureau: Local Authorities. PRB Report 2016. [online], p. 1. Available at: http://prb.pmo.govmu.org/English/PRB%20Reports/report2016/Pages/vol232016. aspx. Accessed 12 Oct 2017
2. La.govmu.org.: Portal of Local Authorities, Mauritius. [online] Available at: https://la.govmu. org/index.jsp. Accessed 12 Oct 2017
3. Kitchenham, B.: Procedures for Performing Systematic Reviews, vol. 33, pp. 1–26. Keele, UK, Keele University (2004)
4. Adelakun, O.: The Role of Business Intelligence in Government: A Case Study of a Swedish Municipality Contact Center (2012)
5. Nycz, M., Polkowski, Z.: Business intelligence in a local government unit. In: Proceedings of Informing Science and IT Education Conference (InSITE), pp. 301–311 (2015)
6. Chandiwana, T., Pather, S.: A citizen benefit perspective of municipal enterprise resource planning systems. Electronic J. Inf. Syst. Eval. **19**(2) (2016)
7. Hartley, K., Seymour, L.F.: Towards a framework for the adoption of business intelligence in public sector organisations: the case of South Africa. In: Proceedings of the South African Institute of Computer Scientists and Information Technologists Conference on Knowledge, Innovation and Leadership in a Diverse, Multidisciplinary Environment, pp. 116–122. ACM (2011)
8. Elragal, A.: ERP and big data: the inept couple. Procedia Technol. **16**, 242–249 (2014)
9. Babu, M.P., Sastry, S.H.: Big data and predictive analytics in ERP systems for automating decision making process. In: 2014 5th IEEE International Conference on Software Engineering and Service Science (ICSESS), pp. 259–262. IEEE (2014)
10. Orosz, I., Orosz, T.: Business process reengineering project in local governments with ERP. In: 2012 7th IEEE International Symposium on Applied Computational Intelligence and Informatics (SACI), pp. 371–376. IEEE (2012)
11. Uwizeyemungu, S.: Motivations for ERP Adoption in the Public Sector: An Analysis From "Success Stories" (2005)
12. Teixeira, R.F.D.S., Afonso, F., Oliveira, B., Portela, F., Santos, M.F.: Business intelligence to improve the quality of local government services: case-study in a local government town hall. In: 6th International Conference on Knowledge Management and Information Sharing (KMIS 2014)
13. Gupta, S., Kumar, V., Chhabra, S.: A Window view of prospects of ERP implementation in municipal corporation of Delhi. Int. J. Emerg. Technol. Adv. Eng. Available at: https://www. ijetae.com (ISSN 2250–2459)
14. Reddick, C.G., Chatfield, A.T., Ojo, A.: A social media text analytics framework for double-loop learning for citizen-centric public services: a case study of a local government Facebook use. Gov. Inf. Q. **34**(1), 110–125 (2017)
15. Anna, K., Nikolay, K.: Survey on big data analytics in public sector of russian federation. Procedia Comput. Sci. **55**, 905–911 (2015)
16. Abai, N.H.Z., Yahaya, J.H., Deraman, A.: Incorporating business intelligence and analytics into performance management for the public sector issues and challenges. In: 2015 International Conference on Electrical Engineering and Informatics (ICEEI), pp. 484–489. IEEE (2015)
17. Fildes, R., Pidd, M.: The Benefits of Analytics in the Public Sector. UK (2011)
18. Schenk, T.J., Burman, B.: Big Data and Analytics: Driving Performance and Improving Decision-Making, Chicago: s.n (2015)
19. Beal, J., Prabhakar, B.: Public Sector ERP (2010)
20. Ramburn Gopaul, H.: A systematic analysis of ERP implementation challenges and coping mechanisms: the case of a large, decentralised, public organisation in South Africa, Doctoral Dissertation, University of Cape Town (2016)
21. Electoral.govmu.org.: Office of the Electoral Commissioner—Election Results. [online] Available at: http://electoral.govmu.org/English/Pages/Election-Results-Mun2015.aspx. Accessed 28 Jan 2018

A Smart Mobile Health Application for Mauritius

Muzammil Aubeeluck, Umar Bucktowar, Nuzhah Gooda Sahib-Kaudeer and Baby Gobin-Rahimbux

Abstract This paper presents a smart health mobile application for Mauritius. One of the main functionalities of the application is a symptom checker and diagnosis tool which provides a bridge between doctors and patients. Once patients input or select their symptoms, the application proposes a list of possible diseases and relevant specialist doctors are recommended through the use of a rule-based expert system. Patients are able to directly book appointments with specialists. The mobile application uses intelligent techniques in order to create a knowledge base of diseases and their related symptoms for Mauritius. The knowledge base also includes information related to the medical specialist for each disease so that relevant doctors can be recommended to patients. The application continues to improve itself as it learns from the symptoms input by users, that is, over time, it learns which symptoms are more likely to occur together. Patients can communicate their symptoms to doctors prior to their appointments, and they can also receive notifications from doctors in case of appointment cancellation.

Keywords Artificial intelligence · Machine learning · Ontology

1 Introduction

In recent years, technology has integrated and improved health care around the world. Artificial Intelligence (AI) in health care has exceeded expectations; IBM Watson

M. Aubeeluck · U. Bucktowar · N. Gooda Sahib-Kaudeer (✉) · B. Gobin-Rahimbux
University of Mauritius, Reduit, Mauritius
e-mail: n.goodasahib@uom.ac.mu

M. Aubeeluck
e-mail: muz.aubee@gmail.com

U. Bucktowar
e-mail: umarbucktowar@gmail.com

B. Gobin-Rahimbux
e-mail: b.gobin@uom.ac.mu

© Springer Nature Singapore Pte Ltd. 2019
S. C. Satapathy et al. (eds.), *Information Systems Design and Intelligent Applications*, Advances in Intelligent Systems and Computing 863,
https://doi.org/10.1007/978-981-13-3338-5_31

for Oncology[1] uses AI combined with data mining to provide treatment plans for cancer patients, and it has recommended the same treatments as doctors for 99% of patients in a study of 1000 cancer diagnoses as reported by [1]. Thus, it is clear that technologies such as AI, data mining and machine learning can be used to create intelligent systems that can support medical practitioners to deliver high-quality health care around the world. Intelligent systems are able to collect data, analyse it and make decisions relevant in real-world contexts, and with mobile devices becoming increasingly popular, they can be integrated in mobile applications to cater to people's healthcare needs 'on the go'.

In Mauritius, technological innovations in health care are scarce as the healthcare systems are still mostly manual and health information is rarely available in a digital format. When a person is sick, he can either go to a public hospital or a private clinic where he will be seen by a general practitioner. If the latter cannot provide the necessary treatment, the person will be redirected to a specialist doctor. In both cases, it can take time and be costly before a patient can be diagnosed and see a relevant specialist doctor. In fact, there are several mobile applications that can help patients in diagnosis depending on the symptoms being experienced for example, Symptomate-Symptom Checker,[2] WebMD[3] and iTriage Health[4], etc. In iTriage Health, patients can also directly book an appointment with a doctor. However, all the applications described above have been developed for the European or North American context and therefore, it is not fully applicable to the Mauritian context. This is so because some diseases are more common in Mauritius or different strains of the same disease may exist in different regions.

Therefore, the main aim of this paper is to describe a smart mobile health application for Mauritius, taking into consideration the local context. The application will provide a bridge between doctors and patients that will allow for quicker diagnosis of patients, and will recommend the relevant specialist through the use of a rule-based expert system. Patients will thus be promptly directed to the correct specialist doctors and can begin their treatment at the earliest. This health application can contribute significantly to improving health care in Mauritius as it can be used in public hospitals in order to help the staff in the triage of patients for their respective departments.

The rest of the paper is structured as follows: Sect. 2 focusses on intelligent systems and existing mobile health applications, Sect. 3 provides a detailed description of the components of the architecture of the application while in Sect. 4, the implementation of the application is discussed. Finally, the user evaluation is reported in Sect. 5 and in Sect. 6, possible future works are outlined.

---

[1] https://www.ibm.com/watson/health/oncology-and-genomics/oncology/.

[2] www.symptomate.com.

[3] https://www.webmd.com/webmdapp.

[4] www.itriagehealth.com.

2 Background Study

2.1 Intelligent Systems

Intelligent systems use techniques from different fields to collect data, analyse it and make inferences in a real-world domain. They employ different technologies such as artificial intelligence, machine learning and data mining to aid users in completing complex tasks. One type of intelligent systems is an expert system which uses artificial intelligence to simulate the judgment and behaviour of a human expert in a specific domain [2]. Expert systems consist of a knowledge base which contains the domain knowledge and an inference engine which considers the user query and derives a conclusion based on the knowledge.

The knowledge in an expert system can be represented using an ontology which is [3] first defined as 'the basic terms and relations comprising the vocabulary of a topic area as well as the rules for combining terms and relations to define extensions to the vocabulary'. Ontologies are versatile and they enable sharing and reuse across applications, thus contributing to sustainable development. There are many ontologies which have been developed for the medical domain for example, the GALEN ontology for medicine [4] and LinkBase®, a biomedical ontology which covers various aspects of medicine, including procedures, anatomy, pharmaceuticals and various disorders and anomalies [5]. These ontologies are complex and cover a wide range of medical concepts and thus reuse becomes difficult as they can be more complex to understand. Simpler ontologies that are more relevant to the scope of this paper can be found in the existing works: a Disease Ontology [6] which is a comprehensive knowledge base attempting to unify all terminology relevant to human diseases and a Symptom Ontology [7], which is an ontology of symptoms including all changed perceived by the patient as indicative of a disease.

2.2 Existing Mobile Health Applications

Given the popularity of smart devices around the world, many mobile health applications have been developed in different countries and a few of them are described in the following:

Symptomate-Symptom Checker: This application has been designed by doctors to diagnose patients using an artificial intelligence algorithm and a medical database of over 1000 symptoms and 500 diseases. Users enter demographic information such as age, gender, height and weight as well as their symptoms. The application asks the user further questions and finally, a list of possible causes of the symptoms is presented to the user. It is available to users worldwide.

WebMD: This is a popular website and mobile application that supply useful health information. It includes a symptom checker which provides medical information about conditions relevant to the user's symptoms. It has several decision-support tools such as drugs and treatments, first-aid information and local health listings. The application allows users to search for physicians, hospitals and pharmacy in European countries and hencem these services are not available in Mauritius. WebMD provides an alphabetically sorted list of diseases and for each disease, useful health information such as symptoms, possible treatments and self-care are provided. The symptom checker is interactive and allows the user to click on different parts of the body to find and choose symptoms.

iTriage Health: iTriage Health is a cross-platform healthcare information application developed by two physicians which gives information on a person's symptoms, health conditions and medical procedures. It has a directory of hospitals, pharmacies, clinics, physicians and emergency care in many countries but it is unavailable in Mauritius. The application allows a user to search for doctors by specialty in European countries. The details and location of the doctor can be viewed on a map and the user can get directions through Global Positioning System (GPS).

Mobile applications discussed previously cannot be fully used in Mauritius for various reasons including the fact that they have been developed for other regions and thus the specificities of the Mauritian healthcare systems are not considered. For example, the details of doctors and hospitals in Mauritius are not included. Thus, there is a need to develop an application that considers the local context and encompasses knowledge about diseases that are most likely to occur in Mauritius. In [8], the previously mentioned Disease Ontology [6] and Symptom Ontology [7] have been merged to create a disease-symptom ontology that can be used as the knowledge base for intelligent systems. However, discussions with medical practitioners in Mauritius revealed that most of the diseases are relevant to the American-Canadian context and many diseases that are common in Mauritius are not represented. As such, there is the need to create a disease-symptom ontology tailored for Mauritius [9].

3   Overview of Architecture

The architectural design (Fig. 1) of the system is a three-tier architecture: The client side will consist of the patient application. Information about doctors and patients is stored in Firebase on the cloud. The server side contains the Disease-Symptom Ontology and the Reasoner which are described in the following.

3.1   Disease and Symptom Ontology

In order to build a disease-symptom ontology for Mauritius, information about diseases was first gathered from the latest Health Statistic Report, published by the Ministry of Health and Quality of Life [10]. From the report, 47 most common diseases in Mauritius were retrieved. Since the report contained only diseases and not symptoms, web scraping was used to find the corresponding symptoms, specialty and

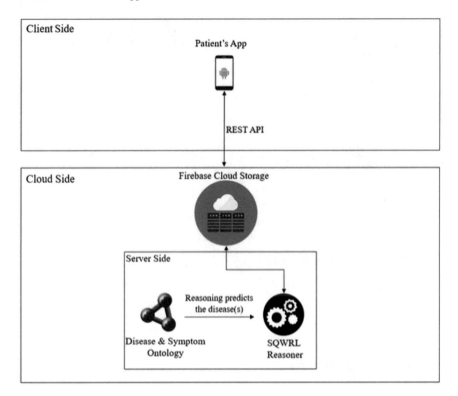

Fig. 1 Architecture diagram of system

description for each disease. From the 47 identified diseases, complete information (symptoms, speciality and description) was obtained for only 27. An expert, that is, a doctor validated the final list of diseases and symptoms. Hence, the ontology contained 27 diseases structured using four top-level classes, namely **Disease** describes the name of the disease, **Symptom** describes the name of the symptoms, **Specialty** describes the specialty of the disease and **Description** briefly describes the disease. After each top-level class has been determined, their individual subclasses were built and instantiated.

After defining the hierarchical structure of the ontology, relationships (referred to as object property) were created between diseases and symptoms as well as diseases and specialty as shown in Fig. 2. The domain, range and object property needs to be defined as constraints conditions.

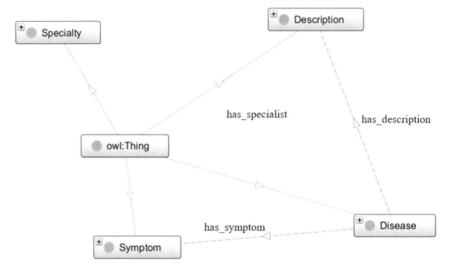

Fig. 2 Architecture of ontology

3.2 SQWRL Reasoner

To be able to infer the appropriate disease Semantic Web Rule Language (SWRL) is used to build the rules. SWRL[5] is based on a combination of OWL DL and OWL Lite which are sublanguages of OWL.[6] The SWRL rule that selects the possible diseases associated with a set of related symptoms is generated each time the person inputs the symptoms. Figure 3 (Panel B) shows how the rule is created through a *for loop* each adding the symptoms to the rule. For instance, if a person inputs 'fever' and 'vomiting', the rule shown in Fig. 3 (Panel A) will be created automatically. The query engine then runs the rule to deduce the possible diseases.

4 Implementation

4.1 The Symptom Checker

As the patient starts typing his symptoms, the application will suggest other symptoms for him to choose from as shown in Fig. 4. The list of suggestions is computed using data mining from other patients, that is, as patients use the application, it learns

---

[5]https://www.w3.org/Submission/SWRL/.

[6]https://www.w3.org/TR/owl-features/.

Symptom(fever) ^ has_symtom(?y,fever) ^ Symptom(vomitting) ^
has_symtom(?y,vomitting) ^ Disease(?y) -> sqwrl:select(?y)

Panel A

```
String [] symptoms = symp.trim().split( regex ",");
String r="" ;

for (int i=0; i<symptoms.length; i++)
{
    String symptom = symptoms[i].trim().toLowerCase().replace(" ","_");

    r +="Symptom("+symptom+") ^ has_symptom(?y,"+symptom+") ^ ";
}

rule = r+"Disease(?y) -> sqwrl:select(?y) ";
```

Panel B

Fig. 3 Rule for diagnosis

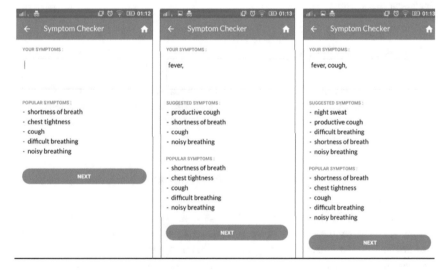

Fig. 4 Symptom checker

about the co-occurrence of symptoms. As a starting point, in order to be able to create suggestion, a doctor was asked to use the application to record the symptoms of his patients for a week.

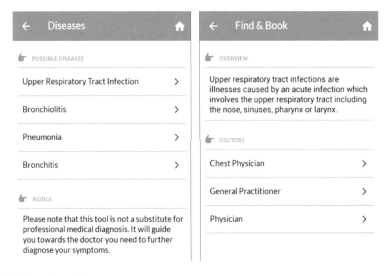

Fig. 5 Diagnosis results

4.2 Diagnosis Result

After patients have input their symptoms, they are given a list of possible diseases. As shown in Fig. 5, the patients are able to view the description of the diseases and they are recommended to the specialist doctors they may see for a specific disease.

4.3 Booking an Appointment with a Doctor

As shown in Fig. 6, the application displays a list of doctors for each speciality. Patients are able to view the location of the doctors and the list of available services with pricing details. Directly through the application, patients can call the doctor or they can also book an appointment, based on the dates and available slots. Patients are given the option to add a note for the doctor during appointment booking.

5 User Evaluation

5.1 Methods

The mobile application was evaluated with 10 users (4 males and 6 females) who were given the following tasks to complete: (1) Think of the last time they were ill

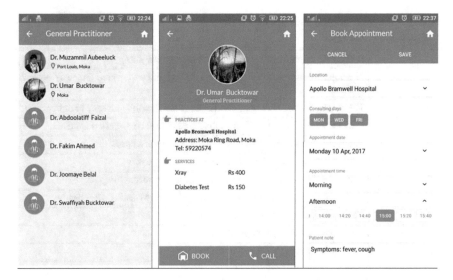

Fig. 6 Booking an appointment with a specialist doctor

and use the symptoms they experienced in order to obtain a possible diagnosis, (2) Book an appointment with a doctor, (3) View the services provided by a doctor and (4) View/Edit appointments.

5.2 Findings and Discussion

After the completion of tasks, participants answered a set of questions which are described in Table 1. Despite 100% of participants finding it easy to input their symptoms in the application, qualitative feedback showed that some users may face difficulty in understanding some symptoms. Additionally, participants would like the application to provide maps and directions to doctors' cabinet as well as nearest hospitals and doctors in case of emergencies. Discussions with participants also highlighted the need to include non-textual descriptions of symptoms to cater for circumstances when it might be difficult for patients to describe their symptoms for example, a skin rash (Fig. 7).

6 Conclusion and Future Work

In this paper, a smart health application for Mauritius has been described. One of the main functions of the application is the symptom checker which provides patients with possible disease diagnoses for the symptoms they are experiencing. The reason-

Table 1 Questions during user feedback

Question No.	Question
Q1	It was easy to input your symptoms in the application
Q2	The application was able to provide a correct diagnosis based on the symptoms you suffered
Q3	The description of each diagnosed disease was helpful and easy to understand
Q4	It was easy to view a list of specialists based on each disease diagnosed
Q5	You were able to book an appointment easily with a specialist
Q6	You were able to search for doctors by specialty easily
Q7	You were able to call a doctor in the application successfully
Q8	You were able to view a list of services provided by each doctor easily
Q9	You were able to view your appointments easily
Q10	It was easy to edit your appointments
Q11	The application is useful and user friendly

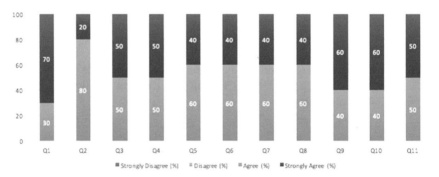

Fig. 7 Findings from user evaluation

ing process to derive the diagnosis is based on a disease-symptom ontology which has been developed from scratch for the Mauritian context. Patients are also able to view description of diseases and are directed to the correct specialist doctors that tackle their respective diseases. Patients can also book appointments directly with the doctors and send their symptoms to the doctor prior to their visit. User evaluation has shown that overall, patients found it easy to input their symptoms and book appointments with doctors following their disease diagnosis. Future work will include the ability for patients to input their symptoms using non-textual descriptions such as images and videos.

Declaration We undertake that we have the required permission to use images/dataset in our work from suitable authority and we shall be solely responsible if any issue in future.

References

1. Galeon, D., Houser, K.: Artificial intelligence. Available at https://futurism.com/ibms-watson-ai-recommends-same-treatment-as-doctors-in-99-of-cancer-cases/ [Online]. Accessed 21 June 2018 (2016)
2. Rouse, M.: Expert system. Available at http://searchhealthit.techtarget.com/definition/expert-system [Online]. Accessed 13 Mar 2017 (2016)
3. Neches, R., et al.: Enabling technology for knowledge sharing. AI Mag. **12**(3), 36–56 (1991)
4. Rector, A., Gangemi, A., Galeazzi, E., Glowinski, A., Rossi-Mori, A: The GALEN CORE model schemata for anatomy: towards a re-usable application-independent model of medical concepts. In: Proceedings of Medical Informatics EuropeMIE94 (1994)
5. Van Gurp, M., Decoene, M., Holvoet, M., dos Santos, M.C.: LinKBase, a philosophically-inspired ontology for NLP/NLU applications. In: Proceedings of the Second International Workshop on Formal Biomedical Knowledge, KR-MED (2006)
6. Schriml, L.M., Arze, C., Nadendla, S., Chang, Y.W.W., Mazaitis, M., Felix, V., Feng, G., Kibbe, W.A.: Disease ontology: a backbone for disease semantic integration. Nucleic Acids Res. **40**(D1), D940–D946 (2012)
7. The OBO Foundry. Available at http://www.obofoundry.org/ontology/symp.html [Online]. Accessed 26 June 2018
8. Mohammed, O.M.: Semantic web system for differential diagnosis recommendations. Thesis, Lakehead University. Available at https://knowledgecommons.lakeheadu.ca/handle/2453/316 [Online]. Accessed 25 June 2018 (2012)
9. Anonymised Authors. A Disease-Symptom Ontology for Mauritius. Manuscript (2018)
10. Health Statistic Unit, Ministry of Health and Quality of Life. Available at http://health.govmu.org/English/Statistics/Health/Mauritius/Pages/default.aspx [Online]. Accessed 27 June 2018 (2016)

A Smart Mobile Application for Complaints in Mauritius

Manass Greedharry, Varun Seewoogobin
and Nuzhah Gooda Sahib-Kaudeer

Abstract Receiving and managing complaints effectively are important for organisations which aim to provide excellent customer service. In order for this to happen, organisations should make it quick and easy for users to report issues. In this paper, a smart mobile application for complaints management in Mauritius is described. Users of this mobile application can report issues for different organisations using a single application on their smartphones. They can register complaints using text, images or videos, and they do not have to specify which authority the complaint is directed to. Instead, the application uses text and image analysis alongside a Convolutional Neural Network (CNN) in order to direct complaints to the correct utility organisations. The classifiers have been trained to identify different categories of complaints for each local utility organisation. Users are notified regarding the status of their complaints and can use the application to directly communicate with the personnel.

Keywords Machine learning · Image recognition · Classification

1 Introduction

In Mauritius, members of the public often have to report issues related to the services they use on a daily basis. For example, citizens often have to report broken pipes, interruption of electricity supply or broken traffic lights and to do so, they have to contact the relevant public utility organisations either by calling in person, by telephone or in writing. Therefore, Mauritians need to know exactly who is responsible

M. Greedharry · V. Seewoogobin · N. Gooda Sahib-Kaudeer (✉)
University of Mauritius, Reduit, Mauritius
e-mail: n.goodasahib@uom.ac.mu

M. Greedharry
e-mail: manass.greedharry@umail.uom.ac.mu

V. Seewoogobin
e-mail: varun.seewoogobin@umail.uom.ac.mu

© Springer Nature Singapore Pte Ltd. 2019
S. C. Satapathy et al. (eds.), *Information Systems Design and Intelligent Applications*, Advances in Intelligent Systems and Computing 863,
https://doi.org/10.1007/978-981-13-3338-5_32

for each public utility and they must know the contact details in order to report a problem. This might deter people from reporting issues as they are not prepared to take time from their busy schedules.

It is argued in [1] that organisations should welcome complaints positively and in a professional manner for many reasons: they represent a way of receiving feedbacks and new complaints can be the starting point for improvement of current practices, they are a way to measure the internal performance of the organisation and they are useful to better understand the consumers [1]. Therefore, an effective complaint management system is important to secure customer satisfaction, loyalty and prevent negative word of mouth [2]. Therefore, public utility organisations in Mauritius should encourage complaint reporting and they should strive to make it simpler, less time-consuming and more effective for consumers. This can be achieved using mobile technology as smart mobile phones are becoming increasingly popular in Mauritius.

The aim of this paper is to create a smart mobile application that allows members of the public to report issues with public utilities organisations using a single mobile application. A smart approach is employed whereby the user does not have to specify which utility organisation the complaint is directed to. Instead, using classification and machine learning techniques, complaints are automatically directed to the correct organisation. The rest of this paper is structured as follows: Sect. 2 describes the existing mobile applications for complaints reporting, Sect. 3 provides a detailed explanation of the classification process, Sect. 4 outlines the architecture of the mobile application and Sect. 5 describes the implementation. Finally, Sect. 6 discusses the potential for future works.

2 Background Study

2.1 Complaints Management Systems

With the rise of network capacity and communication facilities, a growing number of users handle their complaints through the internet [3]. The online environment provides customers with a greater sense of control and encourages them to show their dissatisfaction differently [4]. Previous research [5] has also shown that taking complaint management to online platforms may decrease the number of complainants voicing out directly to the organisation. Consequently, IBM has saved $1.5 billion by handling its consumer complaints and queries electronically in the year 2000 [6]. Recently, a number of services such as self-service machines and touch phones have become popular for complaints management. Thus, customers are increasingly using these technologies to interact independently with organisations in order to report issues [7]. Therefore, there is evidence that online complaint management can contribute significantly to the efficiency of public utilities organisations and if

Mauritius is to pursue a smart island initiative, the local population will benefit from systems that deliver such services seamlessly.

2.2 Complaints Management in Mauritius

There is already a move in Mauritius towards online complaints management. In 2017, the Government of Mauritius launched the Citizen Support Unit,[1] an online portal that allows citizens to submit requests, report complaints and suggest ideas to different Ministries, departments and local authorities. Statistics show that the population has welcomed this approach as in the period of May 2017–March 2018, 20,464 complaints have been made through the portal [8].

Prior to this, eCitizen,[2] a mobile application was launched by Mauritius Telecom (MT) in June 2016 to allow the local inhabitants to take a picture of an issue and report it to the relevant authority. MT handles the complaint and directs it to the organisation concerned. Through eCitizen, users are able to take pictures and add text to the complaint. They are also notified when the issue has been resolved. Users can share their complaints on social media and can view their complaint history.

Both existing mechanisms for complaints reporting in Mauritius require a level of manual assignment to the correct local authority for the issue to be solved. Given that each local authority already has a set of predefined problems that are under their responsibility, smart technologies can be used to automatically classify complaints and direct them to the correct authority without human intervention. In the following: the approaches used for smart complaints management are described.

2.3 Approaches for Smart Complaints Management

Smart applications go beyond traditional data processing as they can sense the environment around them and they adapt themselves to the context and intentions of their users [9]. In order to do so, smart applications have to cope with increasing amounts of data and hence the need for data mining and machine learning algorithms to perform selection, processing and pattern detection. A smart application should also require less effort from the user, that is, technologies should be employed in order to reduce the demands on the user in terms of interaction. In this paper, the smart approaches that are used to achieve the latter are the following:

1. **Text and Image Analysis**. Text analysis is the process of extracting useful information from unstructured text, whereas image analysis is the retrieval of useful patterns from mainly digitally processed image. A typical image mining process

---

[1]https://www.csu.mu.

[2]https://play.google.com/store/apps/details?id=com.wearemobimove.ecitizen&hl=en.

includes preprocessing, transformation and feature extraction, mining and evaluation. Once text or an image has been processed, they can be used for different purposes including clustering, classification or summarisation. The classification approaches employed in this paper are described in Sect. 3.

2. **Deep machine learning**. Deep machine learning [10] is a subfield that includes computational models that aim to mimic the neocortex of the brain in that sensory signals are not preprocessed. Instead, they are allowed to propagate and over time, the models learn to represent observations. Deep learning, therefore, focuses on multi-layered artificial neural networks models through which data is fed, and the data, in turn, is used to make a decision about other data. Deep machine learning can be applied to any format of data: signals, speech, image or text. Some well-established techniques in deep machine learning include Convolutional Neural Network (CNN) and Deep Belief Networks (DBN) [10].

For the scope of this paper, the approaches described previously have been employed to develop a smart complaints management mobile application. The application allows users to submit a complaint or report an issue using text, images, videos or a combination of media. One of the main contributions is that users do not have to specify which local organisation this issue is directed to as the application uses text and image analysis and classification to correctly direct the complaint. For example, an image showing water overflowing will be directed to the Central Water Authority (CWA), and the public utility is responsible for water distribution. Therefore, the classification process is critical for the correct allocation of complaints. In the following section, the classification approach used in this paper is described including details about the training datasets.

3   Classification

A Convolutional Neural Network (CNN) approach has been used both for the text and image classification. CNN architecture has achieved good classification performance and is the baseline for new classification architectures. Model classifiers for text and image were implemented using TensorFlow,[3] an open-source framework for deep learning from Google. The classifiers have been trained with new datasets relevant to the domain and the set of derived categories, representing classes for different complaints, were established for the local organisations included in this paper as shown in Table 1. As each class defines a unique authority for this system, supervised machine learning has been used for the assignment of complaints into discrete categories. Due to unavailability of open datasets on consumer complaints and unauthorised access to database of local authorities, training datasets for both models have been created.

---

[3]https://www.tensorflow.org.

Table 1 Classes for each local organisations

Organisation	Category/class
Central Water Authority (CWA)	Water pipe broken
Central Electricity Board (CEB)	Electricity pole
	Power outage
	Streetlight repair
Road Development Authority (RDA)	Road repair
	Traffic signal
	Road marking
Municipal councils	Scavenging
	Non asphalted road

3.1 Text Classification

Training Data. For the text classifier model, a relatively small dataset was manually created in a Comma Separated Values (CSV) file format. For each category, around 150 sentences were hand-picked from the websites of different local authorities. Each instance in the dataset consisted of the following attributes: textual descriptions and their respective category. An excerpt from the training dataset for the text classifier is shown in Fig. 1.

Data Load and Preprocessing. The sentences are loaded from the raw data files and the text data is then cleaned for unwanted symbols. As neural networks need numeric inputs, the labels are one-hot encoded for generalisation. Longer sentences are trimmed and shorter ones are padded with blank spaces as all labels must be of

520	damaged road barriers	Road repair
521	potholes repair	Road repair
522	no action on water pipe broken	water pipe broken
523	water leakage and bill is high	water pipe broken
524	water leakage before water meter	water pipe broken
525	sound of running water on the street	water pipe broken
526	there is wet areas on the main road	water pipe broken
527	running water on the main road	water pipe broken
528	broken damaged road guard rails	Road repair
529	damaged road barriers	Road repair
530	power loss to my area	Power outrage
531	power cut long enough range in length	Power outrage
532	experiencing frequent and frustrating power cuts	Power outrage
533	power cut everytime	Power outrage
534	power loss to my area	Power outrage
535	power cut long enough range in length	Power outrage

Fig. 1 Dataset for text classifier model

equal size. A vocabulary, a unique set of words in the training dataset, is then built. Therefore, each word in a sentence maps onto a unique integer.

3.2 Image Classification

For the image classification, an image recognition model called Inception-V3 [11] is used which has been shown to yield substantial accuracy on the ImageNet dataset.[4] Thus, the final layer of the model is then retrained using transfer learning with the derived set of classes for the complaints management mobile application. The TensorFlow image classifier Inception-V3 was downloaded and a new final layer, which will do the classification for the new classes of images was added using the retrain.py[5] script.

To create the training dataset, images relevant to the categories mentioned in Table 1 were downloaded from the internet and for some of them, pictures were manually taken using mobile phones. All images were classified in their folders and each folder contained over 100 pictures, since large amount of datasets were needed to increase the accuracy of the image model. The folders were named after the categories of the complaint, which become the class labels for the image classifier. Prior to the model being trained, the bottleneck values[6] of the images are calculated and stored in text files cached in the disk for each image, since these values are used by the classifier to differentiate between the nine classes of images as described in Table 1.

The second step will start training the model with 500 training steps, where each step chooses 10 images from the training batch along with their bottleneck values from the cache. The bottleneck values are fed into the new final layer of the model to get the predictions. The backpropagation process [12] is used, where the predicted categories are compared against the actual class labels of the images. The results of the comparison are used to update the weights of the final layer. At each training step of the classifier, the values of the training accuracy, cross entropy and validation accuracy are calculated, which show how the learning process of classifier is progressing. Finally, when the model was fully trained, the final test accuracy was 86.3%, which indicates the percentage of the images that have correct labels in the test set.

---

[4]http://www.image-net.org.

[5]https://github.com/tensorflow/hub/blob/master/examples/image_retraining/retrain.py.

[6]https://www.tensorflow.org/tutorials/image_retraining.

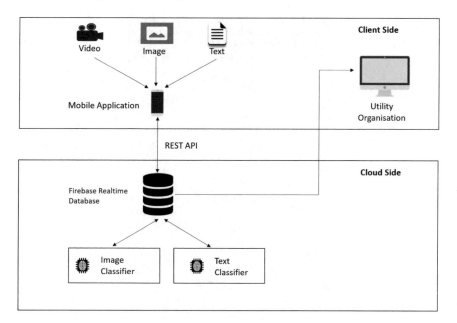

Fig. 2 Architecture diagram of smart mobile application for complaints

4 Architecture

In this section, the design of the smart mobile application for complaints is described according to the architecture diagram as shown in Fig. 2. A user has the option to report a complaint from his or her smartphone using either one of the following: textual description, an image, a video or a combination of media. The video length is limited to 5 s. The Firebase database on the cloud stores information about users, complaints and local utilities organisations. The REST API[7] is used to listen to the database for new complaints and redirect them to the classifiers based on their descriptions.

---

[7]https://github.com/thisbejim/Pyrebase/.

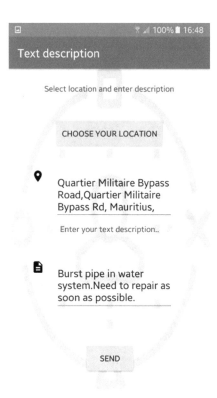

Fig. 3 Complaint in text

4.1 Text Classifier

The text description of the complaint serves as input to the text classification model (Fig. 3). The text is processed and the text classifier determines the class of the complaint in order to direct it to the correct local utility organisation. The text classification process has been described in detail in Sect. 3.1.

4.2 Image Classifier

Complaints including image and video files are both fed to the image classification model. The script label_image.py[8] is used, in which the trained image model is called and it takes the data of an image as input. The video description, limited to 5 s, uses the image classifier model. Due to the relatively small number of frames in the video, there is high probability that the content of the middle one is related to the

---

[8]https://github.com/llSourcell/tensorflow_image_classifier/blob/master/src/label_image.py.

Fig. 4 Complaint in image and text

other frames. It is thus extracted in a proper image format from the average duration of the video and passed to the image classifier model, from which the class of the complaint is predicted. In Sect. 3.2, the image classification process is described.

Both the text and image classifiers output the probabilities, also called scores, of all class labels. This is to demonstrate the confidence level in which category the complaint falls in. The class label of the highest score is retrieved, which eventually become the category of the complaint. The predicted category is then mapped to the respective organisation. Consequently, the user is immediately notified upon the redirection of the reported complaint to the relevant authority.

5 Implementation

The mobile application allows the user to choose the media through which they want to report an issue. Users can submit complaints by using text description, taking picture and video from camera or selecting picture and video from their phone gallery. The user can also use a combination of media to better describe the complaint, for example, use text to accompany a picture of a burst pipe (Fig. 4).

Fig. 5 Complaints around the island

Users are also required to input the location of the complaint, that is, the exact location where the problem is. This allows the local organisation to swiftly address the problem as they know exactly where the problem is occurring. If the Global Positioning System (GPS) is activated on the user's mobile device while the picture is being taken, the mobile application will easily retrieve the information about the location of the complaint. Otherwise, the user has the option of choosing the location of the complaint via Google Map for complaints expressed in all types of media (Fig. 5).

The mobile application also stores the history of the complaints made by the user and can be viewed (Fig. 6). The application offers the possibility to view complaints made by other complainants around the island via place markers on a map, which can be filtered according to the name of authorities. This can help decision makers in identifying problem-prone areas. Users of the mobile application are continuously updated about the status of their complaints, through a notification system and they can also communicate with the relevant authorities via the application itself.

Fig. 6 Complaint details

6 Conclusion

In this paper, a mobile application for smart complaints management is described. The mobile application uses techniques such as text and image analysis and machine learning in order to automatically direct complaints to the relevant authorities. Users of the application can send complaints using text, images, videos or a combination of media together with the location of the complaint. The mobile application uses convolutional neural networks in order to classify all types of complaints so that additional efforts are not required from the user to specify the local authority responsible for different problems. Thus, complaints management is more user-friendly and less time consuming as users can literally send complaints in 'one click'. Local organisations are also set to benefit from this application as they can improve procedures in order to maximise consumer satisfaction. Future works will improve the prediction of the classifiers by enhancing the datasets and the training process.

References

1. Zairi, M.: Managing customer dissatisfaction through effective complaints management systems. TQM Mag. **12**(5), 331–337 (2000)
2. Dong, B., Evans, K.R., Zou, S.: The effects of customer participation in cocreated service recovery. J. Acad. Mark. Sci. **36**(1), 123–137 (2008)
3. Mattila, A.S., Mount, D.J.: The impact of selected customer characteristics and response time on e-complaint satisfaction and return intent. Int. J. Hosp. Manag. **22**(2), 135–145 (2003)
4. Wind, J., Rangaswamy, A.: Customerization: the next revolution in mass 139 customization. J. Interact. Mark. **15**(1), 13–32 (2001)
5. Voorhees, C.M., Brady, M.K., Horowitz, D.M.: A voice from the silent masses: an exploratory and comparative analysis of noncomplainers. J. Acad. Mark. Sci. **34**(4), 514–527 (2006)
6. Agnihorthri, S., Sivasubramanium, N., Simmons, D.: Leveraging technology to improve field service. Int. J. Serv. Ind. Manag. **13**(1), 47–68 (2002)
7. Meuter, M.L., Ostrom, A.L., Roundtree, R.I., Bitner, M.J.: Self-service technologies: understanding customer satisfaction with technology based service encounters. J. Mark. **64**, 50–64 (2000)
8. Citizen Support Unit: Available at https://www.csu.mu/statistics-details/8/statistics.html. Accessed 20 June 2018
9. Preuveneers, D., Novais, P.: A survey of software engineering best practices for the development of smart applications in ambient Intelligence. J. Ambient Intell. Smart Environ. **4**(3), 149–162 (2012)
10. Arel, I., Rose, D.C., Karnowski, T.P.: Deep machine learning-a new frontier in artificial intelligence research [research frontier]. IEEE Comput. Intell. Mag. **5**(4), 13–18 (2010)
11. Szegedy, C., Vanhoucke, V., Ioffe, S., Shlens, J., Wojna, Z.: Rethinking the inception architecture for computer vision. In: Proceedings of the IEEE Conference on Computer Vision and Pattern Recognition, pp. 2818–2826 (2016)
12. Goodfellow, I., Bengio, Y., Courville, A., Bengio, Y.: Deep Learning (vol. 1) Cambridge: MIT press (2016)

An ERP Adoption Evaluation Framework

Sanmurga Soobrayen, Mohammad Zawaad Ali Khan Jaffur
and Bibi Zarine Cadersaib

Abstract Small and Medium Enterprises (SMEs) are confronted to highly competitive markets and need to have recourse to more efficient and effective solutions. An Enterprise Resource Planning (ERP) system could be one solution. A review was performed to identify factors affecting ERP adoption by SMEs in developing countries including Mauritius. The benefits and challenges regarding ERP implementation together with the Balance Scorecard (BSC) have been used to derive dimensions to be considered and a cause and effect analysis diagram is presented. An ERP evaluation framework is then proposed which consists of an ERP dimension weightage matrix and quantitative ERP implementation assessments. Key Performance Indicators (KPIs) are derived for each dimension and the latter is used to propose a matrix for defining the weightage against each dimension. The dimension weightage together with the KPI impact (positive or negative) is then used to derive a formula to be used to evaluate an ERP implementation for SMEs.

Keywords ERP · ERP adoption · ERP adoption evaluation framework · SMEs

1 Introduction

SMEs have an essential part in Mauritius both as a facilitator and as a leading factor for a balanced development [1, 2]. In a competitive market, SMEs are often faced with a lot of difficulties to compete among themselves so as to become more efficient and creative. Currently, there is a low technology utilization among SMEs in Mauritius [3]. To overcome such difficulties and to improve competitiveness, one way might

S. Soobrayen · M. Z. A. Khan Jaffur · B. Z. Cadersaib (✉)
University of Mauritius, Reduit, Mauritius
e-mail: z.cadersaib@uom.ac.mu

S. Soobrayen
e-mail: sanmurga.soobrayen1@umail.uom.ac.mu

M. Z. A. Khan Jaffur
e-mail: mohammad.khan1@umail.uom.ac.mu

© Springer Nature Singapore Pte Ltd. 2019
S. C. Satapathy et al. (eds.), *Information Systems Design and Intelligent Applications*, Advances in Intelligent Systems and Computing 863,
https://doi.org/10.1007/978-981-13-3338-5_33

be the embracement and usage of a suitable Information System [4]. An Enterprise Resource Planning (ERP) can be adopted by SMEs to enhance their business processes as well as to become more competitive, profitable and to achieve efficiency. ERP improves data processing which helps businesses in making better decision to increase productivity [5]. However, the decision to embrace an ERP system is challenging due to the complexity, risks involved, costs and time factor [6, 7] thus ERP adoption rate is still low in developing countries [1, 8–10]. The purpose of this research is to identify the current status of ERP adoption in Mauritius and to propose and evaluate an ERP adoption evaluation framework. It is expected that this study will be beneficial for the research community since there is a lack of study for ERP adoption in Mauritius and a limited number of research carried out on ERP adoption evaluation. The research question of the study is set as follows and is addressed in Sect. 3: *RQ1: Can an ERP evaluation framework be derived and used to assess the ERP implementation results for SMEs?*

2 ERP Adoption in Developing Countries Including Mauritius

A Systematic Literature Review (SLR) was carried out making use of Kitchenham and Charters [11] practical guidelines. It was found that there is a lack of potential research work carried out on ERP adoption in Mauritius, therefore the study was extended over developing countries. Out of 856 papers considered, only 15 were found to be relevant to ERP adoption in developing countries and Mauritius. Due to space restriction, only the major findings are discussed in this paper. It was observed that the adoption of ERP system in developing economy is relatively low. For example, some of the findings showed that there is a low usage of ERP by SMEs due to the nature of business, lack of reliance, lack of knowledge about ERP, time constraint, financial constraint, user resistance, perceived ease of use, Internet connectivity issues and data security issues [3, 8, 9]. The benefits of adopting ERP included: operational benefit, strategic benefit, IT infrastructure benefit, managerial benefit and organizational benefit. The challenges included: inadequate organizational fit, inadequate management structure and strategy, inadequate user involvement and training, inadequate technical skills, lack of knowledge, security issues and inadequate software system designs [3, 5, 7–10, 12–20].

3 Proposal and Evaluation

This section considers the benefits and challenges regarding ERP implementation together with Balance Scorecard (BSC) [21] to derive dimension to be considered and proposes the cause and effect analysis using the Ishikawa diagram. The proposed ERP evaluation framework is then derived based on the cause and effect analysis.

3.1 Deriving Dimensions

The BSC consists of four different perspectives: financial perspective, internal business process perspective, customer perspective and learning and growth perspective. In the financial perspective, the Total Cost of Ownership (TCO) associated with ERP implementation will be considered as a KPI to be evaluated. For the Internal Business Process Perspective, the KPI that will be used for evaluation is speed of service. Customer Perspective deals with the internal users using the system. It is very important to have the point of view of the end users since they are the one to be using the system once it has been adopted. Thus, customer satisfaction measure will be very useful to have a clear indication of whether the user is satisfied with the system. KPIs are used to measure the Innovation (Growth) and Learning Perspective that includes number of training hours and the adaptation time to new system. Dimensions were derived by regrouping factors from the SLR with the perspectives from the BSC to simplify the process and to remove duplications. The dimensions IT infrastructure, management involvement and software system design dimensions were not regrouped in the BSC as these did not fit with the BSC perspectives. The compiled dimensions include (1) Financial Perspective, (2) Customer Perspective, (3) Innovation and Growth, (4) Internal Business Process, (5) IT Infrastructure, (6) Management Involvement and (7) Software System Design.

3.2 Cause and Effect Analysis

A cause and effect analysis was performed to diagrammatically represent all the possible causes of ERP implementation based on the different dimensions and the factors from each dimension as shown in Fig. 1.

3.3 Proposed ERP Adoption Evaluation Framework

The proposed ERP adoption evaluation framework consists of an ERP dimension weightage matrix and quantitative ERP implementation assessments. Key Performance Indicators (KPIs) are derived for each dimension. The dimensions are used to propose the matrix for defining the weightage of each dimension. This KPI weightage together with the KPI impact (positive or negative) are then used to derive a formula to be used to evaluate an ERP implementation for SMEs. The following sec-

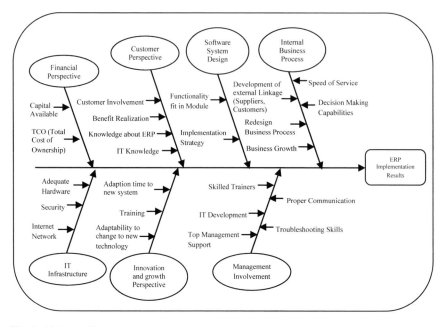

Fig. 1 Ishikawa diagram

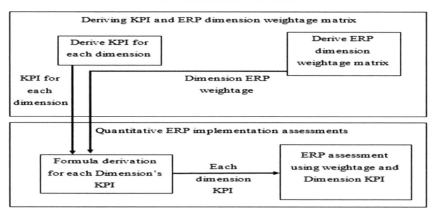

Fig. 2 ERP adoption evaluation framework

tions provide justifications and discussion of these different components. Figure 2 provides a visual presentation of the framework.

Deriving KPI and ERP Dimension Weightage Matrix

KPIs should be SMART that is Specific, Measurable, Attainable, Relevant and Time-bound [22, 23]. It was important to identify the key feature of each dimension and

Table 1 Allocate ID to dimension

Dimension ID	Dimension	KPI
D1	Financial perspective	• Capital available • TCO—Total cost of ownership
D2	Customer perspective	• Benefit realization
D3	Software system design	• Functionality fit in module
D4	Internal business process perspective	• Development of external linkage (suppliers, customers) • Speed of service • Redesign business process • Business growth
D5	IT Infrastructure	• Adequate hardware • Internet network • Security
D6	Innovation and growth perspective	• Adaptability to change to new technology • Adaption time to new system • Training
D7	Management involvement	• Skilled trainers • Top management support • IT development capabilities • Proper communication • Troubleshooting skills

Table 2 ERP dimension weightage matrix

	D1	D2	D3	D4	D5	D6	D7	Dimension Criteria Weight	Percentage of Total Criteria
D1									
D2									
D3									
D4									
D5									
D6									
D7									

to consider only those which were SMART as shown in Table 1. SMART KPIs to be evaluated for each dimension together with the dimension ID is provided. Table 2 illustrates the ERP dimension weightage matrix constructed based on the dimensions identified in Table 1 and weightage importance will be assigned against each dimension. This matrix derivation has been adapted from a previous study [24]. A weightage of 0.2 (Less important), 1.0 (Equal Importance), 5.0 (More Important) and 10.0 (Much More Important) has been used to rate one dimension against another. So if $D1$ against $D2$ is rated as 10.0, then $D2$ against $D1$ will be rated as 0.1 (being 1/10).

Quantitative ERP Implementation Assessments

The Weightage Matrix will provide the 'Percentage of Total Criteria' for each dimension which will be used as the Weightage for each dimension, DN_w, where N represents each dimension set in Table 1. KPIs included in each dimension can represent a positive outcome, a negative outcome or null. If the KPI is found to be a positive outcome, the total positive outcome of the dimension will be increased by 1 point else if it is a negative outcome, the total negative outcome of the dimension will be increased by 1 point. A null means that the KPI has not been used in the case evaluated. A zero (0) point is given to both the positive outcome and negative outcome to denote a null KPI.

For example, if the KPI 'capital available' from the financial dimension is considered as a positive outcome, the financial dimension positive outcome increases by 1 point else if it is a negative outcome, the financial dimension negative outcome increases by 1 point. This weightage is applied with each dimension and the sum of the KPIs (positive or negative) is used to provide a quantitative assessment per dimension. The formula below will be used to compare DN positive outcome(s) to DN negative outcome(s).

$$DN_{\mathbf{KPI}} = \left(\sum DN_{\mathbf{KPI+ve}} \times DN_w\right) - \left(\sum DN_{\mathbf{KPI-ve}} \times DN_w\right)$$

where

$DN_{\mathbf{KPI}+ve}$ KPI with positive outcome
DN_w percentage of total criteria of DN
$DN_{\mathbf{KPI}-ve}$ KPI with negative outcome

Interpretation:

1. $DN_{\mathbf{KPI}} = 0 \rightarrow DN$ positives outcomes are equal to the negatives of DN.
2. $DN_{\mathbf{KPI}} > 0 \rightarrow DN$ positive outcomes are greater than the negative outcomes of DN.
3. $DN_{\mathbf{KPI}} < 0 \rightarrow DN$ negative outcomes are greater than the positive outcomes of DN.

Lastly to identify whether the SME will adopt the ERP system, the following formula has been derived which sums up all the $DN_{\mathbf{KPI}}$ for each dimension.

$$\mathbf{ERP}_{\mathrm{KPI}} = \sum_{N=1}^{7} DN_{\mathrm{KPI}}$$

Interpretation:

$\mathbf{ERP}_{\mathbf{KPI}} > 0 \rightarrow$ The possibility for the SME to adopt the ERP system is high.
$\mathbf{ERP}_{\mathbf{KPI}} = 0 \rightarrow$ The ERP adoption by the SME is neutral.
$\mathbf{ERP}_{\mathbf{KPI}} < 0 \rightarrow$ The possibility for the SME to adopt the ERP system is low.

Evaluation of Framework

A case study approach was used to evaluate two SMEs, SMEA and SMEB from the clothing industry. The name of the SME is not disclosed to maintain confidentiality. An ERP prototype was implemented using an open source ERP (ODOO) where all business processes of the SMEs were integrated in the software and the evaluation was carried out using the proposed ERP evaluation framework. Both SMEs were provided with an explanation to fill in the Weightage Matrix properly and were assisted in filling them. For SMEA, $D1$, $D4$, $D5$ and $D6$ had more importance compared to other dimensions as their corresponding Dimension Criteria Weight is greater compared to the other remaining dimensions. For SMEB, $D1$, $D2$, $D4$, $D5$ have more importance compared to the other remaining dimensions. During the evaluation of the two SMEs, it was found that both SMEs were not ready to embrace ERP system. This can be supported by the figure received for both SMEs according to the ERP_{KPI} calculation. SMEA score -1.384 [$(-0.524) + (-0.066) + 0.058 + (-0.214) + (-0.446) + (-0.36) + 0.168$)] and SMEB scored -0.527 [$(0 + (-0.169) + 0.093 + (-0.294) + 0 + (-0.186) + 0.029$)]. Based on these results, both SMEs have less probability to adopt an ERP system. Moreover, the main hurdles identified were lack of finance, lack of adequate hardware and Internet connectivity, insufficient training, high cost involve for ERP system and customer not realizing the benefits of the ERP system. In addition, during the interview of both SMEs, it was observed that the owners did not have any idea about ERP, meaning that there was a lack of awareness as well.

4 Limitations, Conclusion and Future Works

The evaluation was carried out within a very limited time period (one month) and it was not possible to have an in-depth evaluation of the two case scenarios. The SME owners were reluctant to provide sensitive information about their business activities. The evaluation has to be carried out with a larger pool of SMEs in the same line of business as well as other lines of business to provide more conclusive results. Surveys were also not conducted with local SMEs and the framework was mostly derived from literature findings. A proposed ERP evaluation framework was designed and a case study approach to evaluate the framework. It was found that major factors affecting ERP adoption by Mauritian SMEs include lack of resources (finance, IT skilled staff and Internet connection), the high cost involved in adopting ERP system, insufficient training, lack of awareness about ERP system and most importantly, owner of the SMEs not realizing the benefits of ERP system. Future work will entail in evaluating the proposed ERP implementation framework using other SMEs in different industries and SMEs in other developing countries to have better insights on factors influencing ERP adoption. The proposed ERP evaluation framework consists of seven dimensions which can be extended.

Acknowledgements The authors gratefully acknowledge the owners of SMEA and SMEB for their participation in using the ERP system, filling the Weightage Matrix as well as sharing their feedbacks.

References

1. Antoo, M.: Cloud computing adoption framework by Mauritian SMEs. Postgraduate Thesis, University of Mauritius, Mauritius (2014)
2. Small and Medium Enterprises Development Authority, SMEDA, http://www.smeda.mu (2014)
3. Gobin-Rahimbux, B., Cadersaib, Z., Sahib, N.G. Khan, M.: Investigating technology awareness and usage in Mauritian SMEs in the handicraft sector (2017)
4. Kambarami, L., Mhlanga, S., Chikowore, T.: Evaluation of enterprise resource planning implementation success: case study in Zimbabwe. In: Computers and Industrial Engineering, vol. 42 (June 2012)
5. Shahawai, S.S., Hashim, K.F., Idrus, R.: Enterprise resource planning adoption among small medium enterprises (SME) in Malaysia (2014)
6. Tome Mr, L., Allan, K., Meadows Mrs, A., Nyemba-Mudenda Ms, M.: Barriers to open source ERP adoption in South Africa. Afr. J. Inf. Syst. **6**(2), 1 (2014)
7. Venkatraman, S., Fahd, K.: Challenges and success factors of ERP systems in Australian SMEs. Systems **4**(2), 20 (2016)
8. Rahman, A.: ERP Adoption in small and medium sized enterprises in Bangladesh. M.B.A. Thesis, Southeast University Bangladesh, Banani, Dhaka (2015)
9. Hasheela, V.T., Smolander, K., Mufeti, T.K.: An investigation of factors leading to the reluctance of SaaS ERP adoption in Namibian SMEs. Afr. J. Inf. Syst. **8**(4), 1 (2016)
10. Khaleel, Y., Sulaiman, R., Ali, N.M., Baharuddin, M.S.: Analysis of enterprise resource planning system (ERP) in small and medium enterprises (SME) of Malaysian manufacturing sectors: current status and practices. Asia Pac. J. Inf. Technol. Multimedia, **10**(1) (2011)
11. Kitchenham, B., Charters, S.: Guidelines for performing systematic literature reviews in software engineering [online]. Userpages.uni-koblenz.de. Available at https://userpages.uni-koblenz.de/~laemmel/esecourse/slides/slr.pdf (2007). Accessed 14 Oct 2017
12. Antoo, M., Cadersaib, Z., Gobin, B.: PEST framework for analysing cloud computing adoption by Mauritian SMEs. Lect. Notes Softw. Eng. **3**(2), 107 (2015)
13. Kamhawi, E.M.: Enterprise resource-planning systems adoption in Bahrain: motives, benefits, and barriers. J. Enterp. Inf. Manag. **21**(3), 310–334 (2008)
14. Lechesa, M., Seymour, L., Schuler, J.: ERP software as service (SaaS): factors affecting adoption in South Africa. In: Re-conceptualizing Enterprise Information Systems, pp. 152–167. Springer, Berlin, Heidelberg (2012)
15. Mukwasi, C.M., Seymour, L.F.: Interdependent enterprise resource planning risks in small and medium-sized enterprises in developing countries. In: *IST-Africa Conference, 2015*, pp. 1–10. IEEE (May 2015)
16. Upadhyay, P., Basu, R., Adhikary, R., Dan, P.K.: A comparative study of issues affecting ERP implementation in large scale and small medium scale enterprises in India: a pareto approach. Int. J. Comput. Appl. **8**(3), 23–28 (2010)
17. Zach, O., Olsen, D.H.: ERP system implementation in make-to-order SMEs: an exploratory case study. In: System Sciences (HICSS), 2011 44th Hawaii International Conference on, pp. 1–10. IEEE (Jan 2011)
18. Amba, S., Abdulla, H.: The impact of enterprise systems on small and medium-sized enterprises in the Kingdom of Bahrain (2014)

19. Mirbagheri, F.A., Khajavi, G.: Impact of ERP implementation at Malaysian SMEs: analysis of five dimensions benefit. Int. J. Enterp. Comput. Bus. Syst. **2**(1) (2013)
20. Wolcott, P., Kamal, M., Qureshi, S.: Meeting the challenges of ICT adoption by micro-enterprises. J. Enterp. Inf. Manag. **21**(6), 616–632 (2008)
21. Rosemann, M., Wiese, J.: Measuring the performance of ERP software–a balanced scorecard approach. In: Proceedings of the 10th Australasian Conference on Information Systems, vol. 8, no. 4. Wellington (Dec 1999)
22. Bauer, K.: KPI identification with fishbone enlightenment. Inf. Manag. **15**(3), 12 (2005)
23. Badawy, M., El-Aziz, A.A., Idress, A.M., Hefny, H., Hossam, S.: A survey on exploring key performance indicators. Future Comput. Inform. J. **1**(1–2), 47–52 (2016)
24. Ramkhelawan, S., Cadersaib, Z., Gobin, B.: Cloud computing as an alternative for on-premise software for Mauritian hotels. Lect. Notes Softw. Eng. **3**(2), 113 (2015)

Agent-Based Modelling for a Smart Parking System for Mauritius

Haidar Dargaye, Baby Gobin-Rahimbux and Nuzhah Gooda Sahib-Kaudeer

Abstract With the increasing number of cars on the road worldwide, many towns and cities face parking problems with drivers often spending significant time before finding somewhere to park. Mauritius is no exception and this paper presents an agent-based model for a smart mobile parking booking system for the Mauritian context. The system considers the position of the driver and the speed at which his car is travelling, and proposes the most appropriate and available parking slot to him based on the place he wants to go, his destination arrival time and also the parking fee he wants to pay. The system is modelled using an agent-oriented approach, whereby each agent tackles a specific problem. As such, the model proposes seven different agents, assigned specific tasks and working independently of each other in order to direct a driver to a parking slot. One of the main contributions of this model is that it considers the driver's destination arrival time among other factors when allocating parking slots. It also takes into account the analytics for past parking slots allocations to proceed with the allocation process.

Keywords Smart parking systems · Agent-based modelling
Agent-based parking systems

1 Introduction

A smart parking system contributes to the overall transport system and reduces traffic enormously. It guides the user towards the most appropriate parking slot and handles contingencies that may arise such as two users having the same parking needs. There

H. Dargaye · B. Gobin-Rahimbux · N. Gooda Sahib-Kaudeer (✉)
University of Mauritius, Reduit, Mauritius
e-mail: n.goodasahib@uom.ac.mu

H. Dargaye
e-mail: hair.dargaye@umail.uom.ac.mu

B. Gobin-Rahimbux
e-mail: b.gobin@uom.ac.mu

© Springer Nature Singapore Pte Ltd. 2019
S. C. Satapathy et al. (eds.), *Information Systems Design and Intelligent Applications*, Advances in Intelligent Systems and Computing 863,
https://doi.org/10.1007/978-981-13-3338-5_34

are many approaches to developing car parking systems and agent technologies have been presented in many works [1–3] as a good option for the development of such applications. Agents are autonomous programs that execute simultaneously and independently of each other and are adapted to these kinds of dynamic environment [1].

In Mauritius, there is paid parking throughout the island and drivers have to display parking coupons when using the parking zones. There are two zones, namely Zone 1 and Zone 2, and each which have different hourly tariffs. Once the driver identifies his parking slot, he has to fill the coupon and he is able to stay in the parking spot for a maximum of 2 h. Some privately owned car parks are also available in city centers and towns. The main problem that Mauritian drivers face is that they do not have any real-time information about the availability of parking slots, that is, they have to physically go to the parking to find whether there are any free slots. Additionally, if drivers manage to get a parking slot, often it may be very far from the place they want to go.

In this paper, we present the work done to model a smart mobile parking booking system for the Mauritian context. The system considers the position and speed at which the driver is driving and proposes the most appropriate and available parking slot to him based on his arrival time, the place he wants to go and also parking fee he wants to pay. The system is modelled using an agent-oriented approach whereby each agent tackles a specific problem. The structure of the paper is as follows: Sect. 2 gives a brief overview of the background study carried out, Sect. 3 presents the architecture and describes the different agents to be implemented. The last section concludes with the future work.

2 Background Study

Many technologies and approaches have been used to develop smart parking systems. Some of them are Wireless Sensor Networks [4–6], GPS [5–8], Vehicular Communication Systems [9, 10] and Vision-Based [11]. The agent-based approach has also been used to deploy mobile application for smart parking. Agents are autonomous software that accomplishes the task on behalf of its host in a dynamic and complex environment. They are defined by the characteristic to be able to think by themselves and derive the best possible results on basis of the different variables involved [12]. An agent interacts with other agents and exhibits adaptive behaviour. These multi-agent systems decentralize and organize themselves in a distributed system. They break their complex task into smaller ones and solve the problem [13].

2.1 Agent-Based Mobile Application Development Platform

The three agent-based mobile application development platforms that were identified are: (1) Andromeda [14], (2) JaCa [15] and (3) JADE [16]. The Andromeda platform presents an Agent Model that is designed using components and abstractions that is implemented on the Android SDK. Furthermore, the Agent Model is based on several Agent Models such as TROPOS and GAIA [14]. The implementation of Andromeda is based on the main API components of Android and therefore, this model is wholly integrated with Android's building block. JaCa is a prototype version of a multi-agent platform and is made up of two programming languages: Jason and Cartago. Jason is an interpreter for AgentSpeak. It implements the operational semantics of that language, and provides a platform for the development of multi-agent systems. Cartago is a general-purpose framework/infrastructure that makes it possible to define and create computational work environments in agent-based applications [15]. Jade provides a powerful task execution and composition model, peer-to-peer agent communication using Foundation for Intelligent Physical Agents (FIPA) compliant messages [16]. Jade is the most used platform for developing agents.

2.2 Related Work on Agent-Based Car Parking System

In this section, we present the two works identified in existing literature which helped us in the modelling process. They are as follows:

- **Park Agent** [1]

The authors proposed an explicit model for parking in a city which simulates the behaviour of each driver. The investigation helped reduce the average search time for a parking slot and the walking distance between the parking slot and the destination. Different driver agents simulated the parking process and the city navigation. The results of the simulation were used to analyse and compare parking and management policies, to improve the existing parking situation. Each driver is an agent executing autonomously and follows a set of pattern for parking a vehicle. They also proposed an algorithm for route selection based on the distance to destination from the current junction.

- **Agent-Based Intelligent Parking and Guidance System** [2]

The authors proposed an Agent-Based Intelligent Parking Negotiation and Guidance System (ABIPNGS), which uses mobile agent technology alongside multi-agent systems which allowed for the searching of parking spaces, negotiation of parking fees, reservation of parking spaces, calculation of best parking spots with respect to desired location and guidance system. The various agents developed were the *Communication Service Agent, GIS Agent, Local Management Agent, Negotiation*

Management Agent, Local Car Agent Base, Local Car Agent Base and *Floating Car Agent Base.* A negotiation algorithm was devised to bargain on the price and settle for an acceptable amount. The algorithm is further explained in [3].

It is to be noted that while both systems use an agent-based approach none of them considered the arrival time to assign parking which is the main challenge addressed in this paper.

3   Modelling the Agent-Based Smart Parking Application

The architectural diagram of the proposed solution is shown in Fig. 1. It is separated into client and server side with each containing their own agents. The communication is done via socket connection in JSON format. Users' details are stored on the Firebase database and accessed via REST API. Each agent is discussed in details in this section.

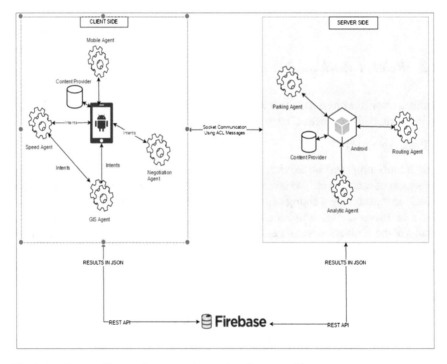

Fig. 1 Architecture diagram for proposed agent-based smart parking system

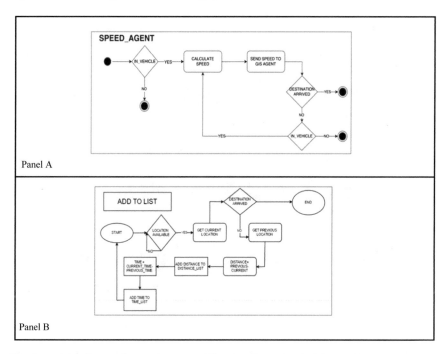

Fig. 2 Activity diagram for speed agent and flowchart for retrieval of time and distance

3.1 Speed Agent

The *Speed Agent* calculates the speed based on the GPS coordinates which are taken at regular intervals. Panel A of Fig. 2 shows the activity diagram for speed agent and Panel B presents the algorithm for the retrieval of the time and distance so as to calculate speed. The input to the agent is the *GPS coordinates* and the output is the *average speed*. The triggering event is when the driver starts moving and the stopping event is when the driver has parked his car. The frequency of GPS reading is increased as the user approaches the parking slot as it becomes more critical.

3.2 Geographic Information System (GIS) Agent

The *GIS Agent* calculates the approximate time of arrival based on the speed obtained from the *Speed Agent*. It also retrieves the location and performs geofencing and routing. Figure 3 shows the activity diagram of the *GIS Agent*. The input to the agent is the GPS coordinates and speed and the output is the time of arrival and the current location. The triggering event is when the speed. The stopping event is speed is obtained from the *Speed Agent* and stopping event is when the driver reaches the

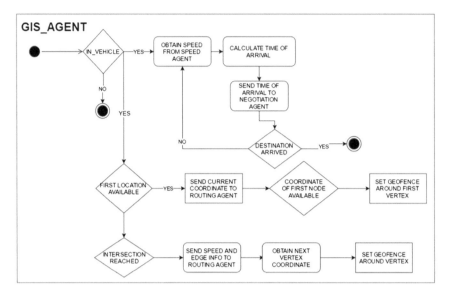

Fig. 3 Activity diagram for GIS agent

parks his vehicle. The agent also sends the time of arrival to the *Parking Agent* and the average speed to the *Routing Agent* and also recuperates the next route to take.

3.3 Mobile Agent

The *Mobile Agent* is responsible to convert user requests into events that will potentially trigger actions of other agents. It also combines events from different agents and convey them to the driver or forwards them to other agents. Figure 4 above shows the activities done by the *Mobile Agent*. When the user enters his destination, the *Mobile Agent* triggers an event that makes the other agents aware of that action and consequently the other agents will follow their own sequence of action. A number of *Parking Agents* will be used and each one will be in charge of a specific area.

3.4 Negotiation Agent

The *Negotiation Agent* negotiates parking spot price which results from a series of steps from the Negotiation Agent and the Parking Agent according to different parameters and goals. The parameters include the place the person wants to go or the amount he is willing to pay. The negotiation does not start until the city radius is reached. Figure 5 is the activity diagram for this agent. The city radius is a special

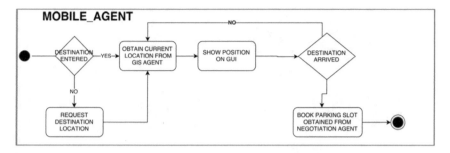

Fig. 4 Activity diagram for mobile agent

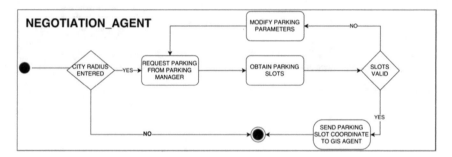

Fig. 5 Activity diagram for negotiation agent

boundary that surrounds the destination of dynamic length adjustable by the analytic agent based on parking miss rates (how many times parking allocation are changing).

3.5 Parking Agent

The *Parking Agent* manages parking slots and assigns them to the driver by communicating with the *Mobile Agent*. Each region will have a specific number of parking slots, if all parking slots are booked, then another parking agent is contacted for parking in another region. Each parking agent needs to have an entry in the database consisting of its ID, Region Boundary and IP address. Whenever the *Parking Agent* is started, it updates its IP in the database. It also ensures that the database is kept updated. The agent uses the user priority list to keep all the drivers in order of their arrival time. Figure 6 is the activity diagram for the *Parking Agent*.

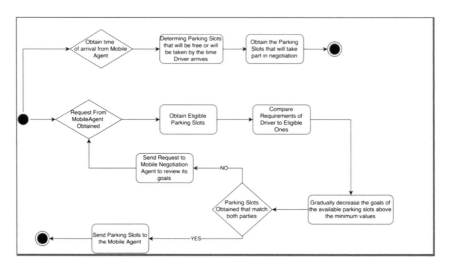

Fig. 6 Activity diagram for parking agent

3.6 Routing Agent

Since the speed is already being calculated by the speed agent, it just needs to send the average speed to the routing agent which holds a repository of all the routes and updates his speed. The routing part consists of two steps: (1) the *GIS Agent* receiving coordinates of the next intersection to follow and informing the server when it reaches a node with the average speed travelled on that edge and (2) the *Routing Agent* (Fig. 7) receiving updates about *GIS Agent*'s position, calculates the next vertex to follow and sends it back to the *GIS Agent*.

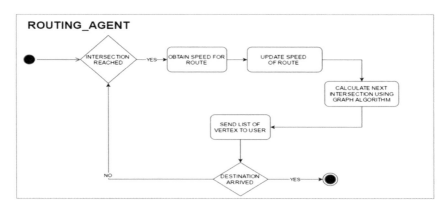

Fig. 7 Activity diagram for routing agent

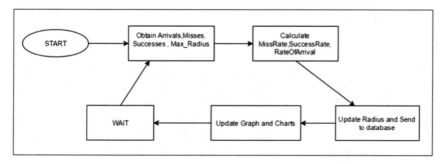

Fig. 8 Activity diagram for analytic agent

3.7 Analytic Agent

The analytic agent (Fig. 8) is used to analyse the different metrics to measure the performance of parking assignment. A radius is set from the destination address to mark when to send the parking locations. If the missed rate (defined as parking assigned and changed) is too high, then the radius may be decreased so that the probability of getting the 'considered' slot is higher. The radius is dynamically controlled by the agent depending on success metrics.

4 Implementation and Testing

Andromeda has been selected as the platform to be used for the development of the application. The Andromeda structure was specifically implemented to embed into the Android system compared to other Agent platforms such as Jade which uses a middleware and thus reducing the agent integrity as remarked by Aguero et al. [14]. Andromeda uses the components of Android operating system to build the components of the Agents. The main components of Andromeda are *Agents*, *Behaviour*, *Capability*, *Task* and *Message*. Interfaces for the mobile application have also been designed.

The system will be simulated for the Port Louis region. A list of scenarios has already been prepared to test the architecture. The next step will be to proceed with the implementation of each agent and complete the first version of the smart parking mobile application.

5   Conclusion

In this paper, an agent-based model for smart mobile parking is proposed for the Mauritian context. The model considers the position of the driver and the speed at which he is travelling and proposes the most appropriate and available parking slot to him based on the following: his arrival time, the place he wants to go and also parking fee he wants to pay. The system is modelled using a total of seven different agents, whereby each agent tackles a specific problem for example, the routing, the negotiation and the geographical positioning among others. Parking allocation is a dynamic environment which responds to changing variable and therefore, it is well suited for an agent-based approach. However, future work will involve the development of a mobile application for Mauritius, based on the proposed model. Additionally, the model can be refined to accommodate other variables such as weather, current traffic and scheduled road repairs etc., in the calculation of arrival time. This additional information will improve the accuracy of arrival time and can eventually lead to more optimal allocation of parking slots.

References

1. Benenson, I., Martens, K., Birfir, S.: ParkAgent: an agent-based model of parking in the city. J. Comput. Environ. Urban Syst. **32**, 431–439 (2008)
2. Longfei, W., Hong, C., Yang, L.: Active parking guidance information system based on dynamic agent negotiation. In: 9th International Conference of Chinese Transportation Professionals, Harbin, China (2009)
3. Longfei, W., Hong, C.: Coorparative parking negotiation and guidance based intelligent agents. In: IEEE International Conference on Computational Intelligence and Natural Computing, pp. 76–79, Wuhan, China (2009)
4. Bhende, M., Wagh, S.: Intelligent car park management system using wireless sensor network. Int. J. Comput. Appl. Res. Cent. Dypiet Univ. Pune **122**, 1–5 (2015)
5. Bagula, A., Castelli, L., Zennaro, M.: On the design of smart parking networks in the smart cities: an optimal sensor placement model. Multi. Digital Publishing Inst. (MDPI) **15**(7), 15443–15467 (2015)
6. Tang, V.W.S., Zheng, Y., Cao, J.: An intelligent car park management system based on wireless sensor networks. In: 1st International Symposium on Pervasive Computing and Applications IEEE, pp. 65–68, The Hong Kong Polytechnic University, P.R. China (2006)
7. Chiang, K.-W., Duong, T.T., Liao, J.-K.: The performance analysis of a real-time integrated INS/GPS vehicle navigation system with abnormal GPS measurement elimination. Sensors. Modelling, Testing and Reliability Issues in MEMS Engineering (2013)
8. Mamta, G., Vinita, M., Dheeraj, K.: GPS based parking system. COMPUSOFT, Int. J. Adv. Comput. Technol. **5**(1), 1–4 (2016)
9. Qureshi, M.A., Noor, R.M., Shamim, A: A lightweight radio propagation model for vehicular communication in road tunnels. PLoS ONE, **11**(3) (2016)
10. Rongxing, L., Xiaodong, L., Xuemin, S.: An intelligent secure and privacy-preserving parking scheme through vehicular communications. IEEE Transactions on Vehicular Technology, Vol. 59, Engineering Research Council of Canada (2006)
11. Yousaf, K., Duraijah, V., Gobee, S.: Smart parking system using vision system for disabled drivers (OKU). ARPN J. Eng. Appl. Sci. **11**(5), 3362–3365

12. Russell, S.: Intelligent Agent (Chap. 2), https://people.eecs.berkeley.edu/~russell/aima1e/chapter02.pdf (1995)
13. Bosse, S.: Design and simulation of material-integrated distributed sensor processing with a code-based agent platform and mobile multi-agent systems. Sensors, Basel **15**(2), 4513–4549 (2015)
14. Aguero, J., Rebollo, M., Carrascosa, C., Julian, V.: Developing intelligent agents on the android platform. In: Sixth European Workshop on Multi-Agent Systems EUMAS, Bath, U.K. (2008)
15. Santi, A., Guidi, M., Ricci, A.: JaCa-android: an agent-based platform for building smart mobile applications. In: Languages, Methodologies, LADS 2010, Lyon, France (2010)
16. Jade http://jade.tilab.com/

Big Data Analytics Life cycle for Social Networks' Posts

Sadiyah Muthy, Thomas Sookram and Baby Gobin-Rahimbux

Abstract Today, the amount and variety of data produced have become more complex from the expansion of new services and new means to produce and share content. Large datasets from different sources and of different types are very helpful to analytics. This has given rise to the concept of Big Data Analytics. This paper focuses on the work done to develop a Big Data Analytics solution for a group of psychologists, whereby the source of data is social network posts. It gives an overview of the proposed life cycle used for the development of the solution and also explains each step through the implementation of the Big Data Analytics solution.

Keywords Big Data analytics · Big Data analytics life cycle
Social network analytics

1 Introduction

The way data is produced and used has changed over the past decades. With the rise of the internet connectivity, the rapid adoption of digital devices, and expansion of social media usage, more and more data are being produced and collected by corporates and service providers. With the decrease in the cost of media storage and cheap cloud storage services, storing large volume of data are no longer constrained by the cost. Nowadays, large volume of data, which exist in different formats, are stored and analysed so as to get valuable insight. This has given birth to the concept of Big Data. According to Gartner [1], 'Big Data is high-volume, high-velocity and/or high-variety information assets that demand cost-effective, innovative forms

S. Muthy · T. Sookram · B. Gobin-Rahimbux (✉)
University of Mauritius, Réduit, Mauritius
e-mail: b.gobin@uom.ac.mu

S. Muthy
e-mail: ummediiyah@gmail.com

T. Sookram
e-mail: thomas.sookram@gmail.com

© Springer Nature Singapore Pte Ltd. 2019
S. C. Satapathy et al. (eds.), *Information Systems Design and Intelligent
Applications*, Advances in Intelligent Systems and Computing 863,
https://doi.org/10.1007/978-981-13-3338-5_35

of information processing that enable enhanced insight, decision-making, and process automation'. The novelty of the field of Big Data and the lack of structure on how to proceed with the latter's analysis became one of the main reasons for the high failure rate of these projects related to data analysis of Big Data. This was, and is still today, due to the lack of a clear detailed approach on how to proceed efficiently with the development of such project [2, 3]. Approaches to proceed with the analysis of Big Data remains at its infancy, with not many research done in the field.

The motivation behind this work was the need for a Big Data solution to analyse Facebook posts, for the psychologists of a Mauritian NGO, Association Kinouete. The biggest challenge was to understand how to build the Big Data Analytics solution for the NGO. Very few methodologies have been proposed in the field, with the first one published only recently in 2015 and until, the preparation of this work, three different approaches on Big Data Analytics were reviewed. Inspired by the three approaches, a Big Data Development Analytics Life Cycle for Social Network Posts was therefore devised and implemented.

2   Background Studies

In 2010, the term 'Big Data' was far from the knowledge of the public, but by mid-2011, it became the new key trend, with all the customary publicity in the IT world. However, analysing and extracting information from 'Big Data' is the actual uprising in technology, since previously unfamiliar chunks of data have now become visible. In short 'Big Data' is the data that derive business value quickly, from a variety of new and developing potential data sources such as location produced by mobile phones, information available on the Web, social media data and real-time machinery equipment [4]. Big Data can be of gigabytes, terabytes, petabytes, exabytes or more, but this solely does not characterize the term Big Data. Big Data is characterized by the 5Vs, namely Volume, Variety, Velocity, Value and Veracity [5].

2.1   Big Data Analytics Life Cycle

Data analytics helps the organization to harvest their data and find new business opportunities. According to Tom Davenport [6], most of the top 50 companies surveyed worldwide implemented data analytics life cycle to reduce cost through data management technologies. This resulted in faster and more accurate decisions as well as the possibility to provide new products and services based on the results obtained from the analysis. The study and representation of data analytic life cycle is often vague and is usually up to the interpretation of the users. Data analytic life cycle has been readapted to cater for Big Data. Three life cycles identified in the literature are presented in this section. Table 1 presents the stages of each life cycle.

Table 1 Big Data analytics life cycles

Dietrich et al. [7]	Erl et al. [8]	Cagle [9]
Data discovery, data preparation, model planning, model building, communication of results, operationalization	Business case evaluation, data identification, data acquisition and filtering, data extraction, data validation and cleansing, data aggregation and representation	Data acquisition, data awareness, data analytics

Dietrich et al. [7] established a six-phase data analytic life cycle which can be used for Big Data. The stages are closely coupled and team members may revert back to uncover new information. The project moves forwards with 'callouts', which are informal criterion planned at the end of each stage, which acts as a means to evaluate if enough information and progress have been made to move forward. The sub-phases can be parallelized if they are mutually exclusive from each other.

Erl et al. [8] proposed the life cycle which consists of nine stages while addressing the 5Vs which characterize Big Data. The methodology provides a step by step guide. The life cycle starts with the *Business Case Evaluation* phase and ends with the *Utilization of the Analysis Results* phase. The emphasis is more on data extraction, analysis and visualization. There is no mention of stakeholder's involvement and hypothesis analysis.

Cagle [9] simplifies the cycle into three main phases which enclose data governance so as to maintain the quality, validity of the project similar to what a quality assurance team would do in a traditional project. He focused on Data Acquisition and Data Analytics and also introduced a new component which is Data Awareness making sure data from various sources and structures are merged in a coherent structure and are relevant to the domain of the project. Formal models to data analysis are also initiated in this phase.

3 Big Data Analytics Solution for Psychologists

'Association Kinouete' is an NGO, which was founded in 2001 which focuses on rehabilitating imprisoned inmates and their reintegration into the society. The psychologists working for the NGO caters for both male and female inmates and a follow-up of their families through psycho-social, emotional and family support programs. The NGO also offers a shelter to ex-inmates without family support a home to live, up to one year, to help them take a new departure in their lives. The work presented in this paper is an extension of an initial research carried out by one of the psychologists at the NGO, on the correlation between drugs and crime from a general population point of view. The project was divided into two phases the first one being an analysis of data based on surveys done in Mauritius and the second phase consisted of the study carried out for the international opinion over the matter to provide a comparative study between the difference of the local and international point of view. International opinion over the matter needs to be analysed so as to provide a comparative study between the local and international point of view.

The solution developed is an automated sentiment analytic tool for analysing posts on Facebook. The posts contents were mined and comments and reactions were analysed so as to evaluate the opinions over the topic of research. Charts were drawn for visualizing the results. The life cycle used and the implementation process are described in the next section.

4 Implementation of the Big Data Analytics Solution

The proposed life cycle used for the development of the Big Data Analytics Solution for psychologist is shown in Fig. 1 and further elaborated in Table 2. The life cycle is inspired from existing Data and Big Data Analytics life cycle presented in this paper as well as some Software Engineering principles. All the nine phases are mutually inclusive to each preceding phase but their sub-phases can be parallelized or skipped depending on the type of data extracted. The whole process is englobed under quality assurance following Cagle's emphasis to ensure minimal error production and more accurate data analysis. Each phase's implementation is further elaborated under this section.

4.1 Domain Study

During the domain study, the stakeholders were identified, and they helped the development team to understand the psychological domain. Members of the development

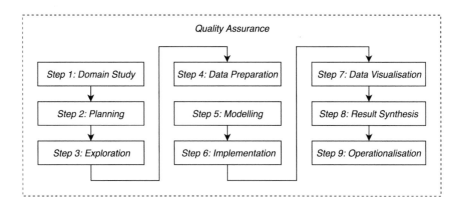

Fig. 1 Proposed Big Data life cycle for analysing social networking posts

Table 2 Big Data life cycle for analysing social networking posts

Phases	Sub-phases	Description of phase/sub-phase
Phase I: Domain study		This phase gives a clear interpretation of the stakeholders' key working area that shall impact on the planning phase and data investigation phase for a resourcefully successful output. Needs, interests, the level of technicality and functionality expertise of the staff, technologies are analysed
Phase II: Planning		This phase ensures that the project meets all the stakeholders needs without excessive expenditures and ensures that the project is assured quality wise
Phase III: Exploration	(1) Data investigation	During this phase potential data sources are identified, and the data is selected and acquired
	(2) Requirement investigation	This phase consists of the problem framing, requirement definition and hypothesis formulation
Phase IV: Data preparation	(1) Prepare analytic sandbox	This phase captures the data at a specific point in time separate from the live data leading to a less risky analysis and enable dry run test for later stages
	(2) Perform ELT	Consistency and quality of data extracted are evaluated in series of process which brings small nuances to the surface enabling data scientist to evaluate its relevance against the analytic hypotheses proposed
	(3) Data conditioning	Extracting, cleaning and normalizing the primary data for pre-analysis
	(4) High-level pattern abstraction	Iterative process where the hypothesis are tested, evaluated and refined against raw data analysis produced
Phase V: Modelling		This phase looks into the development of the model based on the datasets available. Modelling is based on the hypothesis formulation validated in High-Level Pattern Abstraction phase
Phase VI: Implementation		The model is implemented through the expansion of datasets for testing purposes and production
Phase VII: Data visualization		This phase involves analysis of huge amount of data to successfully derive useful patterns. The patterns are better understood and provide a feedback based on the analysis performed
Phase VIII: Result synthesis		The results obtained in the previous phase are illustrated on decision-making platforms such as dashboards
Phase IX: Operationalization		This phase involves the related activities required to enable the end-users to get acquainted with the analytical platform, understand the purpose of the graphical representations

team were acquainted with the work of the psychologists and also looked into the findings of the survey carried out in Mauritius. The initial goal was to leverage data from Facebook as a starting point to generate distinct graphical analysis interpretation charts. These patterns were to evaluate the link between the use of crimes and drugs keywords and the sentiment of the users in correlation with a recent psychology study on the subject and their own research. The development team also investigated the potential tools and technologies that could be used.

4.2 Planning Phase

The planning phase defined with the approved plans and schedule for the implementation project. The plans prepared were the Schedule Management Plan, Scope Management Plan, Cost Management Plans and above all Project Management Plan.

4.3 Exploration Phase

Twitter, Instagram and Facebook were the main sources identified but following discussions with the team of psychologist, Facebook was selected as the main data source. Further investigation on the Facebook platform went on to study and analyse the data. The hypothesis was also derived and discussed with the psychologist at the NGO. A group discussion was performed where everyone could share their opinion and a conclusive hypothesis was agreed upon. In parallel with the data investigation, requirements were then elicited, gathered, specified and validated using through interviews to better specify and validate the requirements. At the end of this phase, the requirements were agreed upon. The non-functional requirement is further broken down in system requirement and data requirement.

4.4 Data Preparation Phase

Sandboxing was used to create a copy of the live data. Graph API [10] was used to extract a copy of the Facebook datasets. After having obtained the datasets, the data had to undergo a format transformation before being loaded in the system. No data cleansing was required as only the required data were extracted. Data profiling was performed to get an in-depth overview of the data extracted and process. Some points identified during this process were: (1) Pagination links in compact and encoded version were sent back and in order to be passed as HTTP request for further retrieval, they needed to be transformed. (2) There is a limit of 5000 comments per post pagination request made, a further request per post is unauthorized and require special tokens and App ID keys to be accessed. During data conditioning, post content,

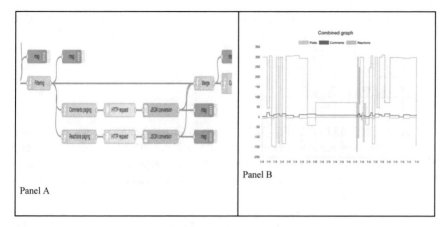

Fig. 2 Big Data life cycle for analysing social networking posts

paginated comments content and paginated reaction content were merged to form a complete set of data and metadata relevant to the post. The raw data analysis pattern was then investigated upon during the high-level pattern phase. Panel A of Fig. 2 shows an extract of the node.js module architecture from Node-RED [11] platform used for the data conditioning, while Panel B shows the raw data analysis pattern obtain during High-Level Abstraction phase.

4.5 Modelling

The analysis of the raw data pattern was used to define the scale of distribution of the sentiment polarity produced after the analysis. The hypothesis was validated based on the data extracted from the sentiment analysis of Facebook post. Metadata was attached to the post object variable. The metadata contained the average score of the polarity of the text, the list of positive and negative words and the sum of all the polarity for the texts evaluated. Different formulas which could be used to derive the sentiments were tested with the help of the team of psychologists.

4.6 Implementation

An investigation was first carried out to study the possible platforms which could be used for the implementation. A cloud-based solution was preferred due to cost and ease of maintenance. Google Cloud [12] and BlueMix [13] were considered. Bluemix was selected due to its flexibility and the possibility to use Node-RED. Sentiment analysis was performed by analysing the texts against the words from the AFFIN lexicon dictionary [14], whereby each word is assigned a polarity value ranges from negative five to positive five, with zero being neutral.

4.7 Data Visualization

The raw graph produced in Fig. 2: Panel 2 was further refined to represent small nuances in the data and balance out the high discrete polarity values of the Facebook post reaction which skewed the rest of the graph for both post content and comments polarity. The final formula used for graph plotting is shown below where s is the raw sentiment value score.

$$10 \log(s/5 + 1)/\ln 10 \tag{1}$$

4.8 Result Synthesis

Figure 3 shows the result after analysing 15,075 posts from the Huffington Post Facebook page between April 2016 and April 2017. From this graph, the psychologists extended their research on drugs and crime. The dashboard consists of three panels. One panel is used for the users' input, the second panel contains the real-time analytical graphs based on the sentiment analysis and the third panel notifies the user of the current status of the analysis, the number of posts, comments and reactions scanned.

4.9 Operationalize

Operationalize phase needs to be carried out. Stakeholders have been presented with the charts. The development team now needs to proceed with the training of all the psychologists who work for the NGO.

5 Conclusion and Future Works

This workgroup together the concepts of systemic data extraction, preparation and analysis of David Dietrich et al., the focus on data visualization and synthesis of Thomas Erl et al. and the quality assurance techniques from Kurt Cagle in what concluded to a complete big data analytic life cycle with a focus on a social media data analysis case study. The real-life case study describes how each phase and process were fit together and help to identify if there were any issues with the cycle.

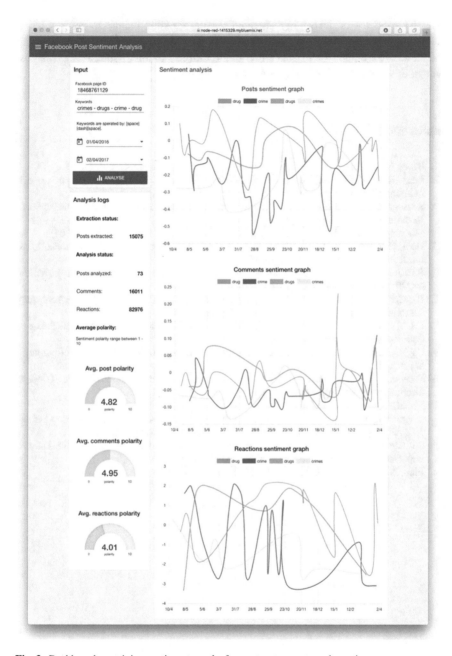

Fig. 3 Dashboard containing sentiment graphs for posts, comments and reactions

The approach proposed was built around a semi-rigid life cycle allowing some sort of parallelization at certain sub-level whenever possible. Future works include how this model would work in a constant changing requirements environment, in a fully integrated agile team environment or in a different domain other than social media where we could have both a mix of structured and unstructured data (images, data nodes from machine learning model).

References

1. Beyer, M.A., Laney, D.: The importance of 'Big Data': a definition Gartner (2012)
2. Reasons: Why data projects fail http://www.kdnuggets.com/2016/11/ten-ways-data-project-fail.html/2 (2016)
3. Reasons: Why most big data projects fail https://www.newgenapps.com/blog/main-reasons-for-failures-of-most-big-data-projects (2017)
4. Mitchell, I., Locke, M., Wilson, M., Fuller, A.: White book of big data, 1st ed. (2008)
5. Hadi, H.J., Shnain, A.S., Hadishaheed, S., Ahmad, A.H.: Big data and 5V's Characteristics. Int. J. Adv. Electron. Comput. Sci. **2**, 16–23 (2015)
6. Harvard Business Review Inc.: Data scientist: The sexiest job of the 21st century. https://hbr.org/2012/10/data-scientist-the-sexiest-job-of-the-21st-century (2017)
7. Dietrich, D., Heller, B., Yang, B.: Data Science and Big Data Analytics: Discovery, Analyzing, Visualising and Presenting Data. Wiley, Indianapolis (2015)
8. Erl, T., Khattak, W., Bulher, P.: Big Data Fundamentals: Concepts Drivers and Techniques. Prentice Hall, New Jersey (2016)
9. Cagle, K.: Understanding the big data life-cycle. https://www.linkedin.com/pulse/four-keys-big-data-life-cycle-kurt-cagle (2015)
10. Facebook Inc.: Facebook for developers - graph API documentation https://developers.facebook.com/docs/apps/changelog (2014)
11. NodeRed, https://nodered.org/
12. Gregor Hope: Google cloud data platform and services https://gotocon.com/dl/jaoo-aarhus-2010/slides/GregorHohpe_DataStorageAndAnalysisInTheCloud.pdf (2010)
13. Stifanie, R.: IBM BlueMix, The cloud platform for creating and delivering applications https://www.redbooks.ibm.com/redpapers/pdfs/redp5242.pdf (2015)
14. Finn, A.N.: AFINN, http://neuro.imm.dtu.dk/wiki/AFINN

A Speech Engine for Mauritian Creole

Wajiha Noormamode, Baby Gobin-Rahimbux and Mohammad Peerboccus

Abstract This paper presents the work done for the creation of a Creole Speech Engine. Speech recognition can improve the experience of a user and can contribute to technology adoption and usage. At present, most speech-based applications are in English or other foreign languages. Having a Creole-based speech engine to support application can be very beneficial. Therefore, a Creole dictionary and grammar file has been developed. Creole recordings with Mauritian accents have also been linked to the Creole dictionary. The success rate has been tested using a particular scenario and the results have been presented.

Keywords Speech engines · Speech recognition
Digital dictionaries Mauritian Creole dictionary

1 Introduction

Speech recognition as part of Natural Language Processing (NLP) by computers is a very complex domain that require thorough analysis of the language being processed but once mastered, can contribute a lot to any speech-based project. Majority of speech engines are English and other foreign languages and therefore any applications based on these engines. Existing Virtual Assistants software are limited to only the languages in which the speech engines are available.

Mauritius being a developing country has not embraced this revolutionary technology yet and still faces challenges in providing efficient customer assistance. Though Mauritius is both an English- and French-speaking nation, Creole is the mother

W. Noormamode · B. Gobin-Rahimbux (✉) · M. Peerboccus
University of Mauritius, Reduit, Mauritius
e-mail: b.gobin@uom.ac.mu

W. Noormamode
e-mail: wajiha.noormamode@umail.uom.ac.mu

M. Peerboccus
e-mail: mohammad.peerboccus@umail.uom.ac.mu

© Springer Nature Singapore Pte Ltd. 2019
S. C. Satapathy et al. (eds.), *Information Systems Design and Intelligent Applications*, Advances in Intelligent Systems and Computing 863,
https://doi.org/10.1007/978-981-13-3338-5_36

tongue of Mauritians and it is this language that is mostly spoken. As at today Creole dictionaries available in hard-copy versions but as at 2018, limited work has been done to digitalise these dictionaries. The motivation behind this work was therefore to bridge the gap by creating a creating a Creole speech engine which can then be used to develop Creole Virtual Assistants which will be a major breakthrough. This will help to have applications with smart interactions with Mauritians who are not at ease with English and French, thereby providing for more user satisfaction and thereby promoting usage of those smart applications.

This paper therefore presents the work done to build a Creole speech engine. It describes the steps used to build the Creole dictionary and grammar file using Sphinx4 [1], which is a Java speech recognition library. An OWL [2] ontology is used to handle the audio files for the speech engine. The speech engine was tested within a specific domain to check the success rate for the identification of the Creole words. The rest of the paper is structured as follows: Sect. 2 focusses on the background study carried out with regards to Natural Language Processing and Speech Recognition. The methodology used to build the Creole speech engine is presented in Sect. 3. Section 4 and the tests carried out to validate the speech engine are discussed. The last section concludes this paper by presenting the future works.

2   Background Study

Complex steps are involved for conversion of vocal speech to on-screen text for processing by a computer. Analog waves created by human voice are converted to digital data using an Analog-To-Digital (ADC) convertor [3]. The signal is then divided into small segments that are matched to known phonemes in the appropriate language. A phoneme is defined as "a sound of a given language, that native speakers agree on in just one segment and which enables them to recognize differences in meaning between words" [4]. There are 44 phonemes in the English language while other languages have more or fewer phonemes giving each one its specificity [5]. The Mauritian Creole has 26 phonemes [6].

The field of NLP involves making the computers perform used tasks through natural language processing. Speech Recognition is being perceived as the key enabler to enhanced user interaction with computer systems [7]. There are various speech recognition and natural language processing tools which have been developed. Some of them are as follows:

1. Apache OpenNLP [8]: The Apache OpenNLP library is a machine learning based toolkit for the processing of natural language text. It supports basic speech recognition.
2. Sphinx4 [1]: Sphinx4 is a purely java and most popular, speech recognition library. It provides quick and easy libraries to convert the speech recordings into text with the help of CMU Sphinx acoustic models. Besides speech recognition, Sphinx4 helps identify speakers, adapt models, align existing transcription to

audio for timestamping and more. Sphinx4 supports US English and many other languages that can be implemented through modification of the in-built dictionary. The Lexicon Tool can be used to generate phonetics from new words upon amendments in the dictionary. Corpus feature allows creating new pronunciations but requires significant NLP knowledge to do so.

3. Java Speech API (JSAPI) [9]: The Java Speech API allows Java applications to incorporate speech technology into their user interfaces. It defines a cross-platform API to support command and control recognizers, dictation systems and speech synthesizers (Oracle.com, 2018).

2.1 Speech Recognition Process

Every word stored in the dictionary has its own set of phonemes which make up a word's full pronunciation. Phonemes captured by the microphones are assembled and compared to those of the words from the dictionary for recognition. The end of a session is marked by silence and this is where the challenge arises: complete silence is hard to achieve in real scenarios. However, a proper microphone sensitivity along with the ability of the API to recognize noise to some degree allows the program to recognize absence of speech and end a session in this manner. Speech recognition sessions are better illustrated using the following scenario.

During a recognition session, the user says the word 'GREETING' and phonemes are recognized as show in Table 1.

After the recognized words are processed, audio is then buffered per results of the system's logic. Audio can be buffered through speech synthesis or audio playback.

Table 1 Recognising the word 'greeting'

Steps	Process	
Step 1	Phonemes are captured and assembled	Capture: {G R IY T IH NG} Assemble: {G R IY T IH NG}
Step 2	Phonemes is then compared to a set of words' translations to phonemes, stored in a dictionary (.dict) file in the library being extended	GREETING consists of phonemes G R IY T IH NG and matches that captured
Step 3	If the set of phonemes captured corresponds to one found in the dictionary, the word is then checked from the grammar file. The grammar file stores words that are pertinent to a part of the speech recognition session. If the word is found both in the dictionary and grammar file outputs the words. Sets of words appearing in a grammar file must be subsets of the dictionary set of words	Output: Greeting

Speech synthesis of a new language involves creation of new pronunciations rules and exceptions and is a very complex process.

3   Building the Creole Speech Engine

A seven-step approach has been used to build the Creole Speech engine. Each step is explained in details in this section.

3.1   Phase 1: Analyze Speech Recognition Platform

The different platforms identified and presented in Sect. 2 were analyzed. The platforms were compared and the Sphinx4 platform was selected as it due to documentation availability, recognition accuracy as well as compatibility with Java and ease of manipulation.

3.2   Phase 2: Understand Speech Recognition Process in Spinnx4

Sphinx4 platform contains libraries, also known as jar files that allow words to be recognized from speech that is input via microphones by users. These libraries contain resources such as dictionaries, acoustic models, variations, grammars, among others. A dictionary of words is needed for the program to browse through it. Just like human beings need to know words before they can start building sentences, Sphinx4 needs to know the set of words expected to be input so that it can process recognition. Words in the dictionary are aligned to the left while their corresponding set(s) of phonemes are separated by tabs on the same line. Elements of the set of phonemes must be separated by spaces. Words do not necessarily need to be sorted in alphabetical order and words can be repeated a few times in cases where different pronunciations exist for a word. It is crucial that no blank lines exist in the dictionary, else the cursor will throw an error message when the program is run. No special characters like number sign, or hash sign, should appear in sets of phonemes. On a side note, phonemes are always written in upper case for English. As part of the external resources that organize words during recognition is the grammar file. A grammar file allows control over recognition for some areas in the project through defined rules that can be applied in coding.

3.3 Phase 3: Understand Particularities of Creole Dialect

US English language has its particularities when it comes to the way certain letters and words are spelled by different people. The same applies for Creole language. Creole language has pronunciations that have a very close resemblance to the French language since Creole is derived from a mixture that includes French in it. Other words include some from Asian cultures mostly Hindi and Urdu. Much care is needed while defining phonetics for Creole words. Letters can have different phonetics when placed in between permutations of vowels and consonants in some cases.

3.4 Phase 4: Build the Dictionary

Starting from a previous work [10], a digital Creole dictionary was obtained in Excel format containing roughly 13,000 words that was obtained from the Ministry of Education of Mauritius. From there, these were transferred over to a text file parsed into the Lexicon tool that did the translation to phonemes automatically systematically. The results were the new Creole dictionary with Creole words' pronunciations based on the English US pronunciation rules. It is to be noted that the French language could have proved to be more appropriate. However, online tools to automatically generate pronunciations based on French rules were not found. Figure 1 (Panel 1) contains the Creole words, left aligned, with corresponding phonemes on the right-hand side of each word. Since the English US pronunciation was used, accuracy of recognition was not of desired level. The set of words was narrowed down to roughly 150 words as per requirements of the test-case for the csu.mu portal. Phonemes of these words were manually rewritten after analysis of trends and patterns of phonemes with corresponding spoken pronunciations. These were then included in the grammar file, a sample of which is depicted in Panel 2. The grammar file being used to only include words pertaining to the test-case, it may be openly manipulated to restrict recognition by the engine. The Creole dictionary and grammar file were embedded in the Sphinx API.

3.5 Phase 5: Validate the Dictionary

Validation of the newly created dictionary inside the sealed jar file was done using a test-scenario from the Mauritian Government Complaints Portal [11]. Speech recognition was used along with guided input following vocal prompts. Updates to the dictionary were done numerous times to test for best pronunciations with different words from the defined set of Creole words. A set of approximately 150 words was tested. Open speech recognition is too prone to errors and will result in garbage-in issue with data input. For this reason, control data structures were implemented in

```
 82 ENN              EH  N
 83 ENN(2)           AH  N
 84 DE               D   EY
 85 DE(2)            D   UW
 86 TRWA             T   R   W   AA
 87 KAT              K   AE  T
 88 KAT(2)           K   AE  T   R
 89 SINK             S   IH  NG  K
 90 SIS              S   IH  S
 91 SET              S   EH  T
 92 WIT              W   IH  T
 93 NEF              N   EH  F
 94 ZERO             Z   EY  R   OW
 95 DIS              D   IY  S
 96 ONZ              AA  N   Z
 97 ONZ(2)           OW  N   Z
 98 DOUZ             D   UW  Z
 99 TREZ             T   R   EY  Z
100 KATORZ           K   AE  T   AH  R   Z
101 KINZ             K   AH  N   Z
102 SEZ              S   EY  Z
103 DISET            D   IY  S   EH  T
104 DIZWIT           D   IY  Z   W   IH  T
105 DIZNEF           D   IY  Z   N   EH  F
106 VIN              V   IH  N
107 VIN(2)           V   IH  N   T
108 VIN(3)           V   IH  N   EH
```

Panel 1: Creole Dictionary

```
 1 #JSGF V1.0;
 2
 3 /**
 4  * JSGF Grammar
 5  */
 6
 7 grammar grammar;
 8
 9
10 public <greet> = (BON-JOUR | BON-ZOOR | BON-ZOUR | BONJOOR | BONJOUR | BONJOUR | BONJOUR | BONZOUR | BONZOUR);
11 public <alpha> = (A | B | C | D | E | F | G | H | I | J | K | L | M | N | O | P | Q | R | S | T | U | V | W | X | Y | Z);
12
13
14 public <0-9> = ( ENN | DE | TRWA | KAT | SINK | SIS | SET | WIT | NEF | ZERO );
15 public <10-20> = ( DIS | ONZ | DOUZ | TREZ | KATORZ | KINZ | SEZ | DISET | DIZWIT | DIZNEF | VIN );
16 public <30-90> = ( TRANT | KARANT | SINKANT | SWASANT | SWASANN-DIS | KATROVIN | KATROVIN-DIS );
17 public <100-1000> = ( SAN | MIL );
18
19 public <command1> = <0-9>;
20 public <command2> = <10-20>;
21 public <command3> = <100-1000>+ [<30-90>] [<10-20>] [<0-9>];
22 public <command4> = <30-90> <0-9>;
23
24 public <makeDecision> = (OUI | NON | ANILER | CONFIRMER | MODIFIER);
25 public <FemaleNom> = (LEEVA);
26 public <MaleNom> = (MAMED);
```

Panel 2 : Creole Grammar File

Fig. 1 Creole dictionary and Creole grammar file

the project that filters possible inputs in different parts of the application. Guided input is favored to eliminate issues of garbage input into the system caused by lack of accuracy of speech recognition.

3.6   Phase 6: Link to Audio Files

An OWL ontology was created to link the audio files to the words in the Creole Grammar file. The ontology consisted of only the 150 words that were validated in Phase 5. Links were done for words having different pronunciations. For example, the word "Bonjour" can be pronounced differently by another person as "Bonzour" or "Bonjoor". Therefore, these three subclasses should become equivalent classes as shown in Fig. 2. The screenshot above shows how different pronunciations of the greeting "bonjour" standing for hello, was grouped together for single access point using Sparql [11] in the program. For different pronunciations of the word "bonjour", the same audio file is then retrieved and played back to the user through this grouping.

Each class and subclass was linked to audio files. For example, if a user says the word "bonjour", the system should reply "Bonjour". To do so, each keyword has its corresponding audio file link, where audio is recorded through a microphone with file format. Instead of going for speech synthesis with English pronunciations, Creole audio recordings were used so as to have better Creole vocal outputs. Our voices were recorded for this purpose. All audio files recorded are stored locally and their file paths are used to playback the audio.

4   Testing the Creole Speech Engine

As already mentioned in Sect. 3, the complaints form from the government portal was considered. This form can be filled by citizens to log complaints regarding issues in their locality. The form is English and quite un-accessible to a set of audience. A simplified version of this form which looks into the reporting of complaints related to pollution was used for testing purposes. The form was in Creole and vocal prompts extracted from the ontology were used to allow users to input their details without using the keyboard. All entries were speech-based. They were then stored in English in a database. The animated Avatar named Lisa acted as an indicator to the user that the system is conversing to them (Fig. 3).

A scenario for recording a complaint for Air Pollution was tested 10 times to check the success rates of the speech recognition. The results are presented in Table 2. For the results it was seen that open speech tends to have a lower success rate as compared to single words. The longer and more complex the words are, the more difficult it is for the system to capture what is vocally input by the user. The reason being that pace of speech affects the silence gaps between phonemes heard by the program. The success rate of 3 represents a successful recognition of 100% in a set of 10 attempts to input the NIC due to the open speech involved. Fields which have 0 success rate did have partial recognition, e.g., Floreal was detected for Address. Other fields like title have a higher success rate due to the keyword recognition involved. Factors like background noise, voice clarity, and microphone quality also play critical roles

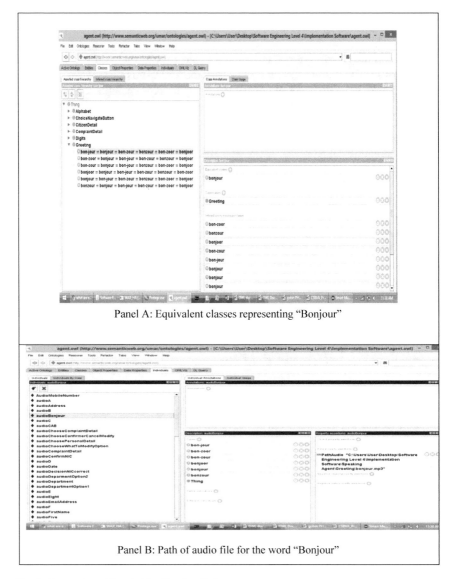

Panel A: Equivalent classes representing "Bonjour"

Panel B: Path of audio file for the word "Bonjour"

Fig. 2 Screenshot of Protégé for "bonjour" class

during recognition. We have used the best and worst available factors during testing to generate the above test results.

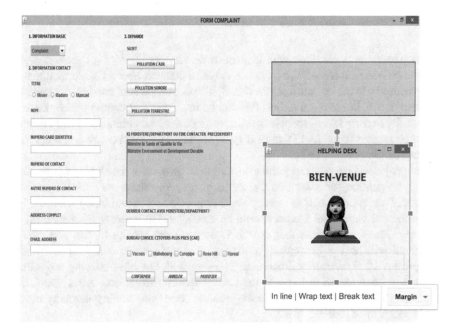

Fig. 3 Screenshot of complaint form

Table 2 Success rates for the various fields recorded

Field	Creole word captured	Success rate	Field	Creole words	Success rate
ID number	H121857295710C	3	Complaint category	Polision L'Air	7
Title	Mamzel	10	Department	Minister Environment ek Development Durable	8
First name	Leeva	10	Date last contacted Ministry	Premier Mars 2018	0
Last name	Maxwell	9	Nearest CAB option	Floreal	7
Phone number	58122449	7	Option	Confirmer	8
Other contact number	42444095	7	Option	Aniler	9
Address	Mangalkhan Floreal	0	Option	Modifier	7
Email address	rolf123@gmail.com	0			

5   Conclusion and Future Works

In this paper, the work done with respect to the development of a Creole speech engine has been described. We have been able to create a new a Creole dictionary and grammar file which was integrated in Sphinx4. The success rate for recognition was also tested. Some observations that can be made are that certain phonemes are more easily captured than others suggesting the nature of their complexity of spelling. Easier phonemes like S and IY are easily captured by the microphone as compared to complex phonemes like R.

Our project is just a starting point to an ambitious project of developing a powerful speech recognition for Mauritian Creole language. As future work further testing like better equipment such as microphones or processing power along with highly controlled environmental factors should be considered to leverage the recognition accuracy. The fact remains that speech is not uniform among human and every individual will have a particular pronunciation signature that computers will not recognise a 100% accuracy. Tokenisation along with machine learning are areas which will be explored to improve the Creole speech recognition. Also, linguistic expertise to allow development of Creole pronunciation rules would potentially allow open speech to be more accurate.

References

1. CMUSphinx: https://cmusphinx.github.io/wiki/tutorialsphinx4/#demos (2018)
2. Web Ontology Language (OWL): https://www.w3.org/OWL/ (2012)
3. Grabianowski, E.: How speech recognition works, https://electronics.howstuffworks.com/gadgets/high-tech-gadgets/speech-recognition1.htm
4. Elgin, S.H.: What is linguistics? 2nd Ed., pp. 85. Prentice Hall, New Jersey (1979)
5. The 44 sounds (phonemes) of English: https://www.dvusd.org/cms/lib/AZ01901092/Centricity/Domain/3795/Sound_Spelling_Chart.pdf
6. Klein, T.B.: Creole pholology typology: Phonem inventory size, vowel quality distinctions and stop consonant series https://www.philhist.uni-augsburg.de/lehrstuehle/germanistik/sprachwissenschaft/Unserdeutsch/publikationen/klein_2004.pdf (2004)
7. Sharma, K., Suryakanthi, T., Prasad, T.V.: Exploration of speech enabled systems for English. In: International Conference on System Modelling & Advancement in Research Trends (SMART), India (2012)
8. Apache OpenNLP: http://opennlp.apache.org (2018)
9. Java Speech Api: https://docs.oracle.com/cd/E17802_01/products/products/java-media/speech/forDevelopers/jsapi-doc/ (2018)
10. Khodabacus, Z., Bissoo, K.: Translation of English to Mauritian Creole and Mauritian Creole to English. B.Sc. (Hons) Software Engineering, University of Mauritius (2017)
11. SPARQL 1.1. Query language: https://www.w3.org/TR/sparql11-query/

A Smart Mobile Application for Learning English Verbs in Mauritian Primary Schools

Bhoovandev Fakooa, Mehnaz Bibi Diouman Banon
and Baby Gobin-Rahimbux

Abstract Learning English verbs in primary schools in Mauritius may sometimes be difficult for students as English is not their mother tongue. Integrating information technology to traditional way of teaching can help to improve the teaching process. In this paper, an AI-enabled mobile application which has been developed to help the learning Mauritian students is discussed. The application provides for personalised learning and an artificial neural network analyses the learning pattern of the students to readapt the type of learning material proposed to them. To bridge the gap due to language issues, the interaction between the students and application is in Creole. The student can say the verbs in Creole and then get the translated version in the various tenses. Instructions on how to use the application is also given in Creole. Quizzes are also given to help the learning process.

Keywords Mobile learning · Artificial neural network · Interactive systems

1 Introduction

During traditional teaching of verbs in primary schools in Mauritius all students are considered at the same level irrespective of their learning style. Not every student has the same pace and capabilities to learn and understand verbs easily specially since English is not the mother tongue of Mauritians. The content may need to be adapted and personalised for the students to help them better comprehend. This process requires a lot of time and teachers in some case cannot afford to spend too much time on particular problem facing students. Some students may have problems

B. Fakooa · M. B. D. Banon · B. Gobin-Rahimbux (✉)
University of Mauritius, Reduit, Mauritius
e-mail: b.gobin@uom.ac.mu

B. Fakooa
e-mail: bhoovandev.fakooa@umail.uom.ac.mu

M. B. D. Banon
e-mail: bibi.diouman@umail.uom.ac.mu

© Springer Nature Singapore Pte Ltd. 2019
S. C. Satapathy et al. (eds.), *Information Systems Design and Intelligent Applications*, Advances in Intelligent Systems and Computing 863,
https://doi.org/10.1007/978-981-13-3338-5_37

understanding a particular verb, tense or even how to use verbs in an English phrase. To help the learning process, the use of tablets has been introduced in primary schools. The benefits of mobile learning have been widely discussed in many publications. Tablets have been introduced in primary schools to support teaching. Therefore a smart mobile application to help students learn English verbs is definitely in line with the set objectives of the Government of Mauritius.

This paper presents an AI-enabled mobile application for primary students of Mauritius to help learn English verbs. Students can interact with the mobile application in Creole, i.e. say the verb in Creole. This verb is then translated to English and the conjugation in the various tenses is given. The application takes into consideration the different learning styles thereby providing text-based, audio and pictorial description of the verb. Quizzes are proposed to the students, and neural network algorithm is used to adapt the quiz based on the learning style that has been identified for the student. Instructions are given in Creole so that the students can easily use the application.

The rest of the paper is structured as follows: Sect. 2 gives a brief overview of the background study carried out. Section 3 presents the mobile application developed while Sect. 4 focuses on the Artificial Neural Network developed for predicting the learning style of the students. The last section concludes with the future work.

2   Background Study

In this section of the background study, research was done on the learning styles of students as well as the tools and technologies that can be used to develop AI-enabled mobile applications for learning.

2.1   Learning Styles

Advanogy stated that there are different learning styles and techniques that each person prefers [1]. Every person has a mix of learning styles in which they may find a dominant style of learning when compared to the utilisation of other styles. The seven learning styles that were stated by Advanogy are as follows: *Visual*, *Aural*, *Verbal*, *Physical*, *Logical*, *Social* and *Solitary*.

Another learning style model is the Felder and Silverman Learner Style Model (FSLSM). The FSLSM characterises learners according to these four dimensions: (1) Active/Reflective, (2) Sensing/Intuitive, (3) Visual/Verbal and (4) Sequential/Global. The FSLSM was used by Graf [2] to identify learning styles in learning management systems such as Moodle and WebCT. There are two ways to know the learning styles of the student which were determined by Brusilovsky [3] which are known as collaborative and automatic modelling. In collaborative modelling, learners give feedback on their learning styles which is then used to determine students' learning styles. In automatic approach, students' learning styles are determined automatically by observing students' behavior while they use the system for learning.

2.2 Mobile Learning

Mobile learning or 'M-Learning' refers to the use of mobile devices such as smartphones, mobile phones and tablet computers to support new and modern ways of learning process for education purposes [4]. Another definition for mobile learning provided by Saccol et al. [5] is as follows: 'Mobile learning is the learning or teaching process that happens with the use of Mobile and Wireless Information Technologies where people are in movement and are apart from each other physically and far from formal education physical spaces such as classroom or training rooms'. Nowadays, it is very usual for any person of all ages to possess any kind of mobile device. Kindergarten kids of aged five and six are always trying to reach for their parent's smartphones at any given opportunity. School age children have easy access to mobile learning technology due to the fact that they have smartphones within their reach for so many hours out of the day [6]. Mauritius is no exception. The use of tablets has been incorporated in the curriculum of primary students. Mobile learning provides the solution for personalised learning which is one of the challenges faced by traditional education. Study shows that touch devices such as mobile tablets, notebook, iPads and Android tablets are very popular devices for mobile learning because of their cost and the app store where countless free and paid education apps are available to download [4].

2.3 Artificial Neural Network and Learning

The need to provide personalised learning environment to learners has prompted the use of machine learning algorithm such as Artificial Neural Network (ANN) in e-learning platforms. By identifying the learning styles of students make sure that they are aware of the strengths and weaknesses. Thus, their learning environment can be personalised to according to their learning styles [7]. The learning experience of the students is then adapted by choosing the best learning path that suits the knowledge level and the acquired competencies of the students [8]. User's actions and usage patterns throughout the application can be monitored through adaptive user interface and this data can be passed through the ANN to infer the student's preference. The system can then adjust the interface components and content accordingly. An adaptable user interface provides the users with guidance to maximise on their potential and learning styles [9]. Identification of learning styles allows more students to gain more insights from more accurate information about their learning styles in a learning environment. Furthermore, teachers can have a better understanding about their students and provide necessary interventions [7].

2.4   Existing Mobile Applications to Learn English Verbs

Two mobile applications for learning English verbs have been identified. They are English Verbs and English Irregular Verb. English Verbs[1] contains around 1000 common English verbs which are either regular and irregular. Tenses such as the present simple tense, past tense and even the future perfect tense are present in the application. It also comes with definitions, pronunciation and pictures illustrations to allow its users to better understand the meaning and the context in which the verb is being used. Additional features such as text-to-speech and verb translation for languages such as French, Spanish and German are also present in the application in order to attract a wider audience around the world. The application is mainly targeted to users of all ages.

English Irregular Verbs[2] is an application which is mostly targeted to users whose weakness is mostly irregular verbs. Verbs such as to be, to buy and to bring have irregular participles and this application helps in learning and assessing its users' understanding of these verbs. The application contains three levels and each level contains a set of irregular verbs that are suitable for that particular level. Beginners should start at Level 1 and pave their way to Level 3. The system will ask the user after each level if they have properly understood the simple past and past participle of the verbs they have just learned through a mini exam. It is after passing the mini exam that a user gets to the next level. The application contains text-to-speech technology to output the conjugations of verbs for users who prefer audio over reading of text. Some aspect of personalization is present in the application such as choosing a preferred language or the look and feel of the entire application.

These applications were downloaded from Google store and tested. However, the main limitations of these applications within the Mauritian context is that these applications lack personalization, that is every user is given the same style of content irrespective of their level of understanding and learning styles. The visual aspect of these applications is not that appealing to young students and that is why these applications fail to attract the younger generation. These applications do not provide any means for students who are weak in the English language to navigate through the app and search for English verbs. That is why a new mobile application must be developed in order to overcome those existing limitations and to provide an effective learning tool to Mauritian primary students.

3   Overview of AI-Enabled Mobile Application

In this section, the features and the tools and technologies that will allow for the development of the AI-Enabled mobile application alongside its architecture are discussed. A description of the system with its user interfaces is also provided.

---

[1] https://play.google.com/store/apps/details?id=com.xengar.android.englishverbs.

[2] https://play.google.com/store/apps/details?id=com.gedev.irregular.

Table 1 Features identified after meeting with the primary school teachers

Feature number	Description
1	The application shall contain different levels for different grades
2	The application can contain a ranking system in order to bring competition among students as a form of motivation
3	The application can display verbs with images, sound, text to increase students' determination to keep using the application
4	A quiz system to test the student's knowledge about certain verb or tense can be implemented
5	Different type of quiz for different level shall be implemented
6	The application shall find the weakness of the student and provide him/her with quizzes
7	The application can adapt to the learning style of each student

3.1 Features Identified Following Meetings with Primary School Teachers

The process and approach of teaching English verbs in a real classroom environment by teachers must be understood and analysed in order to produce an application capable of teaching students. Several meetings were organised with teachers from different types of primary schools including teachers from Education Priority Zone (ZEP) school. Table 1 summarises the features that were requested by the teachers.

3.2 Tools and Technologies Selected

E-learning platform use ontologies as knowledge base. Gruber originally defined the notion of an ontology as an 'explicit specification of a conceptualization' [10]. Ontologies are being recognised in multiplicity of research fields and application areas, now gaining a specific role in Artificial Intelligence [11]. Ontologies are used to keep learning content and this allows for better searching [12]. The Ontology Web Language (OWL) [13] which is recommended by World Wide Web Consortium (W3C) is used to create ontologies. Protégé was selected to build the ontologies. Protégé comes with built-in plugins to extend its abilities to integrate features such as Reasoners and Semantic Web Rule Language on ontologies.[3] The Apache Jena Framework is a free and open source Java framework was used to manipulate the ontologies. Jsoup [14] was selected for the extraction and manipulation data from websites and populate the ontologies Finally, the Microsoft Azure [15] cloud computing platform was selected to host Fuseki server and perform other operations. The mobile application was developed using the Android studio.

---

[3]https://protege.stanford.edu/.

3.3   Description of the System

Primary students can use the application to search English verbs in specific tenses. The application synchronises with the remote verb ontology to the offline verb ontology found in the mobile application and checks whether the verb or tense requested exists in verb ontology. If the verb does not exist, then the application will scrap websites for verb conjugations and verb images and populate the online and offline verb ontology and respond to the student's request. This makes sure that the students will always have the most updated data offline. Students are also able to use to query a verb in Creole language.

The application uses an Artificial Neural Network to learn the student learning style and then present the conjugation in a way that best suits the student's learning style. After every conjugation, student is given a quiz according to their levels. Students will have to clear ten questions in the quiz and earn points to compete with friends. The application makes use of rule-base query to recommend similar and opposite verbs after each completed quiz. The application generates and corrects the quiz. The admin panel monitors the integrity of the verbs in the ontology. Task such as updating verb images, adding quiz template and adding similar and opposite action verbs of a particular verb is carried out by the administrator of the system. Furthermore, teachers will be able to monitor only his/her student's progress in the admin panel.

Figure 1 shows the architecture of the system which is a three-tier architecture. The client side consists of the mobile app and the admin panel. AndroJena API [16] provides the android application with tools to manipulate the ontologies. It also comes with built-in OWL-based Reasoner for inference. The admin panel will be responsible for monitoring the integrity of both remote ontologies. The server side will consist of Apache Fuseki Server. Fuseki is a SPARQL server that provides REST-style SPARQL HTTP Update, SPARQL Query and SPARQL Update of ontologies using the SPARQL protocol over HTTP. The cloud side consists of Azure Virtual Machine that will be used to run the Apache Fuseki Server remotely which will hold both the English Verbs Ontology and the Learner Ontology.

Figure 2 shows the various panels developed. Panel A is the homescreen. Panel B and Panel C show the various instructions given to the students in Creole. The same is said as audio instructions. Panel D shows a more visual oriented style for the presentation of the text while Panel E presents a text-based version. Panel F is an example of a Level 2 quiz.

4   Using Neural Networks to Predict the Learning Style of the Student

In this section, the design of the multilayered artificial neural network (ANN) will be discussed and how it will be used to suggest a learning style based on a student

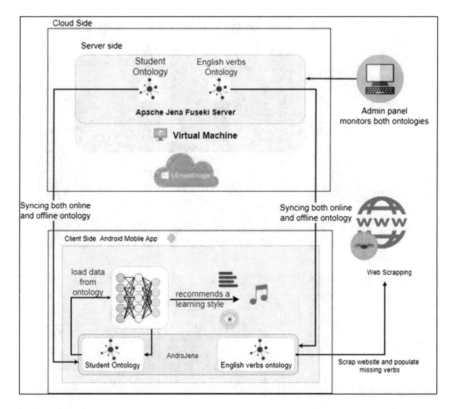

Fig. 1 Architecture diagram for AI enable mobile application for learning English verbs

behavior. The ANN is composed of an input layer to receive the signal, an output layer that makes a decision or prediction about the input, and in between those two, a hidden layer that is the true computational engine of the ANN. Figure 3 (Panel A) shows the design of the ANN. The input layer will be responsible to take in past data about the performance and behavior of all students. Table 2 shows the different inputs the input layer will take. The data from the input layer is multiplied by the hidden layer's weights and then added together. The sum is then passed through the activation function before passing it on. The perceptron makes use of the Sigmoid function as the activation function.

Figure 3 (Panel B) shows one perceptron from the hidden layer. Depending on the activation function, one of the output neurons will fire. The fired neuron will indicate which learning style is suitable for the student. Table 2 shows the different output neuron. The Weka API will be used as it provides the ability to build multilayered perception (ANN).

To build the trained model for the ANN, data were collected from 15 primary students. First, the students were given a questionnaire to fill with assistance from an elder to identify their most dominant learning style. The students were separated into

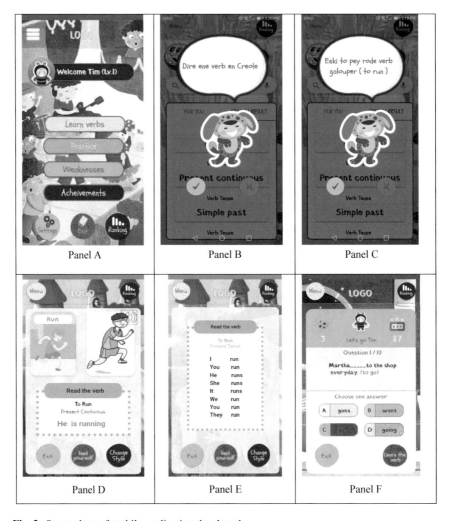

Fig. 2 Screenshots of mobile application developed

three groups based on their learning style and were given the application to perform 10 quizzes so that their behavior and points can be used to train the ANN model. It was then used to predict the learning style of another student and propose quiz based on the output.

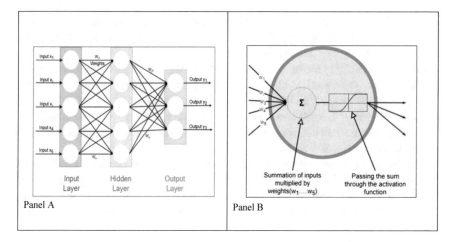

Fig. 3 Design of the ANN on Panel A and one perception function on Panel B

Table 2 Inputs and outputs of artificial neural network

Input Neuron	Data	Output Neuron	Data
x_1	The amount of time spent on visual content	y_1	More visually oriented
x_2	The amount of time spent on text content	y_2	Likes more text-based contents
x_3	The amount of time spent on text and visual content	y_3	Likes both text and visual content
x_4	The amount of time spent on the quiz		
x_5	The point scored after each quiz		

5 Conclusion and Future Works

The use of mobile application to teach English verbs allowed the possibility to build personalised and appealing e-learning tool. By using user interfaces that are friendlier toward primary students, attracted and kept them engaged when learning particular verb. The text-to-speech technology in the mobile phone allows the students to listen to the verb conjugation and repeat after them to better remember the conjugations. The quiz alongside the ranking system maintained a friendly competition among classmates in the quest to attain the first place in their respective class. The ANN captured the time the students were spending on different style of conjugation and the score they get in the quiz and based on that a preferred style was selected for a particular student. Future work includes the following:

- **Expanding the English Verb Ontology**. Based on the feedback obtained from the primary teacher other types of conjugation such as negative and interrogative conjugations will be added to the ontology.
- **Adding more quiz games to the application**. More quiz games can be added to increase the way a student might want to test himself/herself. This can keep the student more engage with the application thus learning more verbs.
- **Improvement of Creole-based interaction**. The Creole-based audio interaction in the application can be improved so that students who have difficulties in English can use the application without depending on teachers.
- **Classification verbs into different groups depending on their conjugation through inferencing**. Different English verbs conjugation have different ending. For example, past tense verbs that end with '-ed' can be grouped together and recommended to the students once the learn a verb with similar conjugation ending.

Declaration The dataset was taken with permission of the participants. All the authors are responsible for any kind of issues in future.

References

1. Overview of Learning Styles: https://www.learning-styles-online.com/overview/
2. Graf, S.: Identifying learning styles in learning management systems by using indications from students' behavior. In: 8th IEEE International Conference on Advanced Learning Technologies, pp. 482–486. IEEE Press, Spain (2008)
3. Brusilovskt, P.: Methods and techniques of adaptive hypermedia. In: User Modeling and User-Adapted Interaction (1996)
4. Mehdipour, Y., Zerehkafi, H.: Mobile learning for education: benefits and challenges. Int. J. Comput. Eng. Res. (2013)
5. Saccol, A.Z., Reinhard, N., Schlemmer, E., Barbosa, J.L.V.: M-Learning (mobile learning) in practice: a training experience with IT professionals. J. Inf. Syst. Technol. Manag. (2010)
6. Passing the Torch: https://issuu.com/abilenechristian/docs/acu_today_summer_2011
7. Bernard, J., Chang, T., Popescu, E., Graf, S.: Learning style identifier: improving the precision of learning style identification through computational intelligence algorithms. Expert Syst. Appl. **75**, 94–108 (2017)
8. Mota, J.: Using learning styles and neural networks as an approach to eLearning content and layout adaptation. In: Doctoral Symposium on Informatics Engineering (2008)
9. Kolekar, S., Bormane, D.S., Sanjeevi, S.G.: Learning style recognition using artificial neural network for adaptive user interface in e-learning. In: Computational Intelligence and Computing Research (ICCIC), pp. 1–5. IEEE Press, (2010)
10. Gruber, R.T.: Toward principles for the design of ontologies used for knowledge sharing. Hum. Comput. Stud. **43**(5–6), 907–928 (2005)
11. Guarino, N.: Formal Ontology and Information Systems. National Research Council (2017)
12. Khozooyi, N., Seyedi, N., Malekhoseini, R.: Ontology-based e-learning. Int. J. Comput. Sci. Inf. Technol. Secur. (2012)
13. Web Ontology Language (OWL): https://www.w3.org/OWL/

14. Handley, Z.: Is text-to-speech synthesis ready for use in computer-assisted language learning? Speech Commun. **51**(10), 906–919 (2009)
15. Microsoft Azure Cloud Computing Platform & Services: https://azure.microsoft.com/en-us/
16. Porting of Jena to Android: https://github.com/lencinhaus/androjena

The Need for Automatic Detection of Uncommon Behaviour in Surveillance Systems: A Short Review

Lalesh Bheechook, Sunilduth Baichoo and Maleika Heenaye-Mamode Khan

Abstract Surveillance system used for monitoring existed since decades. However, these surveillance systems required human intervention for the monitoring of suspicious behaviors. With the rapid revolution in technology, automatic detection of uncommon behaviors is gaining much attention from researchers. Being rapidly accepted in most public places to ensure transparency and security, surveillance systems are contributing in many applications like live traffic monitoring, crime scenes, and old people care. The deployment of automatic detection is very complex since it requires complex algorithms that should accurately detect uncommon behaviors, which is context-sensitive. In addition, it involves a lot of incoming data from various cameras, making it more challenging. In this perspective, the factors affecting surveillance systems and the techniques devised so far to detect uncommon behavior from these systems are analyzed and discussed. Robust automatic detection applications may yield to proactive decisions to be taken to prevent any harms/damages that would be caused by any uncommon/suspicious behaviors. Thus, it is important to explore techniques that can be used to implement automatic surveillance systems.

Keywords Surveillance · Uncommon behavior · Automatic detection

1 Introduction

Surveillance System was introduced back in 1942 in strategic places as a means of security. Surveillance systems are now being introduced in most public places as society is rapidly accepting the use of cameras in a wide variety of locations and applications: live traffic monitoring, parking lot surveillance, inside vehicles, and

---

Please note that the LNCS Editorial assumes that all authors have used the western naming convention, with given names preceding surnames. This determines the structure of the names in the running heads and the author index.

L. Bheechook (✉) · S. Baichoo · M. Heenaye-Mamode Khan
Department of Software and Information Systems, University of Mauritius, Moka, Mauritius
e-mail: lovishbheechook@gmail.com

© Springer Nature Singapore Pte Ltd. 2019
S. C. Satapathy et al. (eds.), *Information Systems Design and Intelligent Applications*, Advances in Intelligent Systems and Computing 863,
https://doi.org/10.1007/978-981-13-3338-5_38

intelligent spaces [1]. These cameras offer data on a daily and real-time basis that needs to be analyzed in an efficient manner. Unfortunately, most visual surveillance still depends on a human operator to monitor these videos, mostly stored passively for record purposes [2]. This is due to the fact that understanding uncommon behavior automatically from videos is a very challenging problem [3, 4].

Uncommon behavior is one which deviates from all behaviors. However uncommon behavior can only be defined in a specific situation, that is, it cannot be subjective but rather context-sensitive [5]. For example, standing in a queue can be categorized as normal behavior whereas standing in a moving crowd can be categorized as uncommon or suspicious. Therefore contextual data is important. Human behaviors have been a studied problem in Computer Vision for a long time. Since decades, a lot of studies have been made on simple human actions such as hand movements, small gestures, and gait [6]. However, more researches have been carried out recently in order to obtain maturity and promising results for automatic detection of uncommon behaviors from Surveillance Systems [7].

Uncommon behavior surveillance is important not only for detection but for prevention in a lot of situations ranging from incidents such as thefts, watching movements of old people in homes, avoiding the need of babysitting every time, to crimes such as spotting a potential terrorist in crowded public places like airports [7, 8]. The reason why automatic detection is being given so much importance in surveillance system recently is because of the monotonous, tiring, and continuous focus required from the human brain. According to [9], different person can find different movements suspicious or uncommon; giving as reason the *gut* feeling of the person.

In order to achieve an effective automatic detection and predict uncommon behavior recognition from video surveillance systems, there are three levels of studies that need to be carried out [10]. The first level is the core technology, which are the different algorithms involved at object segmentation, feature extraction, and object detection or classification levels. The second level is human activity recognition level, that is the level where human kinetics or movements are taken into consideration and lastly, the highest level is the applications or situations where human activity and related anomalies occur [7] explained that further studies are still required despite several promising results have been obtained till now. This is due to the fact that studies have been mainly focused on improving the recognition rate from video analysis or human motion [11–13]. Therefore, little reflection has been given to the prediction of uncommon human behaviors. In surveillance systems, multiple cameras with high resolution and quality produce huge amount of data. To be able to automate this process, a framework needs to be devised so that this data is stored and processed in a reasonable amount of time, to be able to minimize the bad consequences of the uncommon behavior. However, since big data management is a concept that requires a lot of maturity, finding an appropriate infrastructure to be able to cope with challenges [14, 15] like privacy and security, storing large amount of information and processing them in a required amount of time is still a prospective area of study.

The organization of the paper is as follows: Sect. 2 describes the different components required in a surveillance system. Section 3 focuses on the review and analysis of different techniques and algorithms that have been deployed so far in this area of study. Finally, a conclusion based on the different techniques identified is being discussed in Sect. 4.

2 Components of a Surveillance System

As mentioned in Sect. 1, surveillance systems comprise of three layers as shown in the block diagram below (Fig. 1).

2.1 Capture Devices and Preprocessing Techniques

The first main layer in a surveillance system is the data capture. Capture devices can be categorized into static and moving cameras. Static cameras are fixed ones that cover a specific and restricted range of view. However, it is easier to process and

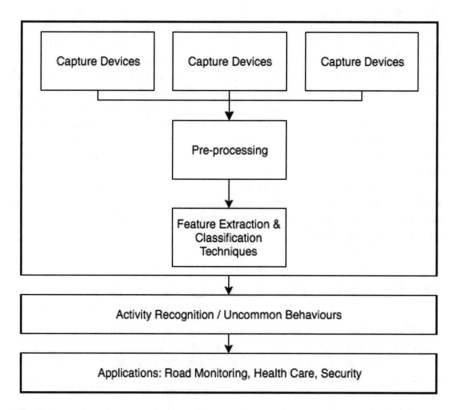

Fig. 1 Layers in a video surveillance system

preprocess images using such cameras. Moving cameras offer a wider range of view but are more complex to manipulate in terms of object tracking and detection. This is because the motion of the camera needs to be differentiated with the motion of the tracking object. As a result, preprocessing techniques are chosen depending on the capture devices being used.

For static cameras, the most commonly used preprocessing technique is background subtraction because of its simplicity and efficiency [16]. However, it is not sensitive to illumination variation. To cater to this issue, more appropriate algorithms like Gaussian Mixture Models [17] or Statistical Models can be chosen, where parameters like brightness and chromaticity can be varied and adapted.

In more complex situations where moving cameras are used, other techniques like temporal difference [18] and optical flow are applied. These techniques compensate for the change in background before segmenting the background from the foreground.

2.2 Feature Extraction Techniques

After obtaining the data, features need to be extracted to implement the automatic surveillance application. It is important to choose a good feature extraction technique since these are the only data that is used to differentiate uncommon behaviors. Global features like Space Time Volume (STV) [19] and Frequency transform, namely, Discrete Fourier Transform (DTW) [20] are good to track human actions but are restricted on moving cameras and occlusion. This is due to the fact that they consider the whole image when extracting a feature. Local descriptors are considered to cater to this issue. Local Descriptors such as Histogram of Oriented Graphs (HOG) [21], Scale-invariant feature transform (SIFT) consider characteristics of an image, making them invariant to occlusions and are resistant to image distortions like rotation and scaling.

However, the above feature extraction techniques were not created in the view of tracking human motions and therefore cannot capture all body actions. Thus, body more dynamic modeling techniques like simple blobs [22], 2D modeling and 3D modeling are proposed [23].

2.3 Human Activity Detection and Classification Techniques

To achieve a good recognition rate, a good classification technique must be used depending on the feature extraction technique used. Classifiers can be categorized as dynamic time warping, static classifiers, and dynamic classifiers. Dynamic time warping compares the similarity between two features when there is a change in motion or time. One common example of a static classifier is the Support Vector Machine (SVM). Static classifiers are very powerful for categorisation, however, dynamic classifiers like Hidden Markov Models (HMM) [24] or Dynamic Bayesian

Network (DBN) consider all variables in a model. This is important in uncommon behavior where all variables other than predefined or target variables need to be taken into consideration [25].

3 Analysis of Related Techniques in Suspicious Behavior Recognition

A series of processes need to be done in order to achieve automatic detection in order to detect uncommon behaviors automatically. An appropriate capturing, preprocessing, processing, and pattern recognition method needs to be implemented so that a good recognition rate is obtained for the detection of uncommon behaviors. Various studies have been done in this perspective to be able to achieve this.

3.1 Tracking Using Gaussian Mixture Models (GMM)

In [26], the authors proposed a simple method to detect if a person leaves a bag in a place and detects if it is an unusual behavior or not. The whole human body was tracked instead of tracking each part. GMM was used to segment the human body from the background. Further noise and shadow reduction were performed to the subject, which was then tracked. Tracking features included center of mass, the size, and bounding box. If the person disposed a bag, both the subject and the bag are tracked in different frames to detect human unusual activity.

However, to be able to track uncommon behavior in real-life situations, a more complex environment should be used to test the rate of recognition. [27] investigated suspicious behavior detection in public transport, mainly in Aircrafts.

3.2 Fusion Approach

Having presented a frequency-based fusion approach to discriminate between unruly and neutral movements in [27], six different behaviors were defined in a database for test-runs, namely, aggressive, cheerful, intoxicated, nervous, neutral and tired behavior. For practical usability, the monitoring system had to be real-time so that reaction can be done without delay. Therefore, low-level features were used since computational power and energy supply in vehicles are limited. In order to extract features robustly, a fusion of these descriptors was done, that is, global motion features were used to extract different parts of the images. To enhance classification, the images were split into several parts. The full frame was used as the first feature set and the face area was the second extracted feature.

3.3 Neural Network

Neural Networks was used and trained with varying parameters for robustness. For recognition, a dynamic classification with Hidden Markov Models was initially dismissed. This was due to the failure of another level of segmentation of behaviors [27]. Therefore a static method was adopted using Support Vector Machine. A recognition rate of up to 67.6% was obtained for a specific behavior. However, out of 222 samples, 204(91.8%) were correctly classified, neglecting which behavior it might be. It was concluded that promising results for the behavior detection task with low-level video features were obtained. However, a high false alarm rate has still to be decreased.

3.4 Clustering Topic Model (EM-MCTM)

Isupova et al. [4] proposed a technique called Expectation—Maximization Markov Clustering Topic Model (EM-MCTM) that improved the recognition rate of the original Hidden Markov Model. The error rate was reduced to less than 0.16. However, scalability is a major problem when detecting uncommon behavior in Surveillance Systems. Isupova et al. [4] came up with a technique called Dynamic Hierarchical Dirichlet Process. This technique observes normal behaviors and stores them so that it can be compared with uncommon behaviors. It updates its own benchmark as the system gets new data. It has the same concept of Dynamic Oriented Graph (DOG) [28].

4 Big Data Analytics in Surveillance Systems

To solve the scalability issue in such systems, a dynamic technique is not enough, that is, proper consideration has to be given on how to handle the huge amount of data. As quality and size of videos from cameras continues to grow, video surveillance data has become the largest source of big data [29]. Big data is not just about the volume of the data but also about the diversity of its data types. Big data analytics have emerged to provide advanced analytic techniques of big data sets using tools based on artificial intelligence, data mining, statistics, and natural language processing [30, 31].

To address the above-mentioned problem, [15] focused on designing a framework to analyze videos, large in size, for an Intelligent Transportation System. They proposed an algorithm that detects moving vehicles that were marked as blobs. A blob is described as a square of pixels and the surrounding squares that can be obtained by moving in a horizontal or vertical position from that square. Therefore a blob is a connected region that can be traversed through the pixels of the image. Blob filtering is carried out in order to remove the stationary blobs, that is, to subtract the background pixels in the input images. In their proposed framework, various systems like the CCTV, IP camera, cloud-based video systems, portable camera, embedded camera systems are merged to acquire a large amount of video data. The video is collected both in centralized and distributed modes. In the centralized mode, the workload of the processing algorithm is done at the server site whereas in distributed mode, the low cost, memory optimized embedded cameras needs to capture and preprocess the videos so that it is ready for further processing on the main server. Archiving of data is also included at the local site. Only keyframes, which summarize the video data, are sent to the cloud storage for further processing. This summary is done using Color Histogramming and a keyframe selection. The local sites use HDFS to archive the data locally and therefore this reduces the workload on the main server.

5 Conclusion

Research in automatic detection in surveillance systems has gained a lot of maturity at the core-level technology, that is, the preprocessing, processing and classifying techniques. Simple preprocessing techniques like background subtraction for simple conditions are most appropriate whereas Gaussian Mixture and Statistical Models are better for complex environments. Most feature extraction techniques have proved to be efficient depending on the type of data being captured. Global features are more efficient in large environments but less accurate to detect precise motions in the presence of occlusions. Classifiers that have proved to be efficient for automatic detection are, firstly, Neural Networks. It learns from data that are input and therefore improves scalability. Also, dynamic classifiers like Hidden Markov Model are very useful for detecting uncommon behaviors due to the fact even non-targeted variables are considered in the model. It can, therefore, consider variables that have not been predefined.

However, very little considerations were made on how to manage all the big data received from the capture devices. Since the objective of surveillance systems is to ultimately prevent bad consequences that might have happened, processing these data in a reasonable time period is of huge importance and require more attention in future studies.

References

1. Morris, B.T., Trivedi, M.M.: A survey of vision-based trajectory learning and analysis for surveillance. IEEE Trans. Circuits Syst. Video Technol. 18(8), 1114–1127 (2008)
2. Emonet, R., Varadarajan, J., Odobez, J.M.: Multi-camera open space human activity discovery for anomaly detection. In: IEEE conference on Advanced Video Signal and Surveillance (AVSS), Klagenfurt, Austria (2011)
3. Remagnino, P., Velastin, S.A., Foresti, G.L., Trivedi, M.: Novel concepts and challenges for the next generation of video surveillance systems. Mach. Vis. Appl. 18(3–4), 135–137 (2007)

4. Isupova, O., Kuzin, D., Mihaylova, L.: Dynamic hierarchical dirichlet process for abnormal behaviour detection in video. In: USES Conference Proceedings. University of Sheffield Sheffield Engineering Symposium, Sheffield, 24 June 2015. https://doi.org/10.15445/02012015

5. Harrigan, J., Rosenthal, R., Scherer, K.: The New Handbook of Methods in Nonverbal Behavior Research, Oxford University Press (2005)

6. Moeslund, T. B., Hilton, A., Kruger, V.: A survey of advances in vision-based human motion capture and analysis. Comput. Vis. Image Underst. Spec. Issue Model. People Vis. based underst. person's shape, appearance, mov. behav. **104**(2–3), 90–126

7. Ibrahim, S.: A comprehensive review on intelligent surveillance systems. Commun. Sci. Technol. **1**(1) (2016)

8. Baig, A. R., Jabeen, H.: Big data analytics for behavior monitoring of students. In: Symposium on Data Mining Applications, SDMA, Riyadh, Saudi Arabia. Procedia Computer Science. 1877–1899, 30 March 2016

9. Wells, H., Allard, T., Wilson, P.: Crime and cctv in Australia: understanding the relationship. Center for Applied Psychology and Criminology, Bond University, Australia Tech. Rep. (2006)

10. Janke, A.T., Overbeek, D.L., Kocher, K.E., et al.: Exploring the potential of predictive analytics and big data in emergency care. Ann. Emerg. Med. **67**, 227–236 (2016)

11. Jung, J., Yoon, I., Paik, J.: Object occlusion detection using automatic camera calibration for a wide-area video surveillance system. Sensors **16**(7), 982 (2016)

12. Murthy, V., Aravind, C., Jayasri, K., Mounika, K. and Akhil, T.: An automatic motion detection system for a camera surveillance video. Indian J. Sci. Technol. **9**(17) (2016)

13. Pojage, P., Gurjar, A.: Automatic fast moving object detection in video of surveillance system. IARJSET **4**(5), 190–195 (2017)

14. Katal, A., Wazid, M., Goudar, R.H.: Big data: issues, challenges, tools and good practices, In: 2013 Sixth International Conference on Contemporary Computing (IC3), IEEE, 404–409 (2013)

15. Ganesh, B.R., Appavu, S.:An intelligent video Surveillance Framework with big data management for Indian road traffic system. Int. J. Comput. Appl. Published by Foundation of Computer Science (FCS), NY, USA. **123**(10), 12–19, August 2015

16. Piccardi, M.: Background subtraction techniques: a review. In: Proceedings of IEEE. International Conference on Systems, Man and Cybernetics, The Hague, The Netherlands, Vol. **4**, pp. 3099–3104, 10–13 October 2004

17. Lin, W., Sun, M., Poovandran, R., Zhang, Z.: Human activity recognition for video surveillance. In: Proceedings of IEEE International Symposium on Circuits and Systems (ISCAS), Seattle, WA, USA, 2737–2740, 18–21 May 2008

18. Chaaraoui, A., Padilla-Lopez, J., Ferrandez-Pastor, F., Nieto-Hidalgo, M., Florez-Revuelta, F.: A vision-based system for intelligent monitoring: human behaviour analysis and privacy by context. Sensors 2014, **14**, 8895–8925 (2014). https://doi.org/10.3390/s140508895

19. Roshtkhari, M.J., Levine, M.D.: Online dominant and anomalous behavior detection in videos. IEEE Conference on Computer Vision and Pattern Recognition (2013)

20. Kumari, S., Mitra, S.K.: Human action recognition using DFT. In: Proceedings of the third IEEE, National Conference on Computer Vision, Pattern Recognition, Image Processing and Graphics (NCVPRIPG), Hubli, India, pp. 239–242, 15–17 December 2011

21. Verma,G.K.: Facial micro-expression recognition using discrete curvelet transform. 2017 Conference on Information and Communication Technology (2017)

22. Beaugendre, A., Miyano, H., Shidera, E., Goto, S.: Human tracking system for automatic video surveillance with particle filters. IEEE Asia Pacific Conference on Circuits and Systems (APCCAS), 152–155 (2010)

23. Hoang, L., Ke, S., Hwang, J., Yoo, J., Choi, K.: Human action recognition based on 3D body modeling from Monocular videos. In: Proceedings of Frontiers of Computer Vision Workshop, Tokyo, Japan, pp. 6–13, 2–4 February 2012

24. Adeleh, F., Rahebeh, N.A.: Recognition and classification of human behavior in intelligent surveillance systems using hidden Markov model. IJIGSP **7**(12), 31–38 (2015). https://doi.org/10.5815/ijigsp.2015.12.05(2015)

25. Moeslund, T., Granum, E.: A survey of computer vision-based human motion capture. Comput. Vis. Image Underst. **81**, 231–268 (2015)
26. Agarwal, A., Triggs, B.: Recovering 3D human pose from monocular images. IEEE Trans. Pattern Anal. Mach. Intell. **28**, 44–58 (2006)
27. Arsic, D., Schuller, B., Rigoll, G.: Suspicious behavior detection in public transport by fusion of low-level video descriptors. In: Proceedings of the 8th IEEE International Conference on Multimedia and Expo. IEEE Computer Society Press, Beijing, China, 218–221, (2007)
28. Duque, D., Santos, H., Cortez, P.: Prediction of abnormal behaviors for intelligent video surveillance systems. In: Proceedings of the 2007 IEEE Symposium on Computational Intelligence and Data Mining (2007)
29. Madden, S.: From databases to big data. IEEE, Internet Comput. **16**(3), 4–6 (2012). https://doi.org/10.1109/mic.2012.50
30. Russom, B.P.:Big data analytics. Tdwi Best Practices Rep. (2011)
31. Tiejun, H.:Surveillance Video: The biggest big data. Computing Now, **7**(2) (2014)

A Smart Parking and Mapping System Using Aerial Imagery

Yuvaraj Purahoo, Gowtamee Gopal and Nitish Chooramun

Abstract In this work, we describe the implementation of a smart parking and mapping system, named Smart Park, for searching and booking parking spaces in uncovered parking areas. This system comprises of a mobile application which allows users to view and select available parking slots on an interactive map based on aerial imagery. The system can suggest parking spaces based on criteria such as vehicle dimensions, distance to destination and user preferences. QR technology has been used to allow users to check-into and out of the parking lot. Moreover, the application integrates a map guidance system to guide users to their designated parking spaces. The application features an administrator interface, with parking data represented in visual format, for analysis and strategic decision making. The features implemented are aimed at alleviating the difficulties in parking identification and allocation, thereby minimising congestion in and around parking lots.

Keywords Smart parking · Parking mapping system · Parking dashboard

1 Introduction

Currently, the number of commuters using road transportation, on daily basis, has increased considerably. In fact, studies show that globally, there are over 40% of commuters who spend more than one hour using road transportation every day [1]. This substantial increase has undeniably led to severe congestion problems, thereby causing delays to drivers and commuters. Another major reason which contributes to

Y. Purahoo · G. Gopal · N. Chooramun (✉)
Faculty of Information, Communication and Digital Technologies, University of Mauritius, Reduit, Mauritius
e-mail: n.chooramun@uom.ac.mu

Y. Purahoo
e-mail: yuvaraj.purahoo@umail.uom.ac.mu

G. Gopal
e-mail: gowtamee.gopal@umail.uom.ac.mu

© Springer Nature Singapore Pte Ltd. 2019
S. C. Satapathy et al. (eds.), *Information Systems Design and Intelligent Applications*, Advances in Intelligent Systems and Computing 863,
https://doi.org/10.1007/978-981-13-3338-5_39

this increase in travel time is the time spent in finding parking spaces [2]. In Mauritius, the number of vehicles registered in between the years 2006–2017 has witnessed an increase of more than 4% annually [3]. However, in spite of this increasing fleet of vehicles, the number of parking spaces available has been relatively constant. Due to the limited number of parking spots, finding a place to park, especially during peak hours, has become increasingly tedious. This process is not only time-consuming and unpractical but also causes fuel wastage and air pollution [4]. Moreover, limited availability of parking spaces may also lead to parking lot accidents which can entail further delays and expenses.

The majority of parking spaces in Mauritius are in the form of outdoor parking spaces, whereby vehicles are parked as per yellow or white guidelines on the road. In this work, we describe the implementation of a smart parking and mapping system to efficiently allocate and manage parking spaces in outdoor parking areas. The proposed solution is technology-based and requires no infrastructural changes, thereby making it cost-effective and faster to deploy. The system, named Smart Park, allows parking locations to be mapped onto the satellite imagery of actual parking lots via a mobile interface. These parking locations can then be booked in advance by customers or during drive-ins. QR technology has been used to allow users to check-into the system and to enable real-time updates of parking status. Furthermore, the system features voice guided navigation to direct users to the parking slots assigned to them whilst minimising detours. In addition, the system can suggest parking spots to users depending on their preferences, destination and vehicle dimensions.

2 Related Work

The problem of parking space identification and management has been investigated in previous studies. For instance, in [5], wireless sensor networks have been used to show the occupancy status of parking spaces. This technique requires sensors to be installed on each parking location, which then communicates the status information wirelessly. However, such systems inculcate considerable costs with regards to the installation and maintenance of the sensors and were therefore not considered financially viable for implementation in the local context.

Existing mobile parking applications were also investigated with a view to identifying different options and features provided to users. For instance, Parking.sg [6, 7] is a parking app developed by the Government Technology Agency of Singapore which allows users to pay for parking using a mobile application, thereby eliminating the need for using parking coupons. Moreover, it allows parking sessions to be extended remotely. Nevertheless, this application can be used once a user has already found a parking slot and as such, it does not allow parking reservations to be made beforehand. Parkopedia [8] is another parking application which facilitates the identification of parking lots. It provides parking information for over 70 countries [8] but does not cater for parking areas in Mauritius. In addition, it does not allow the user to select a particular parking spot but rather shows the location of the entire parking area. Furthermore, another example of a parking application is JustPark which covers more than 20,000 parking locations within the UK [9]. This application allows parking reservations and extensions but does not allow choice of an actual parking slot within the parking areas. Besides, it can be noted that the applications presented herein [6–9] do not provide guidance mechanisms to the users until they reach their parking slots.

The system proposed in this work differs from other studies in the sense that it provides users with the flexibility of selecting a parking location from a real interactive map based on satellite imagery. The added convenience of this feature implies that users are able to select available parking locations which are closest to their intended destination. In addition, the ability to preview the map even before reaching the park lot implies that users have a heightened sense of familiarity with the parking area and its vicinity. This awareness is conducive towards minimising confusion and reducing delays, especially for drivers who are new to the area.

3   System Description

This section depicts the interactions between the core components of the system. Figure 1 illustrates the workflow between a user and the parking system during the parking booking process.

The workflow in Fig. 1 is described as follows:

1. A user logs into the system using the appropriate credentials. These credentials are obtained following the user registration process, during which a user can add personal and vehicle details. The user identity is verified via Firebase authentication.
2. The user initiates the booking process for a parking slot. The system retrieves the booking times from the Firebase database and verifies the availability of the time slot. Smart Park displays only the slots available during the time slot entered by the user. The user can then choose a suitable parking spot on the interactive map, which is discussed in more detail in Sect. 3.1. Alternatively, the user can select the suggestion option, which allows the system to assess all the available parking slots and suggest an appropriate one for the user. This procedure is described in more detail in Sect. 3.2. Following the selection of the parking slot (self or suggested), the user is prompted for effecting the payment for the booking.
3. Once the booking has been confirmed, the database is updated and a QR code is generated. The resulting QR image is saved to Firebase storage and displayed to the user. The size of the QR code is 200 × 200 pixels which accounts for its readability by a QR scanner at approximately 25 cm [10].

Figure 2 depicts the workflow between the user and the parking system during the check-in process, whereby the user reaches the entrance of the car park.

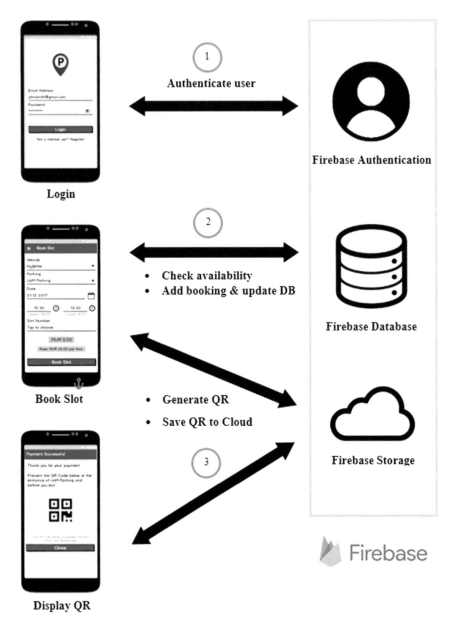

Fig. 1 Workflow between driver and system during booking process

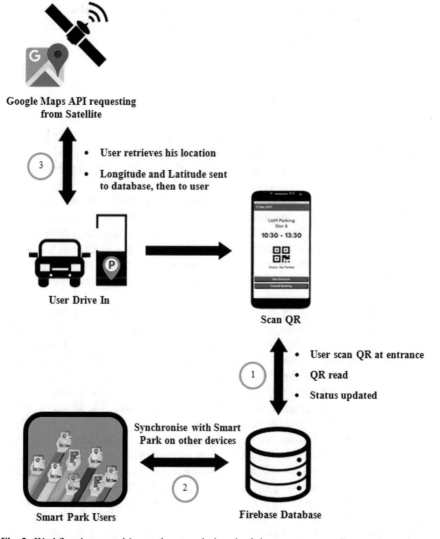

Fig. 2 Workflow between driver and system during check-in process

The workflow in Fig. 2 is described as follows:

1. The user drives in and scans the QR code at the entrance. The QR code is validated resulting in the status update of the parking slot from 'Not Parked' to 'Parked' in Firebase database.
2. Following the update in the database, all Smart Park users are automatically synchronised.

3. The user requests for directions to the designated parking slot. Smart Park retrieves the user's location and computes the directions to the parking slot. This feature is detailed in Sect. 3.3.

3.1 Mapping the Parking Lot

An important consideration during the design of the system was the flexibility to enable users to select parking locations within parking lots. One way of achieving this would be by displaying the available parking slots as a list of parking slot numbers and then making selections accordingly. However, this listing technique would not be intuitive as it would limit the user's knowledge with regards to the structure of the parking lot. Knowing the structure of the parking lot prior to reaching there has its benefits. For instance, drivers are less likely to be confused once they have reached the car park entrance which in turn implies less likelihood of congestions within and around the parking lot. In this respect, a mapping system was implemented such that individual parking locations could be superimposed on an actual aerial image of the parking lot.

To implement the map of the parking lot, a satellite image of the location was retrieved using Google Maps API. The individual parking slots within that parking area were then located. Four pins were dropped on the satellite image of the parking lot in such a way that a rectangle was formed to represent each parking slot as shown in Fig. 3. The coordinates (longitude, latitude) of every pin, for each parking slot were saved to Firebase database. For each slot, a fifth coordinate named Marker Position, representing the centre point of the parking slot was computed. The Marker Position was utilised for determining the shortest distance for effective parking allocation.

Figure 3 shows individual parking slots mapped onto aerial image of a parking location in Reduit, Mauritius. Smark Park also allows dedicated parking bays to be allocated to people with disabilities. In this example, three disabled parking bays (that are aside) have been assigned in such a way that they are closer to the entrance. Moreover, designating parking bays as disabled implies that these can be configured in order to benefit from lower parking rates or exemptions.

In Smart Park, parking slots can also be assigned or suggested according to the size of the user's vehicle. Upon adding a vehicle, the size class (small, medium, large) of the vehicle are automatically determined via web crawling according to the make and model input by the user. Users with large vehicles are typically assigned parking bays that are far from a fleet of parked vehicles as drivers may find it difficult to manoeuvre large vehicles in constrained parking spaces.

Fig. 3 Individual parking slots mapped onto aerial image of parking area

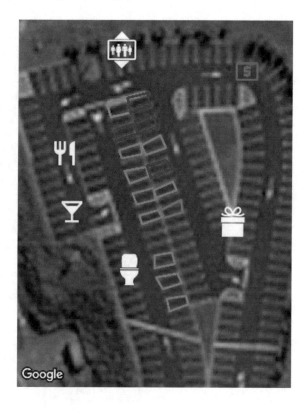

3.2   Suggesting a Slot

Smart Park also has the facility to suggest parking slots to users. In order to use the suggestion feature, users are required to specify their intended destination. In Fig. 3, six fictional destinations (for demonstration purposes) have been provided to the users and are represented by icons around the parking slots. When a destination has been selected, the application suggests a free parking slot that is nearest to the destination. The parking suggestion algorithm takes into account the Marker Positions of the parking bays, the coordinates of the destination, the availability of the parking based on the time, the user type (disabled or non-disabled) and the vehicle size class.

3.3   Determining Directions

When a user arrives at the parking lot, Smart Park provides directions to guide the user to the designated slot. The Dijkstra's algorithm [11] has been used to determine the shortest path from the entrance of the car park to the user's designated slot. The Dijkstra algorithm operates by taking as parameter the slot number assigned to the user in during the booking process. Each direction point on the map is treated as a

Fig. 4 Screenshot of Smart Park showing directions indicated by the solid (yellow) line

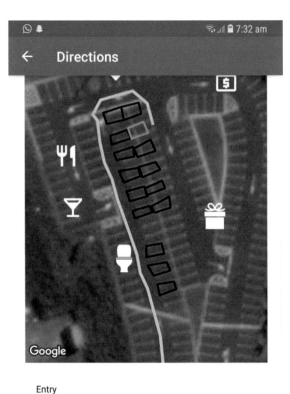

Vertex (Node) and each path between the Nodes is treated as an Edge. By taking into consideration the entrance of the parking lot and the Marker Position of the slot in concern, the algorithm is able to determine the shortest path from the entrance to the slot. Moreover, polylines are drawn on the map to guide the user accordingly as shown in Fig. 4. As the user advances forward, the user's current location is updated as a blue dot on the map. As it may be inconvenient for a driver to read directions or

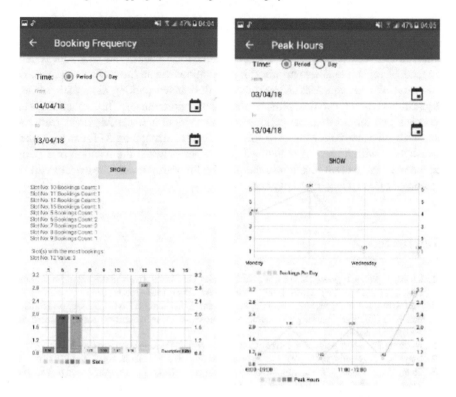

Fig. 5 Administrative interface showing booking frequency and peak hours

look at polylines on a mobile device while driving in a parking lot, a voice assistant has been integrated in the application.

3.4 Administrative Interface

Smart Park also features an administrative interface to facilitate the management of the parking area. This interface is in the form of a dashboard displaying current as well as historical booking data in graphical form as shown in Fig. 5. This data can be used to identify peak hours and booking frequencies of users.

4 Conclusion and Future Works

In this work, we have described the implementation of a smart parking and mapping system using aerial imagery. This system differs from other works by providing a low cost, technology-based parking solution without requiring costly infrastructural deployments such as sensors. The flexibility of the solution proposed implies that it can be extended to other open car parks in the local context. As such, the system presents benefits including parking slot reservations via an interactive map, remote extension of parking sessions, guidance to designated parking slots and parking suggestions based on vehicle dimensions and user preferences. This technology can be developed further to allow parking owners to input new parking areas from the administrative interface. Moreover, the zoom feature in mapping APIs can be further investigated with a view to reducing inaccuracies between the location of a point on the map and its real location in the parking lot. Furthermore, the system will be demonstrated on a larger parking area.

References

1. Zhang, J., Wang, F., Wang, K., Lin, W., Xu, X., Chen, C.J.: Data-driven intelligent transportation systems: a survey. IEEE Trans. Intell. Transp. Syst. **12**(4), 1624–1639 (2011)
2. Vera-Gómez, J.A., Quesada-Arencibia, A., García, C.R., Suárez Moreno, R., Guerra Hernández, F.: An intelligent parking management system for urban areas. Sensors (Basel, Switzerland) **16**(6), 931 (2016)
3. National Transport Authority. http://nta.govmu.org/English/Statistics/Pages/Archive.aspx
4. Elavarasi, R., Senthilkumar, P.K.: Smart car parking. Int. J. Sci. Tech. **10**(9), 1–4 (2017)
5. Tang, V.W.S., Zheng, Y., Cao, J.: An intelligent car park management system based on wireless sensor networks. In: First International Symposium on Pervasive Computing and Applications, pp. 65–70 (2006)
6. Parking.Sg. https://www.parking.sg/
7. Urban Redevelopment Authority. https://www.ura.gov.sg/Corporate/Car-Parks/Short-Term-Parking/Parkingsg-Mobile-Application
8. Parkopedia. https://en.parkopedia.com/
9. Intelligent Parking Management. https://www.justpark.com/car-park-management/
10. QRstuff. https://blog.qrstuff.com/2011/01/18/what-size-should-a-qr-code-be
11. Wang, J., Niu, H.: A distributed dynamic route guidance approach based on short-term forecasts in cooperative infrastructure-vehicle systems. Transp. Res. Part D: Transp. Environ. (2018)

Transform-Based Text Detection Approach in Images

C. Naveena, B. N. Ajay and V. N. Manjunath Aradhya

Abstract Nowadays, every document is very essential to be digitized. Increase in the gadgets where everyone likes to take the information in the form of images, but these images contains important information and necessary to be digitize. To do the digitization text detection in an image is one of the important stage in any field of document image analysis. But its not an easy task due to some of the challenges like complex background, varying light condition, low resolution etc. Hence, this work proposed detection of text in images. The proposed methodology consists of three steps. Initially the gabor filter is applied to extract the uncertainty features of the images. Then, 2D wavelet transform is applied to decompose the text information. Finally non-text information is removed using textual features based on the edge information. The proposed method is tested on MRRC and MSRA-TD500 standard dataset and obtained encouraging results.

Keywords Text detection · Gabor filter · Wavelet transform · Edge detection
Localization

1 Introduction

In recent years, most of the information is captured in the form of image. Electronic gadgets becoming much cheaper also contribute to the gathering of information in the form of image. This tends to generate complexity in gathering the information

C. Naveena
Department of CSE, SJBIT, Bengaluru, India
e-mail: naveena.cse@gmail.com

B. N. Ajay (✉)
Department of CSE, VTU-RRC, Belgaum, India
e-mail: ajaybn30@gmail.com

V. N. Manjunath Aradhya
Department of MCA, JSS Science and Technological University, Mysore, India
e-mail: aradhya.mysore@gmail.com

© Springer Nature Singapore Pte Ltd. 2019
S. C. Satapathy et al. (eds.), *Information Systems Design and Intelligent Applications*, Advances in Intelligent Systems and Computing 863,
https://doi.org/10.1007/978-981-13-3338-5_40

431

because of the varying configuration of the gadgets. However complex background, low resolution, varying text size, font, colors, styles makes text detection more challenging.

Many Optical Character Recognitions (OCRs) have been developed to recognize text by taking character corners or junctions as feature points of learning and matching scanned documents. But, OCR has some limitation where it gives very low recognition result for images which have complex background and low resolution.

Gabor representation of a signal that treats both the frequency and time domain on an equal basis, recommended a Gaussian-derived expansion that has come to be known as Gaussian function or Gabor function. Gabor function have been of particular interest in understanding and analysis of an image because they offer the means by which an image can be represented in an invariant form in both time and frequency domain.

Many researchers used transformation function for text detection in images. Among many transformation gabor is one of the promising techniques for extraction of text features. In [1], proposes a combination of wavelet transform and gabor filter to extract texture feature and sharpening respectively. The basic nature of Chinese characters possess stroke variation. To extract the different stroke effectively gabor filter is used in [2]. Another important features of gabor is it will give different orientation features of images by the help of wavelength parameter. The orientation feature are very important to detect text in images. This characteristic is explored in [3].Gabor features and collaborative representation based classification is used in [4]. Gabor energy and maximum extracted difference are extracted for expiration code detection. Gabor wavelet are employed to capture directional energy features of the word image. In this paper words from the kannada documents are retrieved based on the gabor feature.

Discrete wavelet transform basic characteristics are producing different orientation of frequency image of original image. By the help of these characteristics many researchers explore this idea in text detection. In [5] the decomposing of the given image into pyramid like structure for segmenting words is explored using the wavelets. With the increase of the decomposing level the gap between the characters decreases due to reduction of the size of the input image. Then the Laplacian wavelet is applied to detect the candidate text region. Text detection in digital images is achieved by combining the region-based- and texture-based methods. Wavelet-based features and edge projections in different directions are used. To disregard certain non-textual Area heuristic filtering is applied in [6]. In [7] text detection of regular and color images is described. This is carried out in two stages. In the first stage sharpened edge and textual feature of an image is obtained by applying the wavelet form and gabor filter. Discrete wavelet transform for scene text detection is proposed in [8]. There are several stages in this method. First, wavelet decomposition LH, HL and HH sub-bands are applied for detecting edges in original scene text image.

From the above survey it is clearly evident that, identifying the varying frequency characteristics of Gabor and orientation characteristics of wavelets when combined leads to identify some of the portion of text image in different clause. Motivated from the above two characteristics we plan to combine both the characteristics and detect the edge feature to remove false alarm. Hence in this paper we propose transform base algorithm for text detection.

2 Proposed Method

This section explains the proposed methodology which consists of three steps. Initially the Gabor filter is applied to extract the uncertainty features of the images. Then, 2D wavelet transform is applied to localize the text information. Finally non-text information is removed using textual features based on the edge information.

2.1 Gabor Filter

Gabor filters are examples of multiresolution techniques that can break down texture statistical complexity [9]. Gabors high sensitivity to local features facilitate the processes of preattentive or subtle texture discrimination as well [10]. Hence we applied gabor filter using following Eq. 1 and obtained the Gabor filter image which is shown in Fig. 1

$$G(x, y, \theta, u, \sigma) = \frac{1}{2\pi\sigma^2} exp \left\{ -\frac{x^2 + y^2}{2\sigma^2} \right\} exp \left\{ 2\pi i (uxcos\theta + uysin\theta) \right\} \quad (1)$$

where i is equal to $\sqrt{-1}$, u represents the frequency of sinusoidal wave θ controls the orientation of the function σ gives the standard deviation of the Gaussian envelope.

2.2 Discrete Wavelet Transform (DWT)

The uncertainty features extracted in the image reduces the low frequency level but due to invariant feature of the image not able to remove completely. Hence we used the decomposition characteristics of DWT effectively using Eqs. 2 and 3 and reduced most of the background position which is shown in Fig. 2. After obtaining different DWT components, we merge HL,LH, and HH for further process which is shown in Fig. 3.

$$W_{\varpi}(j_0, k_1, k_2) = \frac{1}{\sqrt{N_1 N_2}} \sum_{n_1=0}^{N_1-1} \sum_{n_2=0}^{N_2-1} s(n_1, n_2) \varpi_{j_0,k_1,k_2}(n_1, n_2) \quad (2)$$

$$W_{\omega}^i(j_0, k_1, k_2) = \frac{1}{\sqrt{N_1 N_2}} \sum_{n_1=0}^{N_1-1} \sum_{n_2=0}^{N_2-1} s(n_1, n_2) \omega_{j_0,k_1,k_2}^i(n_1, n_2) \quad (3)$$

(a)   (b)

Fig. 1 a Input Image b Image after applying gabor filter

Fig. 2 Results obtained after
discrete wavelet transform

3 Edge Features

After applying a DWT most of the text information in the image is highlighted along
with some unwanted information. To localize the text information properly used
vertical and horizontal edge detection approach which is shown in Fig. 4a. Then by
using connected component techniques draw the bounding box for connected area

Fig. 3 Results obtained after merging DWT components

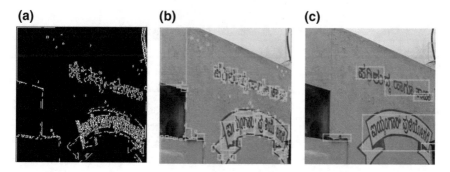

Fig. 4 **a** Results obtained after edge detection **b** Results obtained after applying connected component technique **c** Results obtained after successful text detection

which is shown in Fig. 4b. To remove non-text region used the heuristic rules such as aspect ratio, size, geometric features which is shown in Fig. 4c.

4 Experimental Results

The performance of the proposed method is analyzed using MATLAB 14 on 3.1 GHz with 8 GB RAM. The two different datasets MRRC [11] and MSRA [12] were used to evaluate the performance of the proposed method. The performance was measured using Recall(R), Precision(P) and f-measure(F).

Multi-script Robust Reading Competition(MRRC) consists of 167 training images 167 testing images with all kinds of challenges. Some of the challenges posed by images in the datasets are shading, slant, glossy, occlusion, shear, engrave, low resolution, depth, night-vision, emboss, artistic, illumination, multi-script, multi-color, handwritten and curve with variety of degradation in languages like Kannada,

Table 1 Comparison between proposed method and existing method (on MRRC dataset)

Methods	R	P	F
Proposed method	83	94	96
Yin et al. [11]	64	42	51

Fig. 5 Results obtained for MRRC dataset

Table 2 Comparison between proposed method and existing method (on MSRA dataset)

Methods	R	P	F
Proposed method	86	80	83
Li et al. [13]	63	29	40
Zhao et al. [14]	69	34	46
Kang et al. [15]	62	71	66
Yao et al. [16]	62	64	61
Yin et al. [17]	63	81	71

Fig. 6 Results obtained for MSRA-TD500 dataset

English, Hindi, and Chinese. The proposed method yields better results compared to the existing method which is shown in the Table 1 (Fig. 5).

MSRA is a publicly available benchmark dataset. The dataset contains 500 images collected from various locations. The images captured contain varieties like sign, doorplates, caution plates, guide boards, bill boards and many more with complex background. Some of the challenges put forward by the images in this datasets are font, size, color, orientation with diversity of text and complexity of the background. The proposed method yields better results compared to the existing method which is shown in Table 2 (Fig. 6).

5   Conclusion

The proposed method deals with text detection in images using transform approach such as gabor filter and DWT. Uncertainty features is highlighted by gabor filter and decomposed the image by DWT. Finally by the help of edge features localized the text portion. The proposed method evaluated on two standard datasets such as MRRC and MSRA-TD500. Also, compared with existing method and performed better. The limitation of our method found to be few false alarms in blurred and more complex background images which will be resolved in the future.

References

1. Manjunath Aradhya, V.N., Pavithra, M.S.: A comprehensive of transforms, Gabor filter and k-means clustering for text detection in images and video. Appl. Comput. Inf. **12**, 109–116 (2016)
2. Yan, J., Li, J., Gao, X.: Chinese text location under complex background using Gabor filter and SVM. Neurocomputing **74**, 2998–3008 (2011)
3. Tadic, V., Popovic, M., Odry, P.: Fuzzified Gabor filter for license plate detection. Eng. Appl. Artif. Intell. **48**, 2998–3008 (2016)
4. Zaafouri, A., Sayadi, M., Fnaiech, F., al Jarrah, O., Wei, W.: A new method for expiration code detection and recognition using Gabor features based collaborative representation. Adv. Eng. Inf. **29**, 1072–1082 (2015)
5. Liang, G., Shivakumara, P., Lu, T., Tan, C.L.: A new wavelet-Laplacian method for arbitrarily-oriented character segmentation in video text lines. In: 13th International Conference on Document Analysis and Recognition (ICDAR), vol. 978, pp. 926–930 (2015)
6. Grzegorzek, M., Li, C., Raskatow, J., Paulus, D., Vassilieva, N.: Texture-based text detection in digital images with wavelet features and support vector machines. In: CORES 2013, vol. 226, pp. 857–866 (2013)
7. Manjunath Aradhya, V.N., Pavithra, M.S., Naveena, C.: A robust multilingual text detection approach based on transforms and wavelet entropy. C3IT-2012 **4**, 232–237 (2012)
8. Ali, S.A., Hashim, A.T.: Wavelet transform based technique for text image localization. Karbala Int. J. Mod. Sci. **2** (2016)
9. Al-Kadi, O.S.: Tumour grading and discrimination based on class assignment and quantitative texture analysis techniques. Department of Informatics, University of Sussex, Brighton, Ph.D. thesis edition (2009)
10. Turner, M.R.: Texture-discrimination by Gabor functions. Biol. Cybern. **55**, 71–82 (1986)
11. Yao, C., Bai, X., Liu, W., Ma, Y., Tu, Z.: Detecting texts of arbitrary orientations in natural images. In: CVPR, pp. 1083–1090 (2012)
12. Shivakumar, P., Basavaraju, H.T., Guru D.S., Tan, C.L.: Detection of curved text in video: quad tree based method. In: 12th International Conference on Document Analysis and Recognition(ICDAR), pp. 594–598 (2013)
13. Lu, C., Wang, C., Dai, R.: Text detection in images based on unsupervised classification of edge-based features. In: Proceedings of the ICDAR, pp. 610–614 (2005)
14. Zhao, X., Lin, K.H., Fu, Y., Hu, Y., Liu, Y., Huang, T.S.: Text from corners: a novel approach to detect text and caption in videos. IEEE Trans. IP **20**, 790–799 (2011)
15. Kang, L., Li, Y., Doermann, D.: Orientation robust text line detection in natural images. In: Proceedings of the CVPR, pp. 4034–4041 (2014)
16. Yao, C., Bai, X., Liu, W.: A unified framework for multi-oriented text detection and recognition. IEEE Trans. IP **23**, 4737–4749 (2014)
17. Yin, X.C., Pei, W.Y., Zuang, J., Hao, H.W.: Multi-orientation scene text detection with adaptive clustering. IEEE Trans. PAMI **37**, 1930–1937 (2015)

Hybrid Spatial Modelling, from Modelling People in the Vicinity to Urban Scale Simulation of Pedestrians

Nitish Chooramun, Peter J. Lawrence and Edwin R. Galea

Abstract Buildings, structures, public venues and urban areas can typically accommodate large numbers of people and behaviours. Consequently, it is a challenge for authorities to demonstrate that these heavily populated structures can be evacuated quickly and efficiently in emergencies such as fires, flash floods amongst others. Computational evacuation models are increasingly being used as decision support tools for evaluating evacuation efficiency. These models typically use either macroscopic or microscopic approaches for simulating pedestrian navigation. However, these approaches represent a trade-off between accuracy and scalability. In this work, we present a hybrid evacuation model, combining both macroscopic and microscopic modelling techniques, which can be scaled to larger structures, while maintaining accuracy. This is demonstrated in this work by applying the model to a building, rail tunnel station and large urban area.

Keywords Hybrid evacuation model · Multi-spatial modelling
Agent-based model · Pedestrian dynamics

1 Introduction

Public areas and structures for instance multi-storey buildings, shopping malls, railway stations, concert venues and urban areas typically comprise of large numbers of people. As such, in order to determine the safety levels of such structures, it is crucial

N. Chooramun (✉)
Faculty of Information, Communication and Digital Technologies, University of Mauritius, Reduit, Mauritius
e-mail: n.chooramun@uom.ac.mu

P. J. Lawrence · E. R. Galea
Fire Safety Engineering Group, University of Greenwich, London, UK
e-mail: P.J.Lawrence@gre.ac.uk

E. R. Galea
e-mail: E.R.Galea@gre.ac.uk

© Springer Nature Singapore Pte Ltd. 2019
S. C. Satapathy et al. (eds.), *Information Systems Design and Intelligent Applications*, Advances in Intelligent Systems and Computing 863,
https://doi.org/10.1007/978-981-13-3338-5_41

to demonstrate that the occupants can be evacuated swiftly and efficiently in the event of hazardous situations such as fires, flash floods and high-density crowd formations. Past incidents have shown the dangers to which occupants can get exposed, e.g. Love Parade music festival disaster in 2010 (Germany, 21 fatalities, more than 500 injured [1]), Station Nightclub disaster in 2003 (Rhode Island, 100 fatalities [2]), King's Cross station fire disaster in 1987 (UK, 31 fatalities [3]), retail store fire disaster in 1979 (UK, 10 fatalities [4]). Conventionally, two techniques which are typically used to assess the safety of buildings and structures include building regulations and evacuation trials. Buildings regulations are often prescriptive (e.g. width of doors, distance to exit location, number of exit locations, dimensions of staircases) and do not necessarily cater for the complexities of various types of building/structure designs. Evacuation trials comprise of real evacuation experiments with actual participants. The downside with these trials is that they can be difficult to set up and can expose the participants to significant danger [5]. The shortcomings of these two techniques have led to an increased adoption of evacuation simulation software for evaluating the safety of structures. These tools present several benefits as they allow architects, engineers and safety personnel to evaluate the efficiency of evacuation routes and detect regions prone to bottlenecks within buildings and structures.

Computational evacuation models typically use either macroscopic or microscopic modelling approaches for representing pedestrian navigation. However, these two approaches represent a trade-off between accuracy and scalability. For instance, coarse models [6] represent people at the macroscopic level (coarse regions) and provide fast computation. However, they are unable to accurately simulate people behaviours and movement. On the other hand, continuous models [7] represent pedestrians at the microscopic level (continuous regions) and are capable of modelling pedestrian movement and behaviours to a high level of detail and accuracy. However, continuous models are impacted by poor computational performance. Fine node models, also operating at the microscopic level (fine node networks) [8], provide a compromise between coarse and continuous models by providing reasonable computational performance and representing individual people movement and behaviours.

In this work, we present a hybrid model, with both macroscopic and microscopic features, that is able to harness the efficiency of coarse models and the high fidelity of fine and continuous models. This allows the hybrid model to scale to larger enclosures (buildings, structures) while still being able to maintain accuracy. Moreover, the flexibility of combining different methods of representing the space (multi-spatial) in the hybrid model makes it possible to address different user needs or simulation requirements. For instance, if evacuation time results are required in the fastest time possible, a higher percentage of coarse regions can be used for representing the enclosure. Whereas if the blueprint of an enclosure (e.g. CAD design) is precise, then the higher resolution fine nodes and continuous regions can be used for more accurate representation. This paper is organised as follows: Sect. 2 presents a review of related evacuation modelling techniques, Sect. 3 presents an overview of the architecture of the hybrid model and Sect. 4 shows the methodology used for hybrid modelling. Section 5 demonstrates the application of the hybrid model to three enclosures, ranging from a small-scale building to a large urban area. Section 6 presents the concluding remarks and insights for future work.

2 Related Work

Increasing the level of sophistication of evacuation models for the purpose of modelling the intricacies of crowd dynamics has been investigated in several studies. For instance, in [9], Lammel et al. describe MATSim model as being a multi-agent transport simulation framework which has been adapted for use in large-scale evacuation. MATSim uses a queue model coupled with a 2D model for agent simulation. The queue model represents agent movement as a flow while the 2D model controls agent movement using forces. In another study [10], a hybrid model comprising of a cellular automata model for representing smoke and fire propagation and a pedestrian model comprising of intelligent goal oriented agents has been described. Although the model is described as hybrid, however, the movement space is represented using only fine nodes. In a study reported by [11], a model using continuous space was implemented to demonstrate hikers walking on trails. This model uses a potential field technique, in conjunction with a graph, to guide agents towards their destination. Moreover, in [12], Xiong et al. presented a hybrid model in which some sections of the geometry can be represented using the coarse approach while others using the fine node approach. However, this model has been demonstrated on a very simple geometry comprising of an entrance, an exit and a corridor. Furthermore, a study reported by Nguyen et al. has shown a hybrid model comprising of a macroscopic and microscopic model [13]. The macroscopic model represents agent movement as a homogeneous flow while the microscopic model simulates agent movement at an individual level. However, in the microscopic model, there are only two agent behaviours (agent following shortest path and agent following others) which have been implemented.

Based on this review of related models, it can be noted that the perspective from which the term 'hybrid' has been used varies across various models. For instance, the models described in [9, 12, 13] use a combination of two spatial representation techniques for representing agent movement whereas the models in [10, 11] use only one spatial type. Our hybrid approach for evacuation simulation differs by integrating all three spatial techniques (coarse, fine and continuous) within a single software tool, thereby providing additional flexibility. When modelling the enclosure using the hybrid approach, the user can decide on the discretisation scheme (i.e. the relative proportions of coarse, fine and continuous) that will be used for representing the entire space. The discretisation schemes can be varied to suit to different user requirements.

3   Software Architecture of Hybrid Model

The hybrid model was implemented using the core architecture of the building EXO-DUS software [14], which uses the fine node approach. This software incorporates a plug-in architecture which made it appropriate for developing the coarse and continuous models.

Coarse model: The coarse model uses Flow to Density equations [6] to control the movement of agents across different segments of geometry. The benefits of representing the agents as a homogeneous ensemble are that the simulations can be executed very rapidly. Moreover, this allows larger geometries to be represented at relatively low computational cost.

Continuous model: In this model, each agent is represented as an autonomous entity that can navigate within its environment by perceiving different cues such as boundary walls, exit locations and surrounding agents. This model also features a behavioural framework which allows different agent traits, e.g. exit familiarity, multidirectional agent avoidance and competitive behaviours to be represented. Moreover, the continuous model uses a navigational graph, in conjunction with sub-goals for optimising the path planning process of agents. The model allows the representation of evacuation dynamics at a high level of realism. Nevertheless, due to the complexity of the agent behavioural and movement mechanisms, this model uses considerable computational resources.

The coarse and continuous models have been described in more detail in our previous work [15].

4   Methodology for Hybrid Modelling

The hybrid model provides an integrated platform combining the efficiency of the coarse model and the accuracy of the fine and continuous models. The methodology for hybrid modelling entails the application of the continuous model in areas of the building/structure which require the highest precision (e.g. constrictions such as narrow openings and exits). On the other hand, areas which require less precision, for example, regions that are far from exit locations are represented using the computationally most efficient coarse models. The remaining parts of the geometry are modelled using the fine node approach to provide reasonable computational performance and accurate agent simulation. In Table 1, we show how various combinations of the coarse, fine and continuous models (i.e. discretisation schemes) within the hybrid model can be used to cater for different simulation and user requirements. In the next section, we demonstrate how these discretisation schemes can be applied to simulate different scenarios, ranging from a small-scale building evacuation to a large urban scale scenario.

Table 1 Discretisation schemes used in hybrid model for different requirements

Requirement	Description	Model recommended
Objective of simulation	Preliminary investigation regarding evacuation scenarios	Coarse
	Detailed analysis of evacuation phenomena and representation of agent movement and behaviours	Fine, Continuous
	Predicting flow rates in narrow openings such as internal and external exits	Continuous
Accuracy of the blue print of the geometry	Detailed plan of the geometry available including details of internal compartments and internal obstacles	Fine, Continuous
	Details of internal compartments not available	Coarse
Time critical results	Results required in fastest time possible to support emergency response systems in real time	Coarse
Size of geometry being modelled	Extremely large urban scale scenarios	Coarse, Hybrid (Coarse/Fine)
	Moderate large scenarios	Hybrid (Coarse/Fine/Continuous)
	Buildings	Fine, Hybrid (Fine/Continuous)
Behavioural analysis	Analysing the impact of different human behaviours on evacuation times	Fine, Continuous
	Emergence of behaviours e.g. lane formation, pushing behaviours	Continuous

5   Application of Hybrid Modelling to Different Scenarios

In order to demonstrate the application of the various discretisation schemes when using the hybrid approach, the model was applied to different types of geometries namely: (1) a building made up of multiple internal rooms (2) a large underground rail tunnel station (3) an urban region. For each geometry, 10 simulations were run and the evacuation time results were averaged. The simulations were run on a 3.6 GHz processor with 8 GB of RAM.

5.1 Multiple Compartment Building

This scenario consisted of multiple compartment building with 300 agents. The different rooms and corridor are represented using fine nodes and continuous regions as shown in Fig. 1. The continuous space was chosen to represent the internal exits and the region surrounding them. The experiments were repeated by setting up the geometry using the only fine nodes (All-Fine) and using only continuous regions (All-Continuous). Ten simulations were run for each configuration and the averaged results of the evacuation times of the pedestrians and the time taken for the simulation to run (computational run time) are as shown in Table 2. It can be seen that the different configurations (All-Fine, All-Continuous, Hybrid) produce comparable evacuation times. The All-Continuous produces the slowest run time due to its computationally expensive movement and behavioural algorithmic procedures, while the All-Fine produces the fastest run time.

Fig. 1 Discretisation scheme used in hybrid model

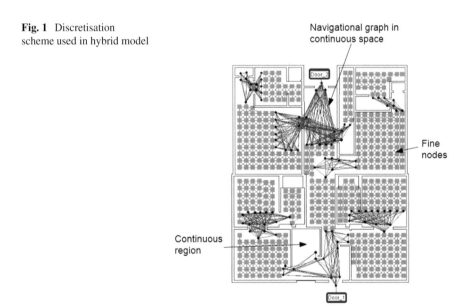

Table 2 Average run times and evacuation times for multiple compartment building scenario

Discretisation scheme	Average total evacuation time (s)	Average run time (s)
All-Fine	58.8	1.5
Hybrid	60.0	8.4
All-Continuous	59.3	14.6

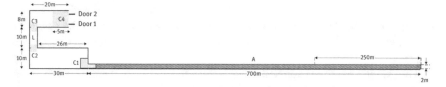

Fig. 2 Large underground rail tunnel station with Section A represented as coarse nodes; sections C1, C2, C3, C4 as continuous space and the remainder of geometry as fine nodes

Table 3 Average run times and evacuation times for large underground tunnel station scenario

Discretisation scheme	Average total evacuation time (s)	Average run time (s)
Hybrid	1023.5	44.2
All-Fine	988.8	80.6

5.2 Large Underground Rail Tunnel Station

To demonstrate the scalability of the hybrid model towards handling moderately large geometries, the hybrid approach was applied to an underground rail tunnel station as shown in Fig. 2. The scenario being represented here entails 2000 agents who have alighted from train carriages as a result of an emergency in an underground rail tunnel. Section A of the geometry is represented using coarse nodes as it is appropriate for modelling the homogeneous movement of the agents going in the same direction. All the internal (C1, C2, C3) and external exits (C4) in the geometry are modelled using continuous space in order to capture the complex behaviours which occur due to the constriction in space at these locations. The locations of the different spatial types in the simulation set up are depicted in Fig. 3.

During the simulations, it was observed that as the agents leave the narrow tunnel to enter the wider platform, they spread out to occupy the available space. The use of the continuous model at this location provided an improved qualitative representation of the agent behaviours. The simulations were also repeated using the same geometry but using only fine nodes to model the entire structure. The average times to run the simulations are shown in Table 3. The results indicate that that the Hybrid model provides an improvement in computational run time by 45.2%, whilst still being able to demonstrate detailed agents behaviours at certain key locations.

5.3 Large Urban Area

The hybrid model can be scaled to include extremely large geometries for example a city block. This is demonstrated in 2 km × 2 km residential area, populated with 10,000 pedestrians. In the scenario, the occupants use the main roads to evacuate towards a 200 m × 200 m safe zone, located 200 m away from the residential area.

The geometrical layout of the city block is as shown in Fig. 4. The discretisation strategy involved the use of coarse regions in areas that are far from the assembly area while the middle artery road and the assembly area were represented using fine nodes.

The average run time results in Table 4 show that the Hybrid model produces an improvement in computational run time by 43.12% when compared to the All-Fine representation.

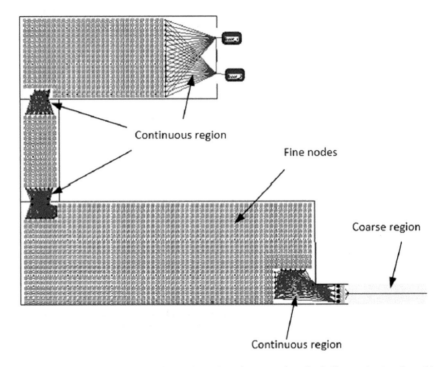

Fig. 3 Location of coarse regions, fine nodes and continuous regions (including navigational graph) in the simulation setup

Table 4 Average run times for large urban area scenario

Discretisation scheme	Average evacuation time (s)	Average run time (s)
Hybrid	2976.82	163.3
All-Fine	3068.40	287.1

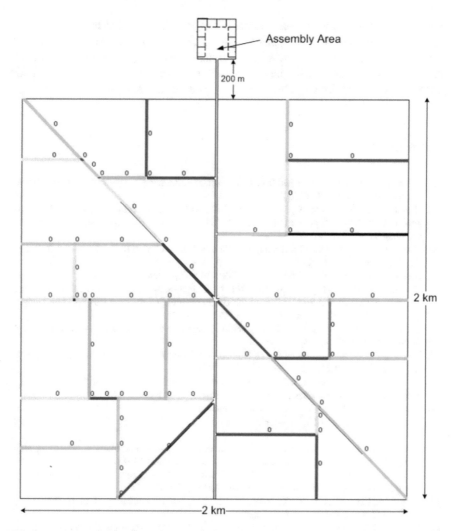

Fig. 4 Road connectivity in urban area

6 Conclusion and Future Work

In this work, we have demonstrated the application of a hybrid model, capable of representing evacuation simulations of pedestrians at different scales, ranging from complex buildings to extremely large urban areas. The flexibility of this approach makes it suitable for different types of scenarios and user requirements. In addition, the hybrid model has been shown to produce results with faster computational run times which accounts for its scalability to handle large-scale geometries with large population sizes, while maintaining accuracy. This is achieved by only applying the appropriate spatial types to suit the modelling objectives or the local behaviours as necessary. Further work will involve the optimisation of the various algorithmic procedures used in order to provide even further improvements in the computational run times of the models. Also, the behavioural framework in the fine and continuous models can be extended to include additional pedestrian behaviours.

References

1. Helbing, D., Mukerji, P.: Crowd disasters as systemic failures: analysis of the Love Parade disaster. EPJ Data Sci. **1**, 7 (2012)
2. Grosshandler, W., Bryner, N., Madrzykowski, D., Kuntz, K.: Report of the Technical Investigation of the Station Nightclub Fire, NIST NCSTAR 2: vol. 1. National Institute of Standard and technology, USA (2005)
3. Cheng, L., Ueng, T., Liu, C.: Simulation of ventilation and fire in the underground facilities. Fire Saf. J. **36**, 597–619 (2001)
4. Beritic, T.: The challenge of fire effluents. BMJ **300**(6726), 696–698 (1990)
5. Marine Accident Investigation Branch.: Report on the Investigation of a Fatal Accident during a Vertical Chute Evacuation Drill from the UK Registered Ro-ro Ferry P&OSL Aquitaine in Dover Harbour on 9 October 2002, Report No 18 (2003)
6. Buckmann, L.T., Leather, J.: Modelling station congestion the PEDROUTE way. Traffic Eng. Control **35**, 373–377 (1994)
7. Pan, X.: Computational Modelling of Human and Social Behaviours for Emergency Egress Analysis. Thesis (Ph.D.). Stanford University (2006)
8. Galea, E.R., Galparsoro, J.M.P.: A computer based simulation model for the prediction of evacuation from mass transport vehicles. Fire Saf. J. **22**, 341–366 (1994)
9. Lammel, G., Grether, D., Nagel, K.: The representation and implementation of time-dependent inundation in large-scale microscopic evacuation simulations. Transp. Res. Part C **18**, 84–98 (2010)
10. Tissera, P.C., Castro, A., Printista, A.M., Luque, E.: Evacuation simulation supporting high level behaviour-based agents. Procedia Comput. Sci. **18**, 1495–1504 (2013)
11. Gloor, C., Stucki, P., Nagel, K.: Hybrid techniques for pedestrian simulations. In: Proceedings of 4th Swiss Transport Research Conference, Ascona (2004)
12. Xiong, M., Lees, M., Cai, W., Zhou, S., Low, M.Y.H.: Hybrid modelling of crowd simulation. Procedia Comput. Sci. **1**(1), 57–65 (2012)
13. Nguyen, N.T., Zucker, J., Nguyen, H., Drogoul, A., Vo, D.A.: A Hybrid macro-micro pedestrians evacuation model to speed up simulation in road networks. In: AAMAS, May 2011, Taipei, Taiwan (2012)
14. Galea, E.R., Lawrence, P.J., Gwynne, S., Filppidis, L., Blackshields, D., Cooney, D.: Building EXODUS v6.2 theory manual, revision 1.1, FSEG, University of Greenwich, Greenwich, London (2016)
15. Chooramun, N., Lawrence, P. J., Galea. E.R.: An agent based evacuation model utilising hybrid space discretisation. Saf. Sci. **50**(8), pp. 1685–1694 (2012)

A Novel Design Architecture to Address the Current Limitations of Fixed and Mobile Phone Data Logging System

Ajit Kumar Gopee, Phaleksha Ujoodha, Vinaye Armoogum
and Yashwantrao Ramma

Abstract Data loggers are devices used to gather data in various environment. However, if they undergo any malfunction or breakdown, data cannot be recovered. Most phone data loggers have a small LCD to display the caller ID only, though a few can also display date and time. Others may not have a display at all and the call information has to be viewed on an external PC. Most of them have limited internal memory of 20 calls only and then are overwritten by the new ones. Some telephone requires an external call recording device since no data logger provides this functionality yet. The EEPROM of the Arduino has limited storage capacity, issues with nature and format of data for executing high load programs. So the present paper attempts to propose a novel design architecture to address these issues. Preliminary output shows that the DTMF decoder could only detect key pressed on a fixed telephone. By using a micro SD card, it is possible to store more data and variables in the Arduino sketch. Unfortunately, for some unknown reasons, the circuit could not detect incoming calls even when the DTMF decoder was replaced with an FSK one. In parallel, an Android app has been being developed that can log the caller ID and indicate the date, time, day, duration, and the type of call such as incoming or outgoing call on an Android mobile phone.

Keywords Data logger · Mobile/fixed phones · DTMF and FSK decoders
Arduino EEPROM · Extended memory and storage

A. K. Gopee (✉) · P. Ujoodha · V. Armoogum
School of Innovative Technologies and Engineering, University of Technology, Port Louis,
Mauritius
e-mail: agopee@umail.utm.ac.mu

P. Ujoodha
e-mail: pujoodha@umail.utm.ac.mu

V. Armoogum
e-mail: varmoogum@umail.utm.ac.mu

Y. Ramma
School of Science and Maths, Mauritius Institute of Education, Reduit, Mauritius
e-mail: y.ramma@mie.ac.mu

© Springer Nature Singapore Pte Ltd. 2019
S. C. Satapathy et al. (eds.), *Information Systems Design and Intelligent
Applications*, Advances in Intelligent Systems and Computing 863,
https://doi.org/10.1007/978-981-13-3338-5_42

449

1 Introduction

Data capture and analysis are essential in the deployment of new innovative products and services, and in the forecasting future events with high probability [1]. Data logger has been widely adopted not only in electronic domains but in all technology-related systems. Data loggers are electronic devices which can also be powered by a battery. They have a microprocessor, some minimal internal memory, sensors, and in some cases an LCD screen as display. Data logging system has three main components: an interface to communicate with a computer, sensors, and the associated software [2].

The function of a data logging system is to automatically collect data toward providing a comprehensive review of conditions upon recording [3]. These are devices which can capture data from a source and take it from the memory to save it as a file or a database object. In general, data logger requires a communication module to interact with the database, a microprocessor or a PC to inspect, transforms and pushes the data to either a database or a log file [4]. Generally, a data logger gets data from measurements in several ways, for example, through an attached analog-to-digital converter (ADC) or sensory data sent via various communication protocols and serial communication port (USART). The data maintained in the data logger collected from monitoring system can be transferred to personal computer for further analysis using software analyzing tools [5]. The unit has to be light, easy to carry, handheld, and cell-operated for capturing "real-world" data which can be easily combined with a PC [6]. A data logger is a surrogate to many data capturing apparatus in a plethora of applications. The former have the capability to accommodate more channels with higher clarity and capture voluminous data. Through programming, the unit can be extended with smart and intelligent features, thus providing the user with diverse unthought-of capabilities so far [7].

The paper is structured as follows: Sect. 1 outlines a brief introduction into the subject. Section 2 covers the literature review, while Sect. 3 describes a comparative study (a technical analysis) of three existing design architectures. Section 4 shows the proposed design. Section 5 briefly describes a first prototype implementation where some preliminary results are also discussed. Section 6 ends with concluding remarks and future works. Section 7 provides a list of references.

2 Literature Review

This section describes different types of telephone data loggers, their specifications, and limitations under three different categories of phone data loggers.

There are many types of data loggers with specific features and different modes of communication. For instance, a general-purpose data logger can be connected to a computer via a USB interface for activation. Such kind of data logger usually comes with a software to record logging variables such as sample frequency, initial time,

Table 1 Some domain examples of using data logger [10]

Field	Applications
Energy management	Temperature, voltage, current, gauge pressure measurement
Environmental	Wind speed, temperature, pressure, humidity and rainfall measurement and follow up, oversee pollution
Life sciences	Monitor refrigerators, freezers, and non-stop cryostats to safeguard invaluable sample and provide advanced alarming and messaging
Oil and gas	Monitoring temperature and pressure of steam in oil wells, water pressure, pumping efficiency
Construction	Monitoring temperature, vibration, pressure, force, along with vibrating wire strain gauge support

and temperature. The former is then operated in a desired place where it records each measurement and stores them in memory along with the time and date. After the data logging process, the device is attached to the PC again and the inherent program allows the measurements to be displayed in human-readable format, graph, or table [8].

Web-based data logging systems are somewhat different from general-purpose data loggers. In such systems, recorded information are sent to the Internet on a server for both wired and wireless accesses. Another type of data logger is the Bluetooth data logger, whereby the data are downloaded directly to mobile devices using the Bluetooth connection. The latter can be used outdoor or indoor or it can be even deployed for use underwater [9]. A brief comparison of the different applications of data logger is tabulated in Table 1.

Using a data logger includes writing programs, deploying these securely somewhere, downloading captured data, and transforming the data obtained from the network. All generic data loggers have those three units and features. Data loggers can be classified as stand-alone, web-based, wireless, or Bluetooth [11].

2.1 The Four Categories Are Briefly Described Below

(a) **Stand-alone data loggers** are usually portable devices for easy setup and deployment. They consist of either internal or external sensors to monitor the surrounding environment. They can be connected to a computer via a USB interface for further data manipulation.
    Example: Temperature and/or humidity data loggers with USB-connected LCD screen, lithium battery, and red and green LEDs for indication.

(b) **Web-based data systems** enable remote and network-based access to information via GSM, Wi-Fi, or Ethernet communications. Just as stand-alone data logger, the web-based system consists of internal and/or external sensors to collect data and send it to a secure web server.

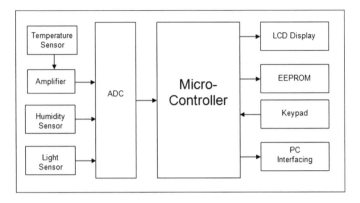

Fig. 1 Generic block diagram of data logger [12]

Example: Web-based voltage, thermocouple, and pulse data logger stores data locally in internal flash memory and remotely on controlled-access servers via Ethernet connection.

(c) **Wireless data loggers** transmit real-time data from multiple locations (very often simultaneously) to a centralized computer. They have a plus-point since it is not necessary to physically deploy these devices in the location and manually take readings.

Example: Wireless node with inherent temperature and humidity sensor propagating real-time humidity/temperature data. Apart from an inherent memory to prevent accidental data loss, they also have an alarm/notification feature via email or messaging.

(d) **Bluetooth data loggers** use minimal energy to measure and transmit data wirelessly to mobile devices via Bluetooth within a coverage range. HOBO MX2302 is an example data logger where temperature and relative humidity sensors can be attached to it. It enables data download directly from a mobile app.

All these different types of data loggers have some components in common. Figure 1 shows the generic block diagram of a typical data logger with either analog or digital sensors. In case of analog readings, an analog-to-digital converter (ADC) is required to transform the signal into digital data which the PC can interpret and process.

(a) **PC-connected telecord**

It is a recording device which is coupled with a PC. The concept describes a recording unit, commonly used in the factory. It can be fixed to a PC. It is referred to as a telephone substation management system. However, the latter has limited storage space and functionalities. Actually, each substation has a PC which acts as a voice recording system. The latter incorporates a customer line identification. The inherent program can generate a list showing missed calls, received calls, actual date, time, and number of calls. The audio files are stored on the PC's hard disk [13].

Table 2 DTMF frequencies [15]

Button	Low DTMF frequency (Hz)	High DTMF frequency (Hz)	Binary coded output			
			Q1	Q2	Q3	Q4
1	697	1209	0	0	0	1
2	697	1336	0	0	1	0
3	697	1477	0	0	1	1
4	770	1209	0	1	0	0
5	770	1336	0	1	0	1
6	770	1477	0	1	1	0
7	852	1209	0	1	1	1
8	852	1336	1	0	0	0
9	852	1477	1	0	0	1
0	941	1336	1	0	1	0
*	941	1209	1	0	1	1
#	941	1477	1	1	0	0

(b) Device to automatically identify DTMF and FSK for caller ID

It is a device that usually identifies dual-tone multifrequency (DTMF) and frequency-shift keying (FSK). A signal detector detects signals on the telephone line. A microprocessor is activated when the signal detector detects a signal on telephone line to activate the DTMF decoder which calculates the frequency spectrum of the detected signal.

For example, the digit "4" translates into a sound with two tones, one at 770 Hz. and the other at 1209 Hz. Each digit is translated into two tones. Table 2 shows the frequencies associated with each character. In our case, a modified version of [14] has been experimented as shown in Fig. 2. Only the DTMF has been used in a first instance.

(c) Incoming call detection and display

This system is used to automatically detect an incoming call and identify the party associated with the latter. The setup consists of a database of telephone numbers and the customers. A display allows the user to see the call and party associated with that number. The latter is compared with those from the database, whereby the name and number are logged for future use [16].

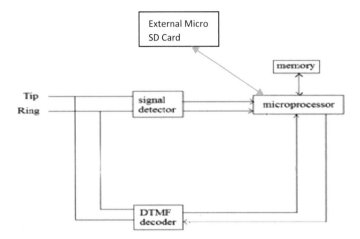

Fig. 2 Block diagram to identify DTMF tones adapted from [15]

3 Comparative Study

Through this current research work, some differences have been identified among the features of three systems. The first work is described as a data logger that is connected to the computer and records calls but has a limited storage space. It has no GUI interface to view any call logs. In the second work, the data logger has neither connected to the PC nor can record calls. It can simply obtain the caller ID, date, time, and duration of the calls. Based on this model, a new novel design architecture is being proposed here in this paper that includes a GUI to view the call logs such as caller ID, date, time, and duration. It can also record calls and provide enough space to store them for external communication. The third work describes an automatic display system to identify the caller name reference to a stored list of directories and caller ID. It includes GUI as compared to the two other designs but still has limited storage features. So the aim of this novel architecture is to build a system that combines the advantages of the three designs, while at the same time addresses the limitations of these models by incorporating extended storage and fully programmable features.

4 Proposed Design

Figures 3 and 4 show an overview of the proposed architecture. The focus of this paper is on the hardware part. It shows the layout and interconnections of the different components used. In contrast to the different works described in Sect. 3, the proposed architecture includes a combination of the features of these three models. It captures a log of all incoming and outgoing calls in terms of their caller ID, duration, date,

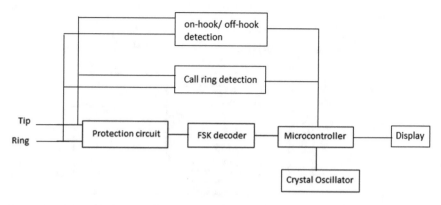

Fig. 3 Proposed block diagram of the circuit

Table 3 Major components used and configuration

No.	Component	Configuration
1	Arduino Uno	16 MHz, AT mega 328, 2 KB SRAM, 32 KB Flash,
2	HT9032D	3.5–5.5 V
3	RJ 11 pin socket	
4	Capacitor	0.2 µF, 275 V
5	Capacitor	0.01 µF, 100 V
6	Diode 1N4007	
7	Resistors	200, 18, 15, 81 kΩ
8	Crystal oscillator	3.58 MHz
9	Capacitor	30 pF, 50 V
10	DTMF MT 8870	8-pin DIP/SOP
11	Jumper cables	
12	Micro SD card	16 GB, class 10

time, and recording. All these information are displayed on a screen. The circuitry is designed to cater for the on-hook and off-hook detection of the telephone and when the phone rings. The circuit also includes a protection circuit which shields from the high voltage fluctuations from the line and to prevent reverse current that may arise. An ATMega 2560 could be an alternative to add more memory when you run out of program memory on an Arduino Uno. However, the use of an external SD card is the better solution since not only the card can be plugged into a PC but it also provides more flexibility in terms of memory virtualization. Figure 3 shows a block diagram of the circuit, while Figs. 3 and 4 show a new circuit modified circuit and a screenshot of the proposed architecture (Table 3).

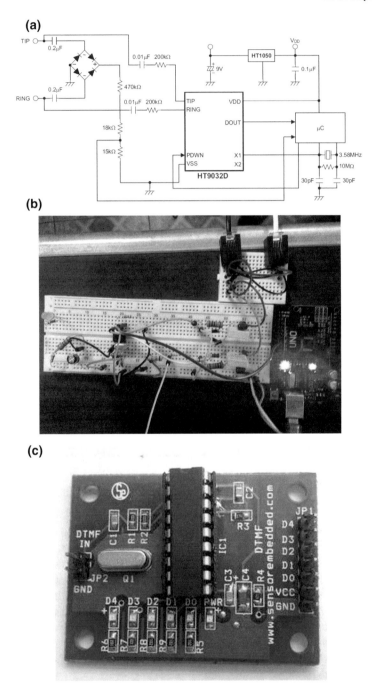

Fig. 4 **a** Circuit diagram (adapted from Holtek datasheet), **b** Setup of the circuit, **c** DTMF MT8878 module, **d** Analog phone—router connection, **e** IP phone— router connection

(d)

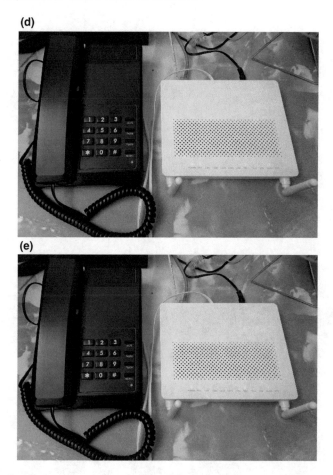

(e)

Fig. 4 (continued)

5 Result and Discussion

Unfortunately, as can be seen in Fig. 5a, the circuit could only detect keys pressed on the phone. For some unknown reasons, it could not detect any incoming calls. Too much time was spent on this issue; so, we decided to use the FSK decoder. Again even with the latter, neither key pressed nor incoming call could be detected. So right now we are trying to address this issue and hopefully to debug the circuit and the Arduino codes soon.

By using an external micro SD card, we have been able to extend the storage of the Arduino to store more data and variables for our program. It also offers more flexibility in terms of memory virtualization. We are already addressing this issue in our next research work and we have come across an Arduino library "*virtmem*" for this purpose. At the same time, in parallel, an Android application (to act as a

(a)

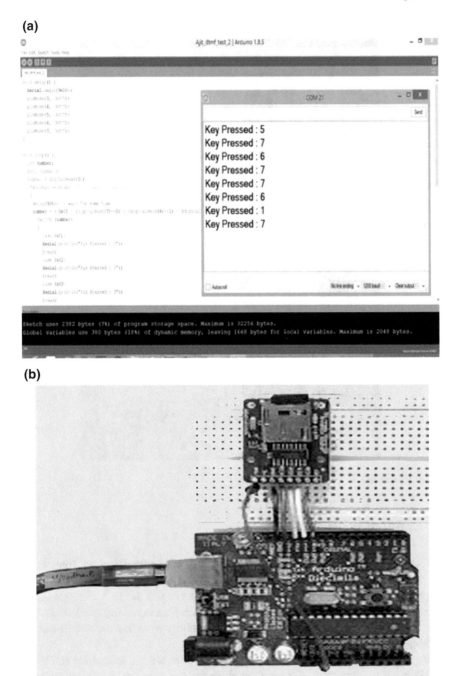

(b)

Fig. 5 **a** Arduino codes and serial monitor output, **b** Arduino with micro SD card, **c** Read/write from SD card

(c)

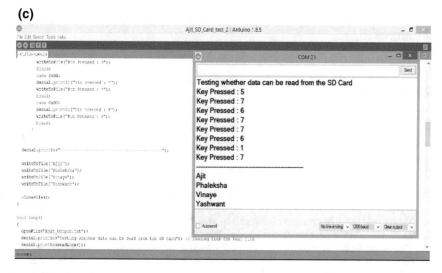

Fig. 5 (continued)

data logger at the software level) has been developed and being tested in the case of mobile phones.

A small sketch was written and uploaded onto Arduino to write and read from the SD card. Preliminary but encouraging results are shown in Fig. 5c.

6 Conclusion and Future Works

Through this novel architecture, the designed circuit has been able to overcome some of the limitations posed by other existing data loggers. Using the DTMF decoder so far, we have been able to detect phone key pressed. Using an external micro SD card, we have been able to read/write from the card, thus extending the storage of the Arduino in a cost-effective way. As future work, sending call information via a wireless communication seems to be the next logical challenge on the design. While testing the Android app incidentally, we came across a problem with WhatsApp messages. The latter is encrypted using the signal protocol. This protocol has been proposed to prevent third parties and WhatsApp from having plaintext access to messages or calls [17]. However, we have managed to transfer the decrypted WhatsApp messages from the mobile to a PC. This research work is being reported in the next paper "*An Android Application as a comprehensive data logging system for a Mobile Phone*" which we are currently working on.

In the short term, future work will focus on, first, how to virtualize memory of external micro SD card connected to the Arduino, and second, a method for assessing the potential interoperability feature of data loggers connected onto different platforms possibly with different operating systems using XML and the implementation of such model.

References

1. Jan, T.: Design and development of a data logger based on IEEE 802.15.4/ZigBee and GSM. In: Proceedings of the Pakistan Academy of Sciences: Pakistan Academy of Sciences A. Physical and Computational Sciences, **53**(1), 37–48, p. 1 (2016)
2. NehaChourasia, S.N.: Review paper on industrial datalogging system. Int. J. Res. Emerg. Sci. Technol. special-Issue-1-JAN-2017, Dahegaon, Nagpur, India (2017)
3. Implementing a data logger with spansion SPI Flash. Zilog Embed. Life Oixys Company, Multimotor Ser. p. 1 (2013)
4. Sio-lin Ao, B.R.: Machine Learning and System Engineering, Springer science + business media B.V, Dordrecht, Heidelberg, London, New York (2010)
5. Thanutong, P.: Implementation of low cost data logger using flash disk with file allocation table
6. Akposionu, K.N.: Design and fabrication of a Low-cost data logger for solar energy parameters. J. Energy Technol. Policy **2**. ISSN 2224-3232, p. 1 (2012)
7. Derek Liu, D.U.: The USB data logger based on the MC9S08JM60. Freescale Semicond. p. 1 (2008)
8. LocalLaw87. Data logging: Online, Available http://www.local-law-87.com/content/data-logging. Accessed 30 05 2018 (2018)
9. Onsetcomp.com. What is a data logger? — onset data loggers: Online, Available http://www.onsetcomp.com/what-is-a-data-logger. Accessed 14/ 04/ 2018 (2017)
10. Evidencia.biz. Types of data loggers mechanical, digital, and wireless: Online, Available http://www.evidencia.biz/what-is/types.htm. Accessed 14/ 04/ 2018 (2010)
11. Hailegiorgis, H.: Implementing a data logger for home automation: Online, Available https://www.theseus.fi/bitstream/handle/10024/93734/Thesis_Final.pdf. Accessed 05 06 2018 (2010)
12. M. Technologies. Microcontroller based data logger: Online, Available https://www.projectsof8051.com/microcontroller-based-data-logger/. Accessed 01/ 06/ 2018 (14/ 06/ 2013)
13. Quingnan, S., Changsuo, Z., Min, L., Cong, Q., Xingwei, B.: Telecord that can be connected with computer: Online, Available https://patents.google.com/patent/CN205681521U/en. Accessed 30/ 05/ 2018 (2016)
14. Hsieh, M.-T.: Device for automatically identifying DTMF and FSK for caller ID: Online, Available https://patents.google.com/patent/US6683948. Accessed 30/ 05/ 2018 (2004)
15. LABS, L. DTMF: Online, Available http://www.lemalabs.com/dtmf/. Accessed 30/ 05/ 2018 (2015)
16. Figa, R., Cohen, J.H., Cohen, S.G.: Automatic incoming telephone call originating number and party display system: Online, Available https://patents.google.com/patent/US4924496A/en. Accessed 31/ 05/ 2018 (08/ 05/ 1990)
17. WhatsApp Encryption Overview. Technical white paper, December 19, 2017, Originally published April 5 (2016)

Evaluation of Quality of Service Parameters for Mobile Cloud Service Using Principal Component Analysis

Rajeshkannan Regunathan, Aramudhan Murugaiyan and K. Lavanya

Abstract The cloud service is communicating between mobile devices through the Internet. They are prepended as platform independent and loosely conjoined. This paper's objective is to obtain effective services with composition of services based on specific QoS parameters using principal component factor analysis method. The Factor analysis, analyzing the interrelationships among variables by characterizing highly correlated known as factors. In factor analysis, particularly statistical principal component analysis is used to consider the interrelationships among variables with a minimal loss of information. This analysis is made from the data collected from the cloud service providers. This result is based on components of cloud service availability, response time, throughput, reliability, successability, compliance, best practices, and web services resource framework. These are few parameters considered among the QoS parameters to find the best mobile cloud services. In this factor analysis based on principal component analysis considers these eight parameters which can be further improved to select the cloud services of QoS.

Keywords Cloud service · Quality of service · Factor analysis · Service selection

1 Introduction

Cloud services are accessed through the Internet by business users. Such that Discover and Integration (UDDI), Web Service Description Language (WSDL), Universal Description, and Simple Object Access Protocol (SOAP). The cloud services comprise application behavior and information systems. Mobile cloud services are supported in business activities to automate on the cloud on a wide scale by auto-

R. Regunathan (✉) · K. Lavanya
School of Computer Science & Engineering,
Vellore Institute of Technology, Vellore, Tamil Nadu, India
e-mail: rajeshkannan.r@vit.ac.in

A. Murugaiyan
Perunthalaivar Kamarajar Institute of Engineering and Technology, Karaikal, India

© Springer Nature Singapore Pte Ltd. 2019
S. C. Satapathy et al. (eds.), *Information Systems Design and Intelligent Applications*, Advances in Intelligent Systems and Computing 863,
https://doi.org/10.1007/978-981-13-3338-5_43

matic discovery and utilization of services. The Business-to-Business incorporation can be extended by clustering multiple service providers into a value-added service [1]. An architectural design which includes the gathering of services in a network which interfaces with each other is called the Service Oriented Architecture (SOA). The impediment of each service is not perceivable to another service. The service is a kind of task which is self-contained, well defined that provides different functionality such as bank statement printing, checking customer account details, etc. and does not depend on the situation of other service [2]. One of the main purposes of this system is the utilization of binding the mechanisms. The feature required in abstract and concrete services [3]. All concrete services correlating to an abstract service which are equivalent operationally and can replace each other. The abstract and concrete service selection can be uttered by Quality of Service (QoS) attributes.

According to Chatfield and Collins [4], factor analysis is basically used to reduce the number of parameters which will help to improve the efficiency selection process. The parameter reduction is a process that aims at evaluating maximum variance from the dataset within each factor. In the experimental work, the factor analysis strongly recommended the Principal Component Analysis (PCA) which is used widely in all types of analysis from neurobiology to computer graphics. The PCA is a straightforward, nonparametric method of extracting pertinent information from confusing sets of data [5]. With little extra effort, PCA returns to limit a complex set of data to a smaller dimension [1].

2 Methodology

2.1 Previous Studies

Cloud services are playing an important role in software engineering with challenge represented by Quality of Service (QoS). This combines the service focused system and run-time with services which provides concerned features, constraints, and optimization. The generic algorithms which can handle generic QoS attributes are proposed to evaluate the service [3]. The QoS parameters are important to consider in selecting the service. Not much research has been done so far to evaluate the quality of service in a fair and open manner. It mostly oriented toward a collection of networking-level criteria [6]. No model gives the complete idea of how many parameters can be used to evaluate QoS. In a QoS model is suggested that all cloud support service providers should exhibit their services using a certifier to evaluate QoS [7]. This model does not allow service providers to improve QoS dynamically and no procedure on how QoS is verified in automatic and cost efficient. The QoS study assists finding the factors most generally used by service investigators in QoS-based selection and configuration of cloud service. Apart from this, the study has also helped to find the factors that need to be taken into account for an individualized selection and configuration QoS based service, thus opening many research opportunities in the same [8].

The flow chart of QoS decision-making analysis in Fig. 1 above which allows decision making in mobile operators combining the technical parameters with information about end user experience. First step is the application of a subjective evaluation in a controlled atmosphere where the samples are biased in periods of time, then next the sample population is selected and QoS parameters are listed, samples go through a statistical process in which the elements out of range are detected and removed. In statistical analysis, the data is collected from the sources for the analysis. Factor analysis depends on the "common factor model" which postulates that experimental measures are affected by primary common factors and unique factors, and the determination of correlation patterns. Maximize possibility attempts to analyze the maximum possibility of sampling the experiential correlation matrix. Factors are rotated for improved interpretation as unrotated factors are unclear. The objective of the rotation is to attain an optimal structure which tries to load each variable. The rotation maximizes the high loadings on each variable as possible. Rotated factor matrix shows the rotated factor loadings. When interpreting the factors, we need to determine the power of the relationships. Factors can be recognized by the largest loadings, but it is also vital to observe the zero and low loadings in order to check the identification of the factors. Figure 1 shows how the users get the processed information and sends recommendations [9]. This is stored in QoS database.

2.2 Parameters and Related Information

QoS parameter is used to characterize the service quality. QoS parameter values differ according to the services. QoS parameters are required to measure the service such as reliability, availability, successability, compliance, throughput, best practices, web services resource framework, response time, and so on [10]. Measuring QoS parameters are essential for how a service or product is sound, therefore selecting a service is a key point for different service providers. If service features are similar, the quality may differ to the users. Looking at a quality perspective of services, the content of the individual with the established service is important. Quality is attained when there are certain provided necessities which public institutions must fulfill. Quality satisfying standards can be only slight assurance of the quality [11].

3 Principal Component Analysis

PCA has many properties, many of which could be used to describe it The first principal component (PC1) is a projection with higher variance. A linear combination of the variables creates a projection with coefficient vector of length one. The sign of PC1 is random as the variance does not depend on the sign. Successive PCs are

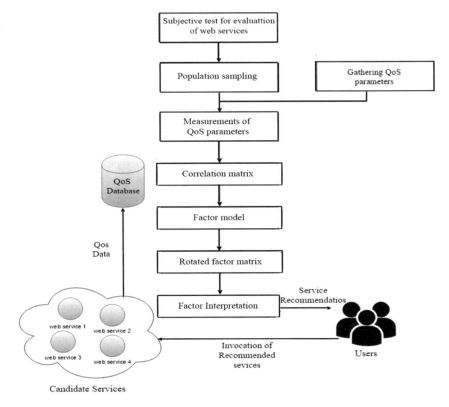

Fig. 1 QoS decision making analysis

described as the projected variable which is not correlated with the earlier PCs and has highest variance, again with random sign. The initial k < p principal components from the set of k dimensions in which the variables are not correlated (and orthogonal if zero), with the highest variance matrix. Consider the projection of first k < p principal components onto the space. To the space, this is the k-dimensional which reduces the sum of squared distances from the cases. The sum of squared distances among all pairs of projected points are maximized by the same projection. The covariance matrix of the data describes the first three properties of the (co)variances of unit-length linear combinations of variables. This allows the use of biased or strong covariance matrices, and also to deliberate about PCs for multivariate distributions. Even though interest is usually in the many variable PCs, seldom the least variable PCs are used to discover which linear combinations of the data are approximately constant.

4 Data Analysis

4.1 QoS Parameter Evaluation

The dataset is comprised of simple random samples of 50 web service providers from 2507 cloud service providers and the data are shown in Table 1 for the specified parameters [12]. Basing on the condition that there are same functionalities for all web service providers the sample size of the study has been fixed. The mean and standard deviation of the QoS parameters are presented in Table 2. QoS parameters are computed to have constant distribution as QoS parameters differ in performing numeric matching. Based on the influence shown on the classification function these parameters are divided into positive and negative [13]. In negative parameters, decrease in objective value when decreases the objective function and in positive criteria increase in attribute value when increases objective. The normalization of positive parameters and negative parameters is done by Eqs. (1) and (2) below [14]:

$$N_{ij}^+ = \frac{q_{ij} - q_j^{\min}}{q_j^{\max} - q_j^{\min}} \quad \text{if } q_j^{\max} - q_j^{\min} \neq 0$$
$$1 \qquad\qquad q_j^{\max} - q_j^{\min} = 0 \tag{1}$$

where N_{ij} is the normalized value of jth parameter of the ith service.

Table 1 Sample QoS raw data

RT	AV	TP	SU	REL	COM	BP	RT
45	83	27.2	50	97.4	89	91	100
71.75	100	14.6	88	85.5	78	80	93
117	100	23.4	83	88	100	87	90
70	100	5.4	83	79.3	100	75	90
105.2	100	18.2	80	92.2	78	84	90
224	100	24.6	83	80	100	87	90
99.2	100	13.7	80	76.3	78	83	89
108.2	100	16.8	80	90.7	78	77	88
125.2	100	16.4	80	89.2	78	84	88
110.3	100	13.9	87	87.5	78	77	87

Table 2 Mean and standard deviation for the QoS

	RT	AV	TP	SU	REL	COM	BP	WsRF
Mean	125.189	98.375	16.677	79.15	84.53	85.975	80.55	85.45
N	40	40	40	40	40	40	40	40
Std.Dev	60.555	8.5	8.45	18	10.85	5.5	5.5	9.5

$$N_{ij}^- = \frac{q_j^{\min} - q_{ij}}{q_j^{\max} - q_j^{\min}} \quad \text{if } q_j^{\max} - q_j^{\min} \neq 0$$

$$1 \qquad\qquad q_j^{\max} - q_j^{\min} = 0 \tag{2}$$

where N_{ij} is the highest value of the jth column of the QoS matrix, $q_j^{\max} = \max(q_{i,j})$, $1 \leq i \leq n$ and N_{ij} is the least value of the jth column of the QoS matrix, if $q_j^{\min} = \min(q_{i,j})$, $1 \leq i \leq n$. $q^{\min}$ denotes least QoS value and $q^{\max}$ denotes the highest QoS value. All the QoS Values lie between a $[0,1]$ interval after normalization [15].

5   Reliability and Validity: KMO and Bartlett's Test

Principal components analysis with Varimax rotation generates the size of discrimination. Bartlet Test of Sphericity and KMO test, In Bartlets test it is observed if the subscales of the scale are inter-independent, KMO (Kaiser-Meyer Olkin Measure of Sampling Adequacy) examines sample sufficiency [1, 16]. The key method to extract factors is studied on main constituents with varimax right-angled rotation, therefore the variance between different loads is improved, on a precise factor, having as an ultimate result less load change into lesser and big loads become even bigger, and at final those values in between are reduced. KMO values lie between 0 and 1. The researchers carried out factor analysis, which accepts the KMO value greater than 0.5. In our analysis, the KMO value is 0.630. KMO and Bartlett's test values are listed in Table 3.

5.1   Other Components

a.   Varimax Rotated Component Matrix:

Streamlining of the expression of a subspace in terms of few major items is done by Varimax rotation. It is being rotated in an orthogonal basis to align with the coordinates without changing the actual coordinate system. Rotation method is more efficient than unrotated method. In unrotated solution set, the parameters show their

Table 3 KMO and Bartlett's test

Kaiser-Meyer-Olkin measure of sampling adequacy		0.630
Bartlett's test of sphericity	Approx. chi-square	152.397
	df	28
	Significance	0.000

impact on more than one-factor component. Orthogonal rotation factors are at best obtained by a Varimax rotation method. The results obtained are shown in Table 4.

6 Results

Varimax rotation set up clearly gives the solution by facilitating the QoS parameters to appear in one direction. These significant loadings depend on the sample size of the dataset. Table 4 presents the thresholds for sufficient/significant factor loadings. Irrespective of sample size, it is better to have loadings greater than 0.500.

QoS parameters are grouped into three factors, namely, trust, performance, and security as given in following Tables 5, 6, and 7 by which cloud service provider performance is measured.

In the final analysis, one has to construct a test battery. The test battery will measure the efficiency of QoS parameters. We have to select one QoS parameter from the QoS parameters which has the highest loadings under each factor group. The results relating to this are presented in Table 8.

7 Conclusion

Mobile cloud service selection can be evaluated by 13 different QoS parameters. With the help of principal component analysis, the results are obtained and identified these eight main factors influence the service selection evaluation extracted by the factor

Table 4 Varimax rotated component matrix

	Component		
	1	2	3
TP	0.915	0.042	−0.036
REL	−0.827	0.193	0.501
BP	−0.705	0.223	0.534
WSRF	0.670	0.418	0.397
COM	0.577	0.440	−0.333
SU	−0.068	0.932	0.029
AV	0.008	0.871	−0.088
RT	0.029	0.941	−0.040

Table 5 Trust factors

QoS	Loadings
TP	0.915
WSRF	0.670
COM	0.577

Table 6 Performance factors

QoS	Loadings
RT	0.941
SU	0.932
AV	0.871

Table 7 Security factors

QoS	Loadings
BP	0.534
REL	0.501

Table 8 Test battery for measuring QoS parameters

Factor group	Items	Loadings
Performance factors	RE	0.941
Trust factors	TP	0.915
Security factors	BP	0.534

analysis method. These eight main factor variables reduced the complexity of the service selection evaluation and indicated that factor analysis had better dimension-reducing effect. From the result obtained, from the data, throughput is the leading QoS parameter as in the trust factors and the loading value is greater for this parameter. Response time parameter is the next best choice as in performance factor has a higher loading factor value. Best practices are the third best choice under trust factors. The authors suggested that consider the eight QoS parameters for service, selection of cloud service for the above analysis.

References

1. Arasi, F.E., Govindarajan, S., et al.: Evaluation of quality of service using factor analysis method [J]. Indian J. Sci. Technol. **9**(41) 2016
2. Newcomer, E., Lomow, G.: Understanding SOA with Web Services(Independent Technology Guides), Addison-Wesley Professional (2004)
3. Canfora, G., et al.: An approach for QoS-aware service composition based on genetic algorithms. In: Proceedings of the 7th annual conference on Genetic and evolutionary computation, pp. 1069–1075. ACM (2005)
4. Chatfield, C., Collins, A.J.: Introduction to Multivariate Analysis. Chapman & Hall, London (1992)
5. Nimako, S.G., et al.: Confirmatory factor analysis of service quality dimensions within mobile telephony industry in Ghana [J]. Electron. J. Inf. Syst. Eval. **15**(2) (2012)
6. Zeng, L., et al.: Quality driven web services composition. In: Proceedings of the 12th international conference on World Wide Web, pp. 411–421. ACM (2003)
7. Ran, S.: A model for web services discovery with QoS. ACM Sigecom exchanges **4**(1), 1–10 (2003)

8. Muthuraman, S., Venkatesan, V.P.: Qualitative and quantitative review of QOS based web services selection and composition techniques. In: Proceedings of the International Conference on Informatics and Analytics, p. 82. ACM (2016)

9. Jinrong, C., et al.: Research on the service industry competitiveness in Liaoning province based on factor analysis. In: IT in Medicine and Education (ITME), 2011 International Symposium on. 1, pp. 441–444. IEEE (2011)

10. Lee, Hyo-Jin, et al.: QoS parameters to network performance metrics mapping for SLA monitoring [J]. KNOM Review 5(2), 42–53 (2002)

11. Kondrotaitė, G.: Evaluation of the quality of public services in Lithuanian municipalities. Intelektine Ekon. 6(3) (2012)

12. Al-Masri, E., Mahmoud, Q.H.: Qos-based discovery and ranking of web services. In: Proceedings of 16th International Conference on Computer Communications and Networks, 2007. ICCCN 2007, pp. 529–534. IEEE (2007)

13. Cardoso, J.: Discovering semantic web services with and without a common ontology commitment. In: Proceedings of the 2006 IEEE Services Computing Workshops. SCW'06, pp. 183–190. IEEE (2006)

14. Al-Masri, E., Mahmoud, Q.H.: Discovering the best web service. In: Proceedings of the 16th international conference on World Wide Web, pp. 1257–1258. ACM (2007)

15. D'Mello, D.A., Ananthanarayana, V.S.: Quality driven web service selection and ranking. In: Proceedings of the 2008 Fifth International Conference on Information Technology: New Generations. ITNG 2008, pp. 1175–1176. IEEE (2008)

16. Anastasiadou, S.D., Anastasiadis, L.: Reliability and validity testing of a new Scale for measuring attitudes toward electronics and electrical constructions subject [J]. Int. J. Appl. Sci. Technol. 1(1) 2011

Author Index

© Springer Nature Singapore Pte Ltd. 2019
S. C. Satapathy et al. (eds.), *Information Systems Design and Intelligent Applications*, Advances in Intelligent Systems and Computing 863,
https://doi.org/10.1007/978-981-13-3338-5

Printed in the United States
By Bookmasters